Simplified Masonry Skills

R.T. KREH SR.

[Van Nost. Reinhold c1982]

Library of Congress Catalog Card Number 81-21879
ISBN 0-442-25337-0

Printed in the United States of America

Designed by Graphics International
50 Wolf Road
Albany, New York 12205

Published in 1982 by Van Nostrand Reinhold Company
135 West 50th Street
New York, NY 10020

Van Nostrand Reinhold Limited
1410 Birchmount Road
Scarborough, Ontario M1P 2E7, Canada

Van Nostrand Reinhold Australia Pty. Ltd.
17 Queen Street
Mitcham, Victoria 3132, Australia

Van Nostrand Reinhold Company Limited
Molly Millars Lane
Wokingham, Berkshire, England

16 15 14 13 12 11 10 9 8 7 6 5 4 3 2 1

Library of Congress Cataloging in Publication Data

Kreh, R. T.
Simplified masonry skills.

Includes index.
1. Masonry—Amateurs' manuals. I. Title.
TH5313.K73 1982 693.1 81-21879
ISBN 0-442-25337-0 AACR2

This book is dedicated to
my wife, Betty, my son,
Ricky, and my daughters,
Patty and Pam
for all their help and encouragement.

Contents

Preface

Masons are highly skilled people who use specialized equipment to mix mortar, bond materials, and lay out and construct masonry walls and other important features of buildings. This requires both study and on-the-job training if the mason is to keep pace with the constant developments of the trade. This completely revised edition of SIMPLIFIED MASONRY SKILLS provides the necessary material to guide you through the study of masonry basics. It also may act as a guide during an apprenticeship program and as a handbook for use on the job.

This edition of SIMPLIFIED MASONRY SKILLS retains the major features that made the former edition a trade standard. These include:

- performance objectives that tell you what information is to be learned through careful study of the material.
- an easy-to-follow format high in illustrations and hands-on application.
- safety features integrated throughout the text.
- skill checklists that explain specific skills that you should have mastered up to that point in study.
- achievement review questions following each unit that allow you to evaluate your comprehension of the unit. Questions after each section similarly allow you to gauge your understanding of section material.
- projects that guide you through actual masonry construction in a step-by-step procedure.
- an extensive Glossary that lists definitions of terms with which you should be familiar.
- technical reviews of text material by field experts, including the Brick Institute of America and the National Lime Association. The material on metrics was reviewed for accuracy by the National Bureau of Standards.

Also included in this edition is new material reflecting the latest developments in masonry, such as the installation of flues for wood-burning stoves.

THE MASONRY APPRENTICESHIP

The study of SIMPLIFIED MASONRY SKILLS develops skills required for work with the masonry contractor. After completion of this study, readers may enter into a masonry apprenticeship to refine their skills and to further their knowledge of trade practices. To review suitable apprenticeship programs, the Bureau of Apprenticeship and Training, U.S. Department of Labor, may be contacted.

A typical apprenticeship program lasts three or four years. Credit for previous experience is given in most programs after evaluation by the apprenticeship committee and a thirty-to-sixty-day tryout on the job. If the apprentice meets contractor requirements, he or she may very well begin at a pay scale that is higher than average. The credit given for previous experience varies according to locality and individual contractor, but previous training is sure to help the apprentice.

Apprentice wage rates, either union or nonunion, are based on a progressively increasing schedule tied to the journeyman rate. This could mean a raise every three or six months, depending upon the agreement. For the pay raise to be in order, of course, the apprentice must produce good work and show constant improvement.

A good program requires the apprentice to complete at least 144 hours of classroom study each year. The training must be done under the supervision of a journeyman bricklayer, who usually teaches in the evening. Attendance is required at all classes.

Union apprentices must have a sponsoring bricklayer before they can enter the program. It is the responsibility of the bricklayer foreman to see that the apprentice receives experience in the various aspects of the trade as the job progresses. A set number of hours' training on the job is required before the apprentice becomes a journeyman bricklayer. The actual number is determined by the apprenticeship committee in that locality.

During the working day, the apprentice is supervised by bricklayers on that particular job. They are responsible for such things as pointing out correct methods of laying units and correcting any errors. This is stated in the agreement between the apprenticeship committee and the contractor. There are usually a limited number of apprentices for each bricklayer on the job to be sure that the apprentice receives adequate instruction. When work becomes scarce, the contractor always attempts to keep the apprentice working as long as possible before any layoffs. This stipulation, stated in the agreement, acts as protection for the apprentice.

RELATED SKILLS

The mason apprentice must develop certain skills in on-the-job training. These include:

- proper care and use of tools.
- study of mortar types and mixing of mortar.
- trowel skill development.
- building of foundations and parging.
- laying of brick, concrete block, tile, and other units.
- cutting of masonry materials and walling around openings.
- laying brick to form arches.
- laying bond patterns.
- construction of fireplaces and chimneys.
- cleaning of masonry work.
- safety practices and accident prevention.
- reading of mason's folding rules.
- construction of various types of walls.

Related technical information taught in the classroom includes plan reading, trade-related math, responsibilities of the apprentice to the contractor, and other information that cannot be covered on the job during the day.

At the completion of the apprenticeship program, the apprentice becomes a journeyman mason and is entitled to full-scale wages. After completing the program and reaching journeyman status, many masons continue their study and become company foremen. This, in turn, means a raise in pay. Still others eventually start a business of their own after a number of years of experience.

Because new techniques are being constantly developed, masons should continually read trade materials, attend trade meetings, and study new methods and products on their own.

THE AUTHOR OF SIMPLIFIED MASONRY SKILLS

The author of SIMPLIFIED MASONRY SKILLS, Richard T. Kreh, Sr., is an instructor of masonry and building trades at Middletown High School in Middletown, Maryland. He is a member of several professional organizations, including the National Education Association, Maryland State Teachers Association, American and Maryland Vocational Associations, Masonry Instructors of Maryland, and the Iota Lambda Sigma Fraternity of Professional Educators. Mr. Kreh is the author of ADVANCED MASONRY SKILLS and SAFETY FOR MASONS. He offers over thirty-three years of experience as a mason and a teacher.

Section 1
Development and Manufacture of Brick and Concrete Block

Unit 1
Development of Clay and Shale Brick

OBJECTIVES

After studying this unit, the student will be able to

- describe the development of brick.
- discuss how brick is used in a modular building system.
- list some of the characteristics of the modern brick.

THE DEVELOPMENT OF BRICK

Brick is one of the oldest manufactured building materials. Recent excavations have uncovered remains of brick walls dating back 6000 years. Considering their age, the walls were in surprisingly good condition.

Greek historians of the fifth century BC relate accounts of the splendid wonders of the city Babylon. These include striking descriptions of immense walls and temples, many of which were built of brick. In Dashur, Egypt, two ancient pyramids of sun-dried brick still stand as monuments to the craft of bricklaying.

Highly respected in early civilizations, masons not only laid bricks but made them, as well. Kings often gave their support to masons by having a royal seal molded into bricks when they were made. This practice continues today as some brick manufacturers mold the name of their company into the bottoms of the bricks.

The recessed panel in the bottom of bricks manufactured today is called a *frog*, Figure 1-1A. It has several purposes:

- To lock the mortar into the depression for a better grip
- To create special effects when the brick is laid on its side as in decorative walls
- To save on the cost of materials

Bricks also sometimes have holes in the bottom. These holes help provide a better bond. Such bricks are called *cored bricks,* Figure 1-1B.

Fig. 1-1A Frogged brick

Fig. 1-1B Cored brick

One of the earliest types of brick and one still used in many countries today is a clay or shale sun-dried brick called *adobe brick.* These bricks contain straw for greater strength, just as steel rods reinforce modern concrete.

After the sun-dried adobe brick had been in use for some time, a discovery was made. It was found that a brick subjected to fire in a closed area such as a *kiln,* or oven, for a definite period of time became very hard and highly fire-resistant. The fired brick resisted erosion far better than unfired bricks. Some of the bricks were coated with a thick enamel or glaze. The glazes were commonly red, yellow, green, or a combination of these colors. When subjected to heat in the kiln, the color hardened and developed a glass-like finish. Some of these glazed bricks, recovered from old buildings, still retain their original color after 2000 years. Glazed bricks are made today but have limited use. This is because they are costly to manufacture.

Brickmaking and bricklaying were regarded by many of the old world craftspersons as secret processes. To keep them secret and confined to their own groups, masons banded together in organizations called *guilds.* These specialized guilds were the forerunners of modern unions.

In 1666, a great fire changed London, England from a city of wooden buildings to a city of brick construction. The manufacture of brick attained a high degree of excellence and dominated the building field in this period of history.

Early records indicate that the first bricks manufactured in the United States were made in Virginia in 1611 and in Massachusetts in 1629. The bricks were made by hand using very simple methods and tools. Many of the bricks used in construction in the early American settlements were brought from England as ballast in sailing ships. Some of these bricks can still be found in the foundations and walls of the remaining original houses in the eastern part of the United States, Figure 1-2.

The invention of the steam engine in 1760, and the subsequent Industrial Revolution, brought a change from manual labor to the use of power-driven machinery to make bricks. This change started the true development of the brick industry in America. The first brickmaking machine was patented in 1800.

THE MODERN BRICK

The term *brick* as used today denotes a solid masonry rectangular unit formed in a plastic state from clay and shale and burned in a kiln. The United States Federal Trade Commission has ruled that no product made from materials other than clay or shale can be called brick. The exception to this is if the name includes the material from which the unit is manufactured, such as cinder brick, sand lime brick, or concrete brick.

Raw Materials

Clay and shale are the principal materials used to make bricks. Usually concentrated in large deposits, these materials are found all over the world.

Clay is a natural product formed by the weathering of rocks. Shale is made in very much the same way and from the same material. However, shale is compressed into layers in the ground. Shale is very dense and is more difficult to remove from the ground than clay. As a result, shale is a more costly raw material.

Two or more kinds of clay and shale may be mixed together to obtain a material having the proper consistency and composition. Good raw material is the backbone of the brick industry. The following are several forms of clay. They have a similar chemical composition but different physical characteristics.

Surface clays are found near the surface of the earth. They may be offshoots of old deposits or the result of more recent weathering of rocks. *Shales* are clays that have been formed by natural conditions under high pressure until they resemble slate. *Fire clays* are mined from a greater depth than are the other

Fig. 1-2 Early brick row houses that have been restored to their original beauty in an urban renewal project

clays. They have fire-resistant qualities (ordinary brick is also fire resistant). They contain fewer impurities and have more uniform chemical and physical properties than shales or surface clays.

Although surface clays and fire clays differ in physical structure from shale, the three types of clay are chemically similar. All three are made of silica and alumina with varying amounts of metallic oxides and other impurities.

Metallic oxides act as fluxes and promote fusion at lower temperatures. The amount of iron, magnesium, and calcium oxides in the clays influences the color of the finished product. The material from each deposit of clay and shale has chemical characteristics which may be uniform for that deposit but may differ from the characteristics of material in other deposits. The changes in characteristics from deposit to deposit are due to differences in the relative amounts of the chemical components. As a result, brick made from the material in one deposit will have one set of characteristics for color, finish, and texture. Brick made from material in another location may look different because the chemical composition of the material varies slightly from that at the first location. In addition, all clay and shale do not react in the same manner to processing methods.

Building Brick and Face Brick

There are many different kinds of brick available today for use in construction. There are several factors to be considered when selecting brick. These factors include composition, the manufacturing method, strength, appearance, color, special effects, and economy. There are two types of bricks: the building brick and the face brick.

Building Brick. The *building* (or *common*) *brick* is formed in a plastic state from clay and shale and then burned in a kiln. Common bricks do not have to meet special standards for color, design, or texture. Because of this, common bricks usually cost less than face brick or select brick.

Common brick is sometimes known as *kiln run brick.* It is used as a filler brick or backing material on many construction jobs. This does not mean that common brick is inferior or lacks durability. Architects often specify common brick for residences. This is because interesting effects are possible due to the various colors and textures resulting from the manufacturing process. This type of brickwork is sometimes called *rustic.* It presents a roughly finished appearance and is meant to resemble old colonial brickwork.

If a brick is overburned because it is placed too close to the highest temperature areas of a kiln, it is known as a *clinker.* The clinker brick is very hard and usually dark in color. Clinker bricks are often warped or twisted as a result of the intense heat. This type of brick can also be used for special effects in construction.

Older styles of kilns did not always provide uniform heat to all of the bricks due to the design of the kiln, differences in the fuel used, and the location of the brick in the kiln. As a result, bricks fired in the same batch often had different characteristics, depending on their location in the kiln.

Because these variations appeared in each batch fired, each variety of brick received a name such as clinker, red, soft, salmon, rough hard, straight hard, and bloat. As brick kilns improved and the control of heat became more uniform in the burning area, many of the old types of common brick were no longer made. The less expensive common brick, however, is still used in masonry work today when face brick is not needed.

Face brick. The term *face brick* comes from the fact that the brick is used in the front or face side of a wall. The material used and the burning of the bricks must meet controlled specifications if the bricks are to be used as face bricks. The size of the face brick must also be within the tolerances established by the American Society for Testing and Materials (ASTM). All face bricks must meet standards for absorption, uniformity, and strength.

The color and texture must meet the specifications of the range number established for the variety of brick being made. The *range number,* an identifying number or letter, means the blend, texture, and color assigned by manufacturers to each of their products. The bricks made for a range number must conform to a sample brick produced earlier for the same coded numbers or letters. The user of the brick is then assured that bricks of the same range number will always match the original sample selected for the job. If it becomes necessary to build an addition to a structure, a match of the brick can be obtained by consulting the range number.

Brick Sizes

One of the most important recent developments in the brick industry is the range of sizes of available brick. The greater number of sizes means that brick laying can be more economical. The mason can cover more area using a brick larger than the standard size. Oversized bricks are popular in construction today since with their use, jobs progress faster and production increases.

Until recently, only three sizes of bricks were available: *Standard, Norman,* and *Roman.* Bricks now can be obtained in the following sizes:

- The thickness or bed depth may range from a nominal 3″ up to 12″.
- The height may range from a nominal 2″ to 8″.
- The length may be up to 16″. An average standard brick weighs about 4 1/2 pounds (lb).

Before starting any job, the mason should consult with the local brick manufacturer or supplier to be sure that the brick selected is available in that area.

The names for the different brick sizes are not the same throughout the industry (with the exception of the Standard, Roman, and Norman sizes). Individual manufacturers often give names to their own lines of brick sizes. To avoid confusion and the risk of getting the wrong size, it is good practice to identify the brick first by its dimensions and then by its name.

The modern brick is made for use in the *modular grid system* of building. The main reason for making a brick on a modular grid is for economy purposes. Standards for modular dimensions have been approved by the American Standards Association for all building materials. These dimensions are based upon a 4″ unit of measure called the *module.* This module is used as a basis for the grid system which must be used when two or more different materials are to be used in a construction job. Any building construction in which the size of the building materials used is based on the 4″ grid system is called *modular design.*

Most masonry materials will tie and level off together at a height of 16″ vertically (16″ is a multiple of the 4″ grid system). For example, 2 *courses* (layers) of block for a wall including the mortar joint will equal 16″ vertically. Six courses of standard brick in mortar will also equal 16″ vertically. As a result, the wall can be tied together at 16″ intervals or at multiples of 16″. Masons should learn early in their work how various building materials tie together on the job.

Most modern bricks are produced in modular sizes. Figures 1-3A and 1-3B indicate the sizes available for both modular and nonmodular bricks. In modular design, the *nominal* dimension of a masonry unit (such as a brick or a block) means the specified or manufactured dimension plus the thickness of the mortar joint to be used. That is, the brick size is designed so that when the size of the mortar joint is added to any of the brick dimensions (thickness, height, and length), the sum will equal a multiple of the 4″ grid. For example, a modular brick whose nominal length is 8″ will have a manufactured dimension of 7 1/2″ if it is

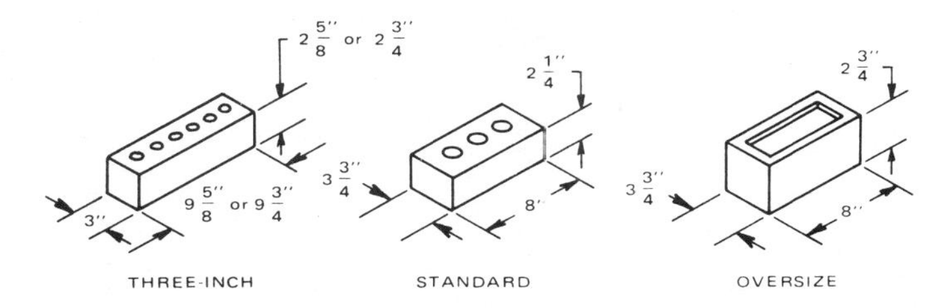

Fig. 1-3A Nonmodular brick with actual dimensions shown

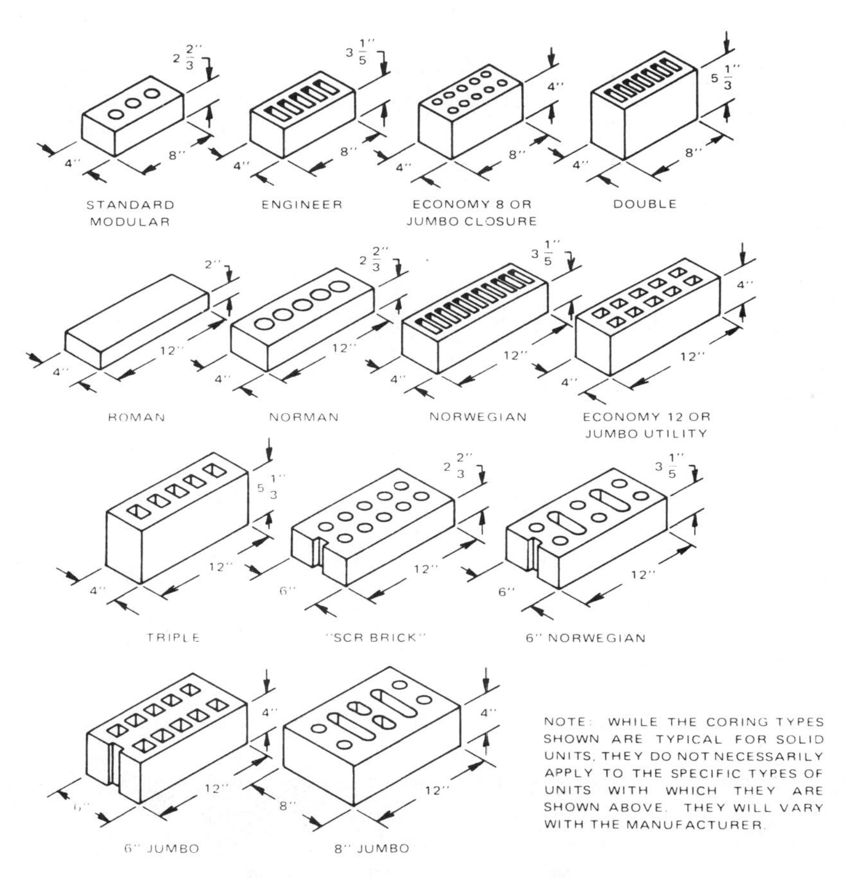

Fig. 1-3B Sizes of modular bricks available

designed to be laid with a 1/2" mortar joint. The same 8" brick will have a manufactured dimension of 7 5/8" if it is designed to be laid with a 3/8" joint.

Figure 1-4 shows nominal and actual manufactured dimensions plus the planned joint thickness for a number of basic modular brick sizes. The last column of the chart indicates the number of courses required for each type of brick to equal a 4" grid unit or a multiple of this unit. For example, for the standard modular brick, 3C = 8. In other words, 3 times the nominal height of 2 2/3" equals 8, or twice the basic 4" grid unit.

Mortar Joints

Masons form mortar joints by using the trowel to place mortar between and under the brick being laid. The two most basic joints are the *head joint* and the *bed joint,* Figure 1-5. Student masons should become familiar with these widely used terms.

SIZES OF MODULAR BRICK								
Unit Designation	Nominal Dimensions, in.			Joint Thickness, in.	Manufactured Dimensions, in.			Modular Coursing, in.
	T	H	L		T	H	L	
Standard Modular	4	2 2/3	8	3/8	3 5/8	2 1/4	7 5/8	3C = 8
				1/2	3 1/2	2 1/4	7 1/2	
Engineer	4	3 1/5	8	3/8	3 5/8	2 13/16	7 5/8	5C = 16
				1/2	3 1/2	2 11/16	7 1/2	
Economy 8 or Jumbo Closure	4	4	8	3/8	3 5/8	3 5/8	7 5/8	1C = 4
				1/2	3 1/2	3 1/2	7 1/2	
Double	4	5 1/3	8	3/8	3 5/8	4 15/16	7 5/8	3C = 16
				1/2	3 1/2	4 13/16	7 1/2	
Roman	4	2	12	3/8	3 5/8	1 5/8	11 5/8	2C = 4
				1/2	3 1/2	2 1/4	11 1/2	
Norman	4	2 2/3	12	3/8	3 5/8	2 1/4	11 5/8	3C = 8
				1/2	3 1/2	2 1/4	11 1/2	
Norwegian	4	3 1/5	12	3/8	3 5/8	2 13/16	11 5/8	5C = 16
				1/2	3 1/2	2 11/16	11 1/2	
Economy 12 or Jumbo Utility	4	4	12	3/8	3 5/8	3 5/8	11 5/8	1C = 4
				1/2	3 1/2	3 1/2	11 1/2	
Triple	4	5 1/3	12	3/8	3 5/8	4 15/16	11 5/8	3C = 16
				1/2	3 1/2	4 13/16	11 1/2	
SCR brick	6	2 2/3	12	3/8	5 5/8	2 1/4	11 5/8	3C = 8
				1/2	5 1/2	2 1/4	11 1/2	
6-in. Norwegian	6	3 1/5	12	3/8	5 5/8	2 13/16	11 5/8	5C = 16
				1/2	5 1/2	2 11/16	11 1/2	
6-in. Jumbo	6	4	12	3/8	5 5/8	3 5/8	11 5/8	1C = 4
				1/2	5 1/2	3 1/2	11 1/2	
8-in. Jumbo	8	4	12	3/8	7 5/8	3 5/8	11 5/8	1C = 4
				1/2	7 1/2	3 1/2	11 1/2	

Fig. 1-4 Table showing the thickness, height, and length of each modular brick unit

Fig. 1-5 Two basic types of mortar joints are (A) the head joint and (B) the bed joint.

ACHIEVEMENT REVIEW

Select the best answer from the choices offered to complete the statement or answer the question. List your choice by letter identification.

1. Many ancient bricks had an unusual feature called a frog. Which of the following describes a frog?
   a. A very hard red brick burned in the center of the kiln
   b. Brick with a recessed panel
   c. Brick with holes to lock the mortar in
   d. Glazed brick made for decorative work
2. One of the earliest types of brick was called
   a. a cinder brick.
   b. a salmon brick.
   c. a sand lime brick.
   d. an adobe brick.
3. Many years ago, masons formed into groups to guard their trade secrets and manufacturing processes. What term was applied to these groups?
   a. Unions
   b. Guilds
   c. Craftspeople
   d. Manufacturers
4. Bricks are
   a. made from cement and clay material.
   b. formed in a plastic state from clay and shale and then burned in a kiln.
   c. formed from a sand and lime mixture.
5. Clay is a natural product which is formed by
   a. earth and sand mixed together.
   b. weathering of rocks.
   c. decomposed vegetation and minerals.
   d. cement and sand mixture pressed into a mold.
6. Fire clays have a high resistance to heat and thereby can be used in fireplaces and smokestacks. When the raw material is taken from the ground, it is always found
   a. near the surface.
   b. on hillsides and in outcroppings of ground.
   c. deeper in the ground than other clays.
   d. on the surface.
7. Kiln run bricks are
   a. bricks that do not meet standards for uniformity, color, and texture.
   b. bricks made in special kilns to withstand heat and moisture.
   c. red bricks made for the front of buildings and face work in walls.
   d. bricks costing much more than common bricks.
8. The brick manufacturer's range number indicates
   a. the hardness of the brick.
   b. the size of the brick.
   c. the blend, texture, and color of the brick.
   d. the degree of water resistance the brick shows under controlled test conditions.
9. The modular system of measurement for the building industry is based on the module. The module unit measures
   a. 16″.
   b. 8″.
   c. 12″.
   d. 4″.
10. The principal reason for making a brick on a modular grid is for
    a. beauty and design.
    b. economy.
    c. ease in manufacturing.

Unit 2
Manufacture of Brick

OBJECTIVES

After studying this unit, the student will be able to

- list the various steps in modern brick making.
- describe briefly each of the basic brick production steps listed.
- describe various kilns used in brick making.

INTRODUCTION

Technological developments during the last century have helped to make the manufacture of brick a very efficient and productive process. More complete knowledge of the characteristics of the raw material, improved kiln designs, controlled heat in the kilns, and extensive mechanization have all played important roles in modernizing brick manufacturing.

There is such a tremendous demand for bricks that they are used faster than they can be manufactured. The modern brick plant meets the challenge of inincreasing production, while retaining a high quality for the final product, by using computerized manufacturing methods and highly skilled workers.

Basically, bricks are made by mixing a specified amount of water with finely ground clay or shale or a combination of both. The mixture is then formed into the desired shape, predried, and burned in a kiln for a predetermined time under carefully controlled conditions.

While the basic steps of brick manufacturing are standard throughout the industry, each brickmaking plant has minor variations to these steps due to local conditions. For example, one brick plant may be near the source of the raw material, while another plant may have to transport the material from a distant source. These two plants will have different ways of obtaining and stocking their raw materials. If possible, the plant should be build on the site where the raw material is obtained.

STEPS IN THE MANUFACTURING PROCESS

The following are the six major steps in the manufacture of brick:

1. Mining the raw clay or shale from the ground
2. Preparing the raw materials for use
3. Forming the raw materials into brocks
4. Predrying the bricks for burning
5. Burning the brick in the kiln under controlled heat
6. Storing and shipping

All other operations stem from these six major steps. Figure 2-1 shows, in schematic form, the basic steps in the manufacture of bricks.

Mining the Material

The removal of the raw material from the ground is called *mining.* Power equipment mines surface clay and shale in open pits. Power shovels or bulldozers work the raw material to the conveyor belts. The conveyor belts then transport the material to the brick plant's storage area where it is deposited on a pile. If the raw material is not mined near the site of the brick plant, trucks or railways bring the material to the storage piles.

Enough raw material is stored to assure plant operations for several days in the event that bad weather halts mining or shipping operations. Several storage areas are provided so that the clay and shale can be blended to yield material with a better composition.

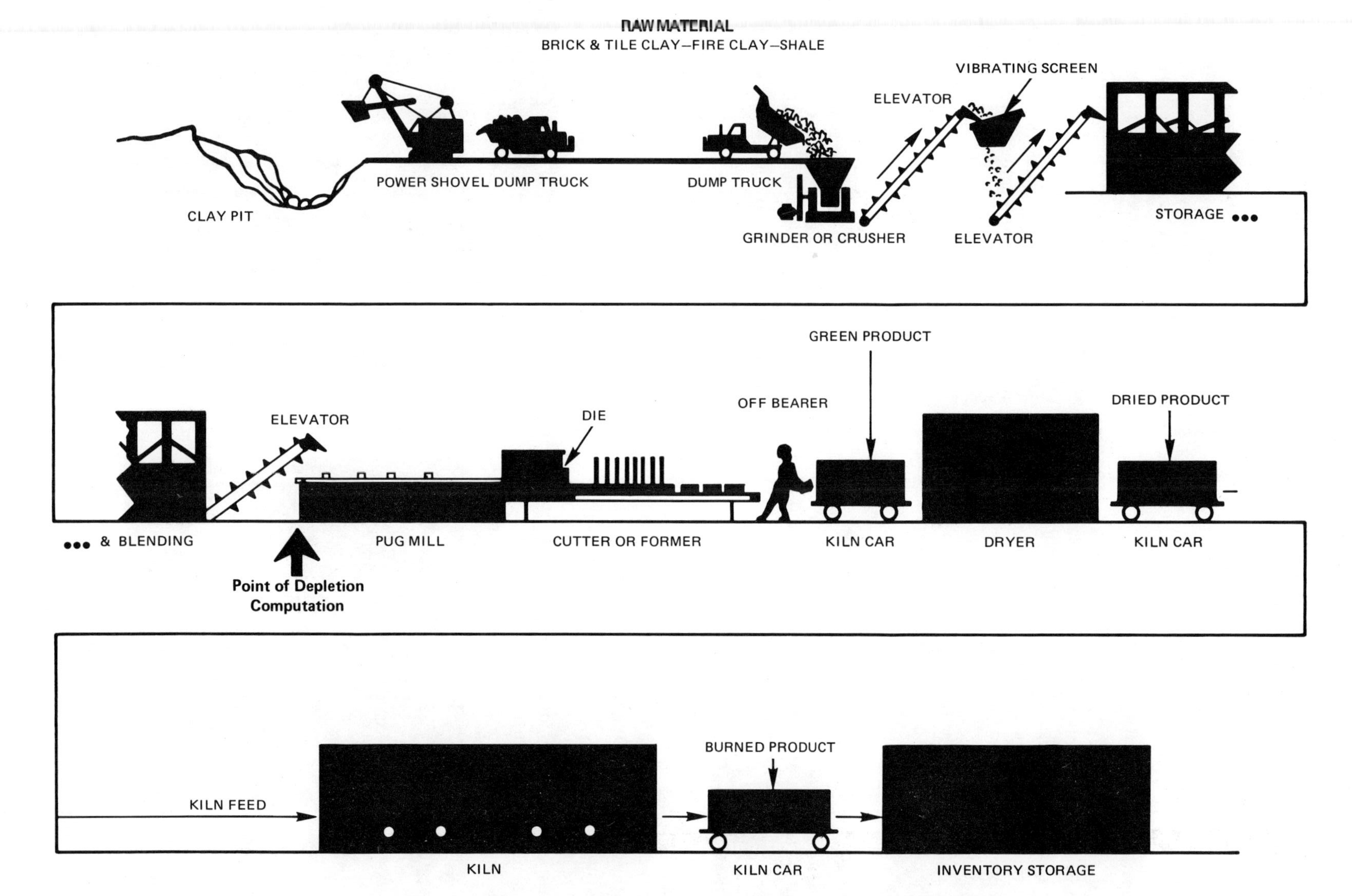

Fig. 2-1 The basic flow of materials in the manufacture of brick in a modern plant

Blending may not be necessary if the shale alone is of sufficient quality that it can be used by itself. Blending produces more uniform raw material, helps control the color of the finished product, and permits some control over providing raw material suitable for manufacturing a given type of brick unit.

Preparing the Material

If the raw material is in large lumps, it may be crushed before it is placed on the storage pile. Crushing breaks up the large pieces and removes the stones. Next, 4-ton to 8-ton grinding wheels revolving in a circular pan grind and mix the material. It then passes through an inclined, vibrating screen which controls the particle sizes. The finely ground material is taken by a conveyor belt to the site where it is formed into single bricks.

Forming the Clay or Shale into a Brick Shape

Three methods of forming are used in the production of bricks: the *stiff-mud process,* the *soft-mud process,* and the *dry-press process.*

Stiff-Mud Process. The most frequently used process at present is the stiff-mud process. It produces a hard dense brick. A greater volume of bricks can be manufactured by this method to meet the growing demands of the construction industry.

The first step in the stiff-mud process is to add water to the raw material to make a plastic, workable mass suitable for molding. This is called *tempering.* The mixing is done in a machine called a *pug mill.* The pug mill has a mixing chamber which contains one or two revolving shafts which thoroughly mix the raw material and a measured amount of water.

After thorough mixing, the tempered clay goes through a machine which removes air bubbles, giving the clay increased workability and plasticity. In addition, a brick made from clay without air bubbles will have greater strength.

The clay is next forced through an opening called a *die,* a process much like toothpaste being forced from a tube. The long, formed ribbon of brick being extruded through the die is called the *column,* Figure 2-2. Texture may be applied to the face of the brick as it leaves the die, if so desired. As the column moves away from the die, it is cut into lengths which are either the height (*side cut*) or length (*end cut*) of the brick. The cutting is done automatically by a large, circular wire cutter which cuts each brick to the same size, Figure 2-3.

Upon leaving the wire-cutting machine, the bricks move to a conveyor for inspection. Bricks passing inspection are placed on the dryer cars. Imperfect bricks are returned to the pug mill for reprocessing. The soft bricks that have not yet been burned hard in the kiln are called *green bricks.*

Soft-Mud Process. This is the oldest way of making brick and was used before brickmaking machines were developed. Automated machinery is now used in this process, Figure 2-4.

Fig. 2-2 A pug mill showing the column of brick after being forced through the die

Fig. 2-3 Bricks are automatically cut to size by a wire cutter.

Fig. 2-4 A soft-mud machine in operation. The soft-mud mixture is placed in wooden molds, formed, removed from the molds, and placed in the kiln. Note that the workers are wearing hardhats to comply with safety rules concerning work near machinery.

The soft-mud process is suited for clays which contain too much natural water for the stiff-mud process. The clay is mixed with twice as much water as in the stiff-mud process and is pressed into wooden molds. The molds are lubricated with sand or water so the clay does not stick to the mold. When sand is used to lubricate the molds, the bricks are *sand-struck* and have a sandy finish. When water is used, the bricks are *water-struck* and have a very smooth finish.

Dry-Press Process. This method is particularly adaptable for clay of low plasticity. In this process, the clay is mixed with a small amount of water and then forced into steel molds under very high pressure. This method is seldom used today due to the high cost and low demand for pressed brick.

Predrying Brick Before Burning in the Kiln

To prevent cracking, excess moisture must be removed from the bricks before they are burned in the kiln. Years ago, bricks were allowed to dry in the open air before they were placed in the kiln. The modern method is to predry the bricks in the forward section of the kiln using the waste heat from the hot section of the kiln. The heat and humidity must be regulated carefully to prevent sudden changes in the temperature. Sudden changes in temperature will cause excessive cracking and deformation of the brick.

Fig. 2-5 Green brick entering the drying chamber. Note the heavy steel doors that close the drying chamber off. The brick can also be seen on the car that will take them on their journey through the kiln.

The drying time in the kiln is greatly reduced from that of the open-air process. In other words, with the brick being dried in the kiln, the weather is not a factor affecting the moisture content of the brick. The bricks are placed in the drying chamber on special, rolling steel cars. They are left there for a predetermined drying time before they move into the hot section of the kiln, Figure 2-5. The temperature in the drying area is 100 to 400 degrees Fahrenheit (°F).

Burning the Brick in the Brick Kiln

The burning process is one of the most specialized steps in the manufacture of brick. Kilns have changed drastically over the years. Since brickmaking emerged as an industry, three kilns have come into use: the *scove kiln,* the *beehive kiln,* and the *tunnel kiln.*

The Scove Kiln. The scove kiln, Figure 2-6, was one of the first type to be used to burn bricks. Scove kilns are now obsolete. The unburned bricks were stacked in piles inside the kiln with air spaces left between the individual bricks to permit the heat to reach all surfaces. The kiln was then plastered with mortar to lock in the heat during the burning process. Openings were left in the bottom walls of the kiln where hardwood fires provided heat for the kiln (in later years, gas and oil were used). This method was not very efficient. The bricks nearest the fires were burned hard, while those near the top of the kiln were soft and could be used only for filler walls or interior construction.

Fig. 2-6 A scove kiln. This type of kiln was one of the first used to burn bricks. Note the openings at the base of the kiln where fires were burned to provide the heat for burning the bricks.

The Beehive Kiln. An improvement upon the scove kiln was the beehive kiln, Figure 2-7. The beehive kiln is a round brick structure wrapped with steel bands which control the expansion caused by the heat in the kiln. The bricks to be burned are stacked in the kiln with narrow spaces between them so the heat will pass completely around them. The kiln is sealed by walling the doors shut with brick and mortar. The heat may be applied either from the bottom or the top. When these kilns first came into use, wood and coke were the main sources of heat. Gradually, these fuels were replaced by gas and oil because they could be controlled and regulated more efficiently. The beehive kiln requires at least 1 week to burn the brick. As a result, this type of kiln yields a limited amount of bricks.

The Tunnel Kiln. The most modern kiln in use today is the tunnel kiln which is built of bricks and lined on the inside with firebricks, Figure 2-8. The bricks to be burned are stacked on flat cars which move very slowly through the long, narrow kiln. The cars move from the predrying section of the kiln into the burning section.

The average time for the unit to pass completely through the kiln is about 36 hours. This is controlled by computers. The number of bricks on a car is rather small compared to the number of bricks stacked in the older types of kilns. The cars, however, pass through the long tunnel kiln continuously and the bricks all receive the same heat treatment, resulting in a more uniform product.

The heat in the kiln is supplied by gas or oil. The heat gradually increases as the bricks pass from the kiln inlet to the zone in the center of the kiln. This is where the greatest temperature is reached, an average of 1950°F. As the bricks move from the center to the outlet, the temperature drops and the bricks cool slowly. Cooling the bricks slowly eliminates cracking, pitting, and other problems which are due to a rapid reduction of temperature, Figure 2-9.

The rate of temperature change in the kiln depends on the raw materials being used. Tunnel kilns are equipped with recording instruments which provide a constant check on the temperatures in the kiln.

While the bricks are still in the kiln, they can be given a treatment called flashing. *Flashing* means that the amount of oxygen used in the burning is reduced. The flame changes from red to blue. Flashing lowers the temperature of the kiln which, in turn, causes the

Fig. 2-7 A beehive kiln. Kilns of this type are still in limited use. Each kiln is wound with steel bands to control expansion. The piping for each kiln supplies the gas or oil fuel used to burn the bricks.

Fig. 2-8 A modern tunnel kiln. The large ductwork on the top of the kiln is used to recycle the hot gases to the preheating chamber of the kiln.

Fig. 2-9 Burned brick at the end of their journey through the kiln. The intense heat of the kiln can be seen in the background.

Fig. 2-10 Bricks coming through the banding machine. This machine places bands around the bricks to form cubes of approximately 500 bricks.

bricks to take on varying shades of color. The type of clay or shale in the brick determines how the brick reacts to flashing.

Storing and Shipping

The process of removing the bricks from the kiln is called *drawing*. The bricks leave the kiln on cars which are placed in a holding area until the bricks are cool. When the bricks are cool enough, they are banded with metal strips to hold them together for shipment and handling. The banded cubes of brick contain approximately 500 standard bricks. The number may vary slightly according to the variation in brick sizes, Figure 2-10.

After the bricks are banded, they are moved to a storage yard and are ready for delivery to a job. During the off-seasons of the year (cold weather), the stock piles in the storage yards are built up to prepare for the construction season ahead, Figure 2-11.

It is true that modern masons no longer manufacture the brick they use as their forerunners did many years ago. However, an understanding of how bricks are manufactured and the materials of which they are made will help masons make better use of the material. The modern brick is a superior product due to careful quality control during manufacturing. Much of the guesswork and trial-and-error methods of the past have been replaced by scientific processes and computerized manufacturing.

Fig. 2-11 Brick cubes being stored for future use. The banding holds the bricks together and simplifies handling.

ACHIEVEMENT REVIEW

The column on the left contains a statement associated with brick making. The column on the right lists terms. Select the correct term from the right-hand list and match it with the proper statement on the left.

1. Coloring brick by reducing the amount of oxygen in kiln
2. Forcing the clay through a die into the shape of a brick
3. Forming the brick in a wooden mold
4. A long brick structure used to burn brick using the most modern controls
5. Clay mixed with only a small amount of water and formed under high pressure in steel molds
6. Round brick kiln surrounded by steel bands
7. The mass of clay being extruded through a die
8. The taking or obtaining of the raw materials from the ground
9. Product made from finely ground shale or clay and burned
10. The process of removing the bricks from the kiln, allowing them to cool off, and then unloading them from the cars
11. The most commonly used process for making bricks
12. A machine which mixes raw material with water to form a workable mass
13. Trade name given the soft condition of brick before burning in the brick kiln
14. Preparatory treatment of the green brick to remove excess moisture before burning
15. Early type of brick kiln which has openings at the bottom of the kiln in which to build the fires; the most inefficient of the 3 kilns discussed in the unit

a. Forming
b. Drawing
c. End cut
d. Tunnel kiln
e. Column
f. Sand-struck
g. Flashing
h. Brick
i. Scove kiln
j. Dry-press process
k. Soft-mud process
l. Mining
m. Beehive kiln
n. Stiff-mud process
o. Humidity
p. Green brick
q. Predrying
r. Pug mill

Unit 3
Properties and Characteristics of Brick

OBJECTIVES

After studying this unit, the student will be able to

- describe how raw materials affect the properties of brick.
- discuss what brick texture is and how it can be used to best advantage.
- explain why absorption is such an important factor in masonry construction.

PROPERTIES OF BRICKS

All properties of structural clay products such as brick are affected by the composition of the raw materials used and the manufacturing processes. The important properties of brick are the following:

- Color
- Texture
- Size variation
- Absorption
- Compressive strength
- Durability

Most burned brick is fireproof regardless of the other properties mentioned in this unit. Due to the technical nature of brick properties, they will be discussed only generally.

Color

The chemical composition of the natural clay and the minerals which may be added to the natural clay determine the color of the finished bricks. Another important factor which affects color is the temperature at which the clay is burned and how well the temperature is controlled in the kiln.

Of all of the natural oxides found in clays, ferrous oxide (iron oxide) has the greatest effect on the finished color of the brick. Regardless of its natural color, clay that has iron in it will burn red due to the formation of ferrous oxide. When burned in a reducing atmosphere (given the flashing treatment), the same clay will take on a purple cast or color.

For bricks made with the same raw materials and methods of manufacture, the darker colors are associated with burning at a higher temperature, with low absorption, and with increased compressive strength. However, for products made from different raw clays, there is no direct relationship between color and either absorption or strength.

Clay with particles of iron mixed through it produces a brick containing black specks in the finish. Iron-spotted bricks may range in color from light gray to mahogany. One popular type is called a *salt and pepper* brick. There may be some differences in the bricks from different burnings. Therefore, it is advisable to buy enough bricks for a job at one time. In this way, the mason has a sufficient quantity of bricks in the proper color to complete the job. Any excess bricks can be used for repairs or additional construction.

Architects and builders can add to the architectural beauty of modern brick buildings by choosing from the range of colors in which bricks are available.

Texture

Texture in bricks is the arrangement of the particles of raw materials in the brick and the appearance and finish of the brick. For example, a hard, smooth finish has a fine finish or texture. Brick that has a sand finish is said to have a coarse texture. These finishes are the result of the type of process used to make the bricks.

In the stiff-mud process, many textures can be obtained. These textures are obtained by using attachments which cut, scratch, roll, brush, or otherwise roughen the clay column as it leaves the die. Figure 3-1 shows several different brick textures commonly available.

Fig. 3-1 Examples of available brick textures

Bark brick, matt face brick, and rug face brick are all bricks that have a rough face of irregular design and texture. The rough face bricks can be found in almost any color or design to suit any job. These bricks are popular for fireplaces and buildings in which the builder wants to show a bold course design in the finished job. Old bricks are often used to obtain this special effect. However, the quality of the old bricks is important since they may deteriorate rapidly when used on an outside wall.

The treatment of mortar joints is a definite form of texture and affects the final appearance of the wall. The same kind of brick can be used to build a wall, but different ways of finishing the mortar joints will create various textures in the wall.

A wall using recessed or projecting bricks achieves a different texture due to the resulting shadow and highlighting effects. The use of colored mortar gives the wall added texture and warmth. The texture of the wall should always be considered so it blends in with the architectural design of the building and its surroundings.

The treatment of mortar joints, commonly called *striking* or *tooling,* creates a definite texture and greatly affects the final appearance of the wall. Figure 3-2 shows a colonial brick laid on an offsetting angle with a rustic tooled joint.

Fig. 3-2 The angled corners and the rustic tooled joints create a textured effect in this brick tower.

Size Variation

Clays shrink during the drying and burning processes. Therefore, allowances for shrinkage must be made in the die sizes and the length of the cut brick. Since air and fire shrinkage vary for different clays, the burning process requires careful control based on testing and experience. The main problem is not so much the total shrinkage of a brick as the variations in the shrinkage of the mass of the clay.

Fire shrinkage increases as the temperature of the burning area increases. Since darker colors are obtained at a higher burning temperature, the mason could expect some differences between the sizes of dark and light bricks. To assure uniform size in their bricks, manufacturers work very hard to control the factors which cause shrinkage. Because of the differences in raw materials and temperature variations in the kilns, absolute uniformity is impossible. However, the *specifications* (written details or a description) for brick types indicate the desired size as a range of permissible sizes to allow economical manufacture.

Size variations may also occur if the green brick is jostled or bumped too hard. This treatment will not change the size so much that it will affect the shape of the brick. That is, the brick may sag or warp. If this happens before the brick is burned, it is possible to reshape it before processing. Otherwise, the brick can only be sold as a second for unexposed work. This situation seldom occurs in modern brick plants, as most of the brick is handled by machinery and not manually.

Absorption

Absorption is the weight or amount of water a masonry unit absorbs at certain conditions for a stated length of time. This weight is expressed as a percentage of the weight of the dry unit.

The rate of absorption of a brick (suction rate), or the amount of water a partially immersed brick can absorb in 1 minute, is important to masons. Masons must know how much time the brick takes to set when it is laid in the wall. *Set* is the length of time required for the mortar to change from a soft mass to a hard mass that will sustain the weight of the masonry placed on it. The action of mortar hardening is known as *setting*. Bricks which contain too much water require more time to set. As a result, the productivity of the mason decreases. Bricks that are too dry absorb the moisture from the mortar quickly, thereby preventing the mortar from curing properly. Hard bricks should be covered on the job so that they do not get too wet. High-suction bricks, on the other hand, must be wet before they are laid in the wall to obtain good masonry construction.

In wetting high-suction brick, sprinkling is not enough. A hose stream should be played on the brick pile until water runs from all sides. The surface of the bricks should then be allowed to dry before the bricks are laid in the wall. Too much water on the surface of the brick causes it to float on the mortar. If this happens, the bricks are difficult to lay, as absorption between the brick and mortar does not occur.

The following is a good field test which can be used to measure the absorption rate of bricks.

1. Draw a circle 1" in diameter on the surface of the brick which will contact the mortar. Use a quarter as the circle pattern. Use a wax pencil or crayon to draw the circle.
2. With a medicine dropper, place 20 drops of water inside the circle. Note the time required for the water to be absorbed. If the time exceeds 1 1/2 minutes, the unit need not be wet. If the brick absorbs the water in less than 1 1/2 minutes, wetting is needed. Figure 3-3 shows this second step in the absorption test.

The water content of bricks must be correct to obtain the best results from the combining of brick and mortar to form a wall. The water content is a factor too often ignored in masonry construction with the result that the strength and durability of the wall are affected.

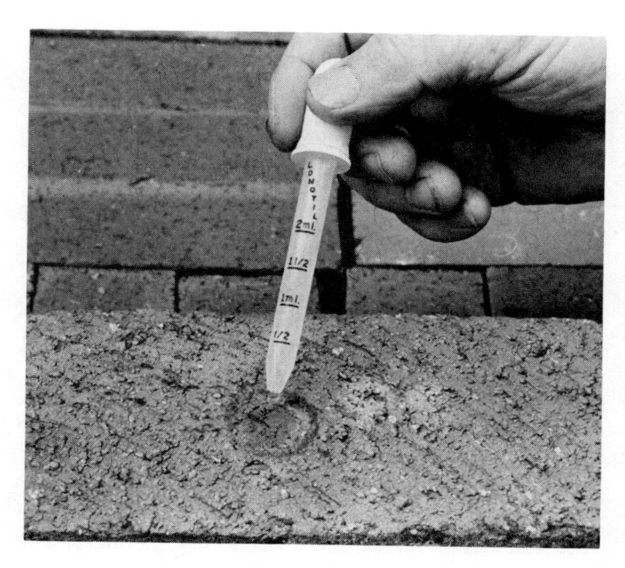

Fig. 3-3 Absorption test to determine if the brick requires wetting before being laid in the wall

Compressive Strength

Strength is defined as the resistance a brick has to increasing loads or stress placed on it before it breaks. Each brick used in construction must first be tested to determine its strength. The testing procedure involves building a test section of a wall. The wall is carefully constructed to assure that it is perfectly true and that all the joints are solid. Following 28 days of curing, the wall is placed in a large press which can apply up to 1 million pounds of pressure to the material, Figure 3-4. The point at which the wall fractures or breaks is recorded on a gauge in the laboratory. The fracture point is expressed in pounds per square inch (psi). Manufacturers submit their products to the Brick Institute of America (BIA) for strength tests. These tests are generally accepted throughout the construction industry.

Both the compressive strength and the absorption of a brick are affected by the properties of the clay, the methods of manufacturing, and the burning process. Although there are exceptions, bricks produced by the stiff-mud process have higher compressive strength ratings and lower absorption rates than units produced by the soft-mud or dry-press methods. These are additional reasons why the stiff-mud process is used more than any other method of manufacturing.

Durability

In general, the characteristics of bricks reflect the way the bricks are manufactured and the type of clay or shale used to make the bricks. Generally, the harder or denser a brick is, the longer lasting and more waterproof it is. Bricks used in construction must endure heat, cold, wetting, drying, and chemical and corrosive actions. Of all of the materials used in construction, brick is probably the most tested by time.

Since kiln-burned bricks are fireproof, walls constructed of brick are given fire-resistant ratings. That is, the wall is rated according to the amount of time it will withstand heat before failing. The ASTM has established tests for determining the safe ratings for masonry walls, Figure 3-4. The fire rating of a wall is usually less than the actual ultimate fire resistance of the wall. As a result, there is a safety margin within which the builders can work. Many buildings are consumed by fire and only the brick walls are left standing. The buildings have been rebuilt to building code requirements without major repairs to the brickwork. This illustrates the fire resistance of bricks as a building material.

Fig. 3-4 A sample brick wall being tested for strength. This test will determine what load in pounds per square inch the wall can withstand.

POINTS TO REMEMBER

- Bricks should meet ASTM requirements for size, color, texture, and durability.
- The weight of bricks varies, but a standard brick weighs about 4 1/2 lb.
- The surface of the brick which contacts the mortar should be rough enough to ensure proper bonding with the mortar.
- Texture, absorption, and compressive strength are all factors contributing to the durability of brick.
- The many properties and the unusual characteristics of brick make it an outstanding choice for lasting construction.

ACHIEVEMENT REVIEW

Write the correct answer for each of the following questions.

1. What causes brick to burn red? The iron becomes ferrous oxide
2. What is the name given to the brick which has iron spots showing as black specks in the face finish?
3. What does texture mean as applied to a brick?
4. Fire shrinkage increases with the hotter temperatures which are necessary to achieve what shade of colors?
5. What does the term *green brick* mean?
6. What process yields bricks which have a higher compressive strength and lower absorption rate than other bricks?
7. What two important factors are affected by laying brick when it is either too wet or too dry?
8. What is the name given to the resistance shown by a brick to increasing loads before it breaks?
9. What does the abbreviation *psi* mean?
10. What does the abbreviation *ASTM* stand for?

Unit 4
Development of Concrete Block

OBJECTIVES

After studying this unit, the student will be able to

- describe the early methods of concrete block manufacturing.
- explain the use of lightweight aggregates used in block.
- list some of the uses of concrete block in construction.

EARLY DEVELOPMENT OF CONCRETE BLOCK

In the early nineteenth century, two American masons named Foster and Van Derburgh sought a method of making a precast building block which would be larger than the brick in use at that time. After a great deal of experimenting, the two masons made a discovery. They combined powdered quick lime and moist sand and placed the mixture in a mold under pressure. The masons found that through this process, a usable building block could be made, provided it was hardened or cured with steam. The natural mechanical heat generated by the pressure formed a silicate of lime. This silicate of lime cemented the block together.

Foster and Van Derburgh realized, however, that their solid block was very heavy and difficult for masons to handle. This was because the block had no hand or finger holes. This block was patented in the early nineteenth century. However, it never enjoyed great popularity.

The next major advance in concrete block making occurred in England in 1850. An English mason named Joseph Gibbs decided to make a hollow block which would be considerably lighter and easier to handle than Foster and Van Derburgh's block. Gibbs also cast his block in a mold, but the mold had dividers, so that the finished block had hollow *cells.* Gibbs intended masons to lay these hollow blocks in the wall and then fill the hollow cells with cement to strengthen the wall. Although masons didn't always follow that procedure, Gibbs' block remains the forerunner of the hollow concrete block we know today.

BLOCK-MAKING MACHINES

The first American block-making machine for commercial use was patented by Harold S. Palmer in 1900. The machine made hollow block. Later machines used the same method of hand-tamping the mixture of cement and aggregate into a mold, Figure 4-1. Using a machine of this type, two workers could make about 80 blocks a day. This hand-tamping process was used to make block from 1904 to 1914. The next 10 years, however, saw a change to power-tamping machines. In power-tamping machines, bars or rods were lifted by mechanical action and pulled down by gravity. The bars compressed the mix into the mold. This operation was controlled by a machine operator. By 1924, the automatic-tamping machine could produce 3000 blocks a day. Block could now be produced in great enough volume to be competitive with other masonry materials. This marked the real beginning of the modern block industry.

The automatic-tamping machine was replaced by the vibration machine around 1938. The vibration machine produced a block with more uniform texture, sharper corners, and denser material composition. Production increased tremendously due to the vibration machine. Today, 15,000 blocks can be made on a single machine in a 10-hour day. New developments and research change this figure each year as methods and machinery are improved.

Since the first hand-tamping machine was used, automation had changed the block industry from a backyard business to the largest masonry products business in the country.

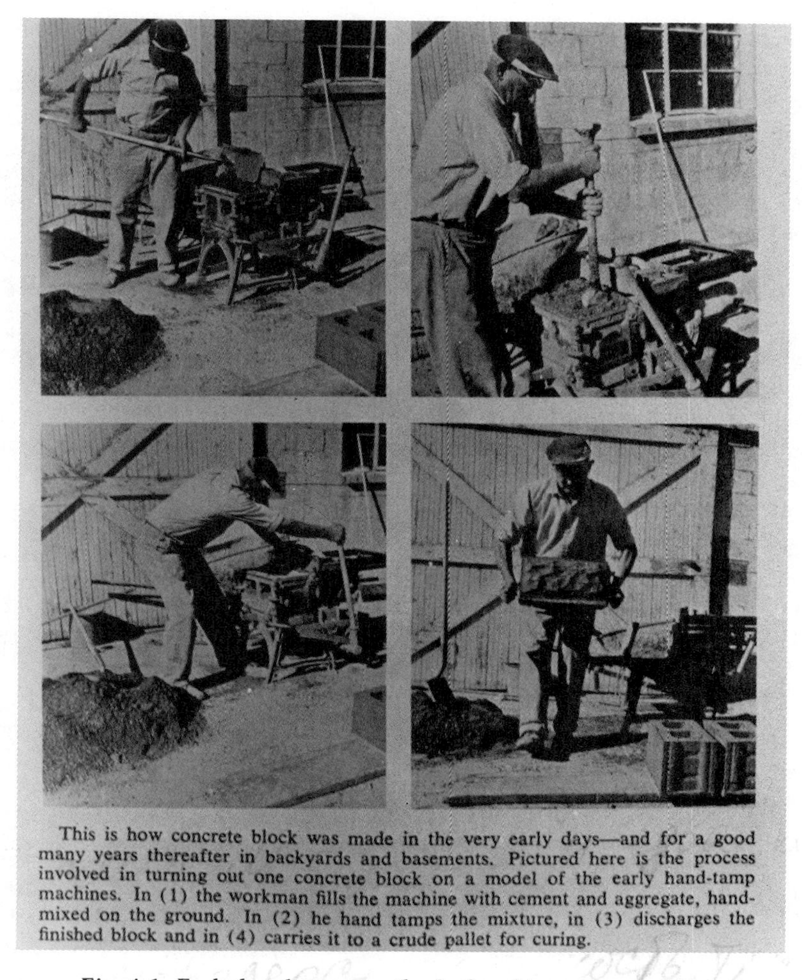

This is how concrete block was made in the very early days—and for a good many years thereafter in backyards and basements. Pictured here is the process involved in turning out one concrete block on a model of the early hand-tamp machines. In (1) the workman fills the machine with cement and aggregate, hand-mixed on the ground. In (2) he hand tamps the mixture, in (3) discharges the finished block and in (4) carries it to a crude pallet for curing.

Fig. 4-1 Early hand-tamp method of making concrete block

LIGHTWEIGHT AGGREGATES

Aggregates are sand and pea gravel or some inorganic substitute particles which are mixed with portland cement to form concrete block. About 2000 years ago Romans used *pumice,* a light, porous volcanic rock. Pumice is still used by some manufacturers to make lightweight concrete block. A problem in the use of pumice is its high absorption rate. This high rate makes it difficult to control the amount of water in the mix.

In 1913, a Pennsylvania bricklayer, Francis J. Straub, invented a system of making block out of waste cinders. The product, known as *cinder block,* was lightweight and was able to hold nails. Cinder block was sold for about the same price as competing products. Since the block was lighter in weight than block made with sand and gravel, masons could lay them with less physical effort, thereby enabling a faster work rate. The savings that it offered made it very popular as a building material. Cinder block was a big factor in the development and growing use of concrete block in the east, since lightweight pumice was only available in the western United States.

Other lightweight aggregates now in use include expanded fly ash, expanded shale, expanded clay, expanded slag, and expanded slate. Different companies give brand names to their own products. The trend in the block industry is to produce more lightweight concrete blocks as the demand increases.

CHARACTERISTICS AND PROPERTIES OF CONCRETE BLOCK

Durability

Durability is the ability of a product to withstand long, hard use. Concrete block is a very durable building material. Block made over 50 years ago is still in good condition. Block made today is even better than that made years ago due to better materials and improved methods of manufacturing.

Strength

Strength is another important characteristic of block. Not all block has the same degree of strength. Concrete block for heavy loads may be either solid or semisolid, depending on where it is used. Block is used in partition walls in a building where the structural frame is concrete or steel. This block is usually of low compressive strength and supports only its own weight. It is meant to provide privacy and fire protection at an economical cost. Generally, but not necessarily, heavier block has greater strength than lightweight block. Block made from stone dust is relatively heavy and is generally used for foundations. Building codes and federal and state regulations refer to ASTM standards for strength standards. Concrete block must meet ASTM standards before it may be used as a building material.

Fire Resistance

Fire resistance is another important factor to consider in the selection of concrete block for a building material. Tests have shown that all concrete block is highly effective in resisting fire and heat. In the test, heat is applied for a predetermined amount of time. The number of hours that the concrete block resists damage and the passage of heat is the fire-resistance rating. Because of its fire resistance, block is used to enclose steel columns and other less resistant materials in construction jobs.

Sound Resistance and Absorption

The sound resistance of concrete block is measured in terms of sound transmission class. *Sound transmission class* is the measure of the loss of sound between two points. Generally, concrete block resists the passage of sound from one room to another very effectively. A heavy, dense block effectively blocks more sound than a lightweight block. The ability of a block to reduce sound which starts inside the room is known as *sound absorption.* Generally, an open-textured block absorbs sound more efficiently than a smooth wall. On some construction jobs, the hollow cores of the block are filled with a sound-absorbing material such as sand, plaster, Styrofoam, or an aggregate. In addition, the cores often contain insulation to reduce energy costs for heating and cooling the building.

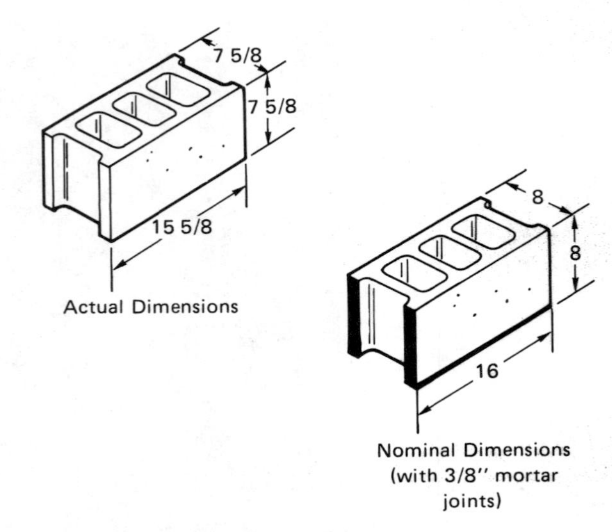

Fig. 4-2 Actual and nominal dimensions of 8-inch block

BLOCK SIZES

Like bricks, concrete blocks are generally made using the modular system. In most modern building construction, blocks are solid or hollow, rectangular shapes. Block sizes are expressed as nominal sizes. This means each dimension includes the mortar joint, which is 3/8". For example, an 8" x 8" x 16" block has an actual measurement of 7 5/8" (height) x 7 5/8" (width) x 15 5/8" (length). The addition of the 3/8" mortar joints gives the block a nominal dimension of 8" x 8" x 16". Figure 4-2 shows actual and nominal dimensions of an 8" x 8" x 16" block. A *solid block* is defined by the ASTM as a unit in which the core (or hollow) area is no more than 25% of the total cross-sectional area. A hollow block has a core area of at least 25% of its total cross-sectional area. Concrete block may contain 2 or 3 hollow cells, depending on the job specifications determined by the architect.

Concrete blocks are produced in many different lengths and widths. The proper dimensions depend upon the requirements of the particular job. In addition, manufacturers produce concrete blocks for special uses, such as for solar screens or for decoration, that vary from normal dimensions. Figure 4-3 shows the various sizes of common concrete masonry units.

BLOCK SHAPES

Block units are made in a variety of patterns and shapes to suit modern construction. In some cases, the name of a shape describes its function in construction such as sash block, pilaster units, control joint block, etc. In other cases, the design itself determines the name, such as bull-nose unit, solid-top block, concrete brick, etc. While terminology has not been formally standardized, the shape names given here are fairly well established by common usage – and some variations are peculiar to certain localities. In most cases, terminology is self-explanatory. Shape design can be more readily understood by referring to Figure 4-4. The drawings are those of units generally available. It is impossible to show all shapes made by all manufacturers. This is because some units are dropped and new ones added. Most block manufacturers have up-to-date catalogs showing the shapes and sizes they stock.

Full Size Units		Supplementary	
Height	Length	Heights	Length
2 2/3	8	4	6, 4
3	8	2	6, 4
4	12		10, 8, 6, 4
5 1/3	12	2 2/3, 4	10, 8, 6, 4
6	12	2, 4	10, 8, 6, 4
8	12	4	10, 8, 6, 4
8	16	4	12, 8, 4

NOTE: Supplementary sizes listed are the sizes required for complete 4-inch flexibility. Such supplementary sizes as are required for a particular job may be cut on the job or furnished by the manufacturer. Maximum economy in construction is achieved when walls are planned and designed on the basis of the 4-inch module.

Fig. 4-3 Concrete masonry nominal dimensions

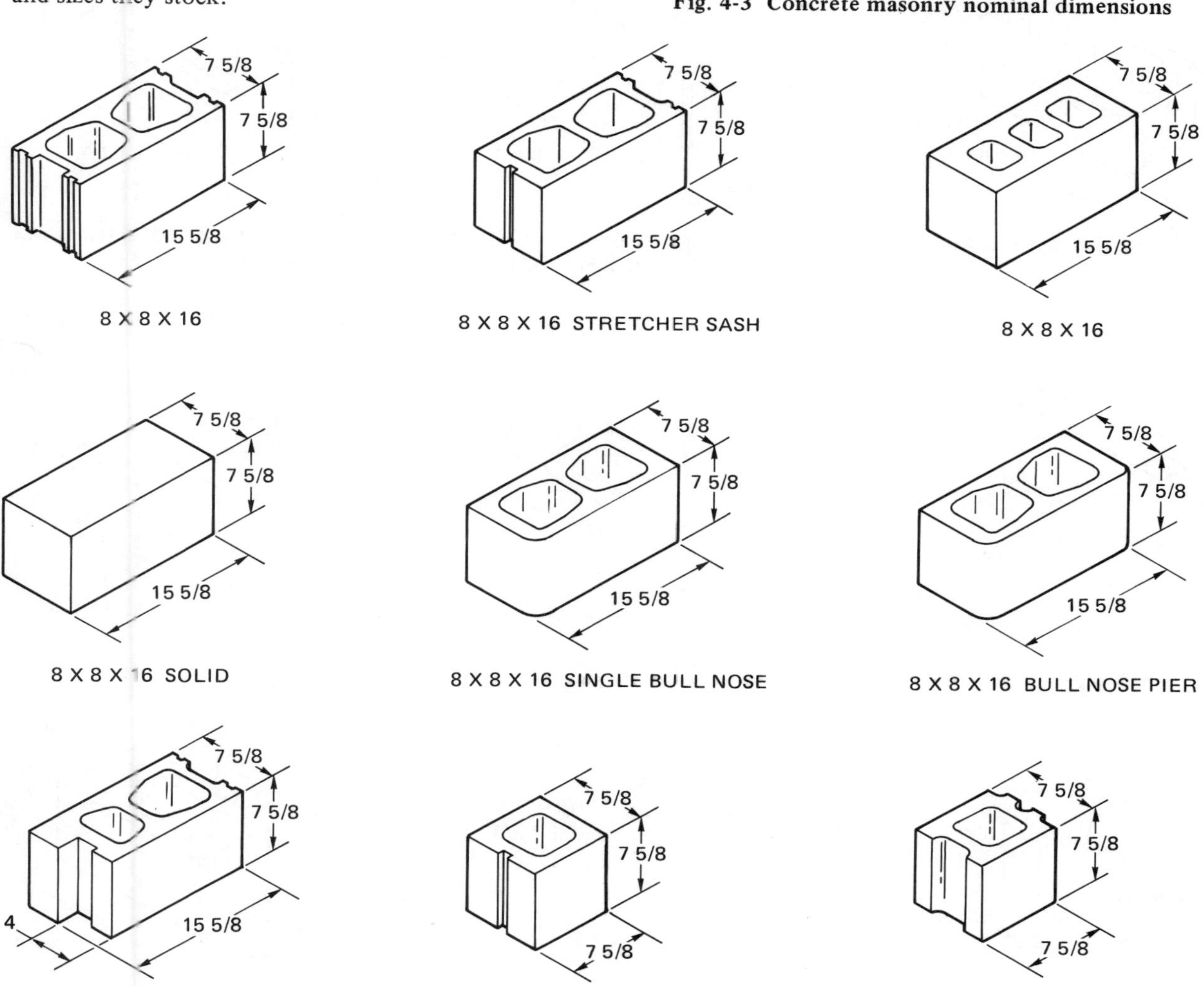

Fig. 4-4 Typical concrete block shapes (continued)

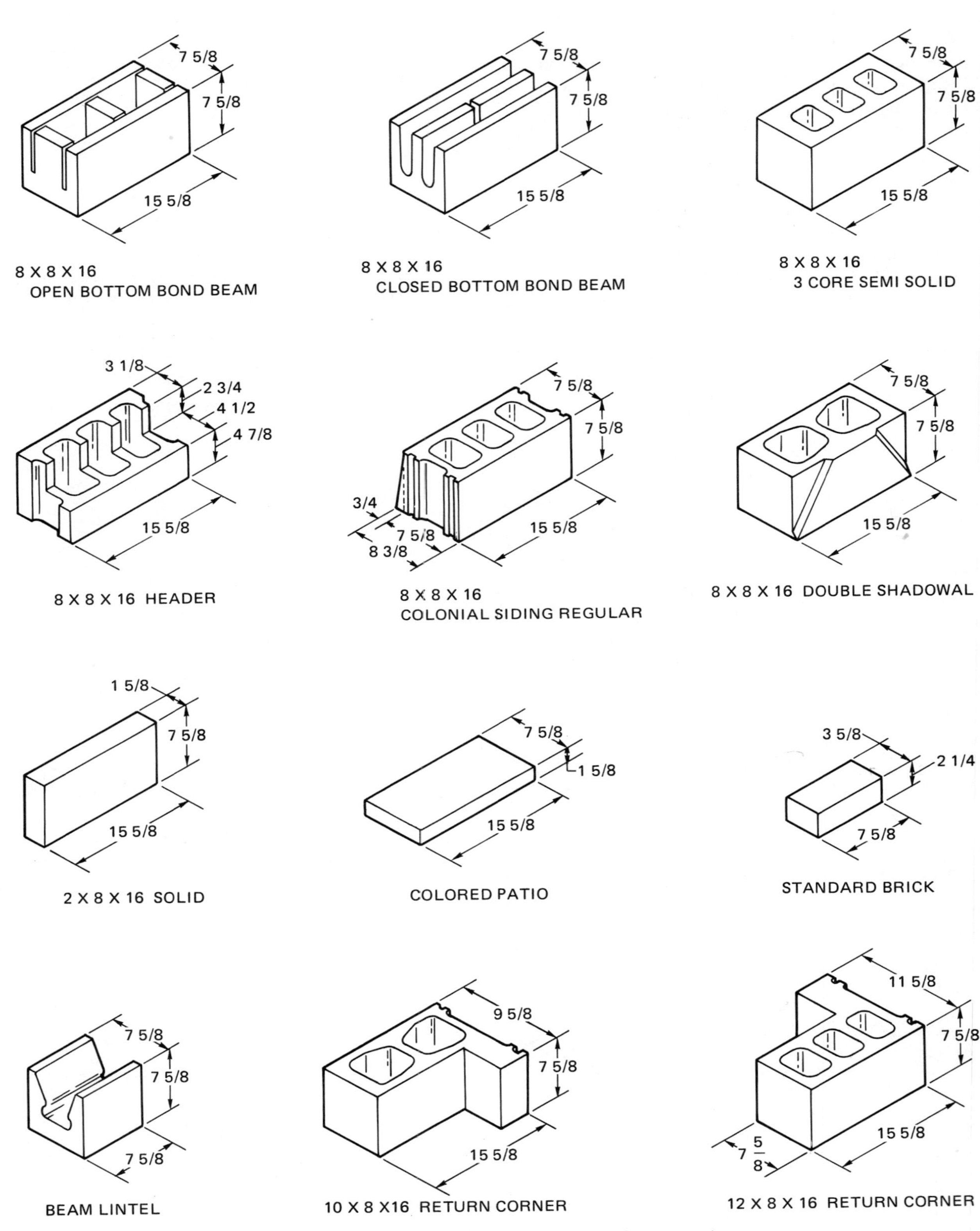

Fig. 4-4 Typical concrete block shapes (continued)

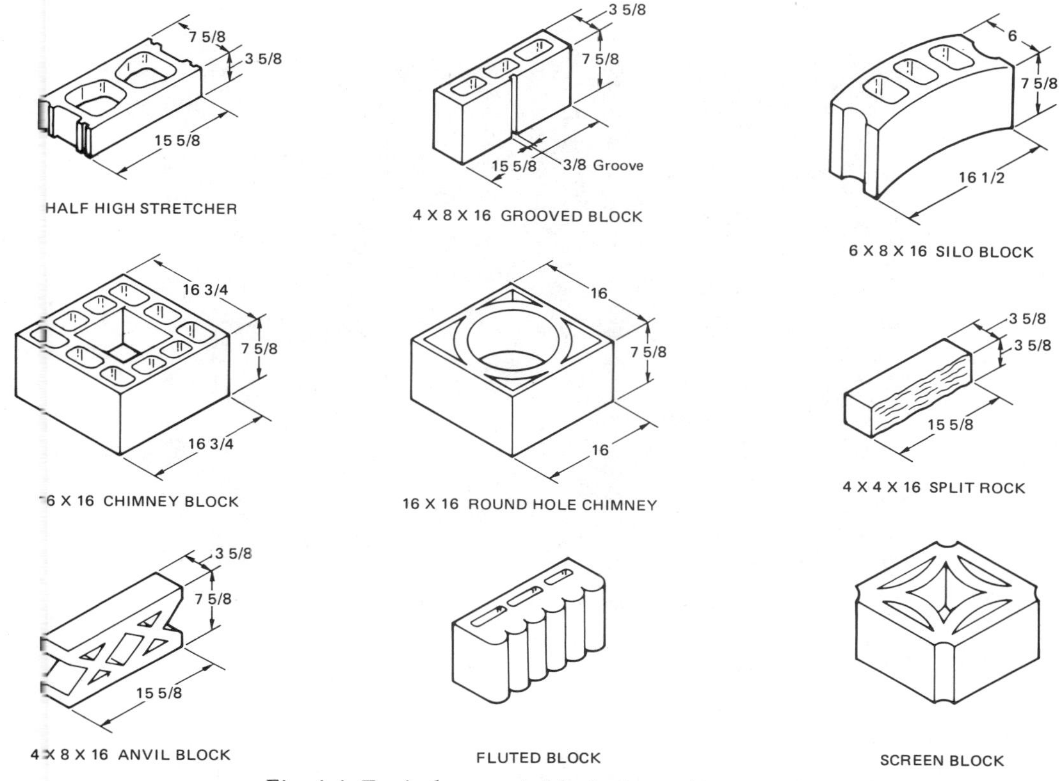

Fig. 4-4 Typical concrete block shapes (continued)

USES OF CONCRETE BLOCK

Concrete blocks are widely used in the basic construction of many buildings because of their low cost and durability. They also provide a variety in design and color. The mason can arrange the blocks to create certain effects. For example, Figure 4-5 shows how recessed and projecting blocks create a shadow effect. Another popular concrete masonry unit used increasingly is *Splitrock. Splitrock,* also called *splitblock* in some areas, is a solid block with a white or gray stone aggregate which gives the appearance of stone, Figure 4-6.

Fluted block, which has projecting vertical ribs, creates another textured effect, Figure 4-7. Over 700 different sizes, shapes, and types of block are available today. The demand for new products to fill new requirements in construction will push this figure even higher.

Fig. 4-5 Concrete block wall showing how recessing and projecting of block creates a shadow effect

Fig. 4-6 Splitrock used for decorative purposes

Fig. 4-7 Close-up of 4-inch fluted block

ACHIEVEMENT REVIEW

Select the best answer to each question from the choices offered. List your choice by letter identification.

1. The solid block is difficult for the mason to work with because
   a. the block is too long.
   b. the block is very fragile.
   ✓c. it does not contain hand or finger holes.
   d. it is too big to handle.

2. Cinder block, an important factor in the development of concrete block, was a popular building material because it was
   a. lightweight.
   b. available in more colors than other materials.
   c. insulated.
   d. much bigger than other materials.

3. Production increase in the concrete block industry since 1900 is mainly due to
   a. the automatic-tamping and vibration machines.
   b. improved cement for the mix.
   c. better management.
   d. highly trained workers.

4. Pumice, used in early times to make masonry units, is still used today in modern block making. One problem with pumice is
   a. it is too dry to use economically in the mix.
   b. it is very heavy.
   c. the amount of water in the mixture is difficult to control.
   d. it is too coarse to pass through a machine

5. In 1913, a Pennsylvania bricklayer named Francis J. Straub devised a new system of making concrete block. The block was called
   a. slag block.
   b. cement block.
   c. cinder block.
   d. pumice block.

6. In masonry, the term *durability* is used in connection with concrete block. To which of the following does the term *durability* refer?
   a. Color
   b. Type of aggregate used
   c. Weight of the block
   d. Ability to withstand long, hard use

7. The term *sound transmission class* is used when discussing building with concrete block. It refers to the ability of concrete block to
   a. reduce the noise in a room.
   b. block the sound from other rooms.
   c. measure the amount of noise in a room.
   d. increase the sound in an area.

8. The concrete masonry business has converted to standard sizes in manufacturing to avoid confusion among builders and architects. This system is called
   a. standardizing.
   b. modular.
   c. nominal.
   d. systematic.

9. To define a hollow block, the ASTM requires that the core area comprise a certain percentage of the block. The percentage required is
   a. 50%.
   b. 75%.
   c. 25%.
   d. 15%.

10. The development of concrete block has brought on the design and construction of many buildings which would have previously been built with wooden frames. The main reason for this is
    a. availability of the material.
    b. improved color arrangements.
    c. durability and economy.
    d. shapes and textures.

Unit 5
Manufacture of Concrete Block

OBJECTIVES

After studying this unit, the student will be able to

- describe the steps in concrete block manufacturing.
- explain how concrete blocks are cured with steam.
- describe safety practices in relation to the manufacturing of concrete block.

The change from the hand method of forming concrete block to the modern, fully-automated block plant is a fairly recent development. Concrete block has become an economical building material due mainly to power-driven machinery and automatic control.

BLOCK PRODUCTION

Block plants may differ in size and types of equipment and methods used, but the basic operation is the same throughout the industry. The operation includes the following:

1. Obtaining and storing the raw materials
2. Batching and mixing the materials
3. Molding the blocks
4. Curing the blocks
5. Cubing and storing the blocks
6. Delivering the units to the job

Raw Materials

The raw materials used to manufacture block are delivered to the plant by truck or railroad cars. Enough material is stockpiled to ensure continuous operation of the plant. A conveyor belt lifts the raw material to storage tanks over the batching and mixing area so that the material feeds to the machines by the flow of gravity. The various aggregates are stored separately to permit proper mixing, Figure 5-1. Portland cement is generally delivered in special cars or trucks and forced through hoses or pipes to overhead storage bins by compressed air.

Batching and Mixing

Materials for concrete mix are measured by weight on a device known as a *weight batcher.* The batcher is located under the storage bin. The proper amount of material is discharged from the bin into the batcher. The materials are then released into the mixer, Figure 5-2, where a controlled amount of water is fed into the mix by an electronic water meter. Each material

Fig. 5-1 Aggregates stored on the roof of the plant directly over the batching and mixing area

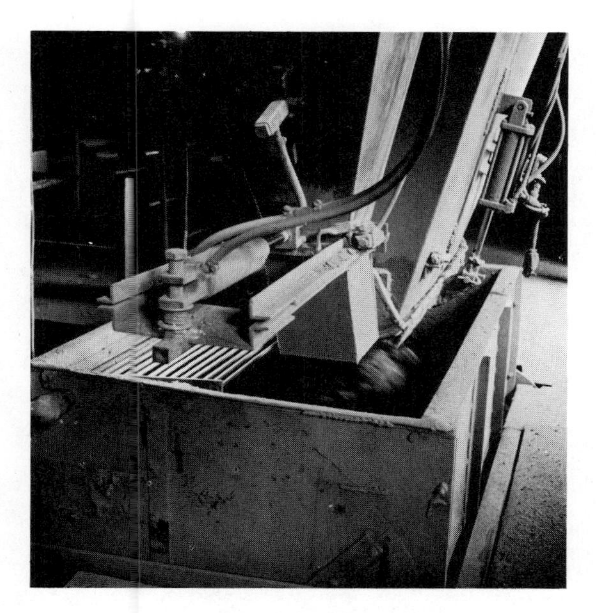

Fig. 5-2 The mixer, which combines the ingredients for concrete block, is located below the batcher.

Fig. 5-3 Mechanical bucket dumping the mixed materials in the block-making machine. The giant hopper is located on the top of the block machine.

that goes into the mix is also electronically regulated so that each batch is exactly the same. To mix a batch, 6 to 8 minutes are required.

After the batch is mixed, it is lifted and dumped above the block machine by a cable-driven winch. The required amount of mix is fed into the block machine for the molding of the block, Figure 5-3.

Molding

The concrete mix is packed into the molds of the machine by vibration and pressure. The blocks emerge from the machine on a belt or rollers. As the block leaves the machine, a brush rolls across the top, removing loose aggregate particles which the block may have collected, Figure 5-4.

Curing

After the concrete blocks leave the block machine, they are placed on curing racks. *Curing* is the process in which concrete block is permanently hardened. When the curing rack is filled with pallets of green block, it is moved to the kiln by forklift tractors, Figure 5-5.

There are two main curing methods used in the production of concrete block, the *low-pressure* or *atmospheric-pressure steam process* and the *high-pressure steam process.* About 80% of all blocks are cured by the low-pressure steam process. Low-

Fig. 5-4 Concrete block emerging from the block machine. Three blocks are produced every ten seconds.

pressure steam curing takes place in steam-curing kilns. The kiln is usually built of concrete blocks and is equipped with sliding metal doors at one end which close off the heated area as the blocks are cured. The maximum curing temperature in this process ranges from 150 to 190°F. During the first hour, the temperature in the kiln remains at 72°. This is known as *presetting* the block. During this time, the cement is *hydrating* (combining with water) and beginning to harden. The heat is then gradually increased to about

Fig. 5-5 Uncured concrete block being placed in the kiln's steam curing chamber

Fig. 5-6 Concrete block leaving the automatic cubing machine

175° and then allowed to remain at this temperature for the rest of the process.

The heat in the kilns is controlled by recording instruments. Blocks require approximately 5 hours in the kiln. In this way, a block with normal compressive strength of 1000 psi can be produced. The total time required for low-pressure steam curing is approximately 24 hours.

High-pressure steam curing is done in *autoclaves* (special metal chambers designed for high-pressure steam) at a temperature of approximately 360°F. The block cured by the high-pressure process can be used on the job within 12 hours after molding.

Cubing and Storing

After the concrete blocks are cured, they are removed from the kiln and taken to the cubing machine. The blocks are assembled in cubes usually comprising 6 layers. Each layer contains 15 to 16 blocks which measure 8″ x 8″. Approximately the same volume should be achieved when using other sizes of units.

When the cubes leave the cubing machine, they are moved to a storage yard where they are stacked 3 to 4 cubes high, Figure 5-6. Depending on the curing method used, they may be left in the storage area from a few days to a few weeks before they are ready for delivery to a job.

Delivery to the Job

Special trucks deliver the blocks to the job, Figure 5-7. The trucks are equipped with a hydraulic-operated fork which is controlled by the driver. The boom of

Fig. 5-7 Block being placed in foundation by truck equipped with hydraulic lifting arm

the forklift has a drop reach of about 12′ and is able to swing in any direction. The truck driver can usually place the block in necessary areas around a building site. This is a labor-saving practice compared to the old method of unloading every block from the truck by hand.

Block Making and Safety

It is the responsibility of the manufacturer to operate a safe place for employees to work. Most workers should wear adequate clothing and steel-toed shoes to minimize the danger of accidents caused by falling material or machinery. Hard hats, required by federal and state law, protect employees from head injuries. Grates on mixers and equipment with guards also help prevent accidents on the job. Most plants employ a safety engineer who supervises inspections of manufacturing operations. Workers should take pride in meeting the safety requirements of the concrete block industry. Many plants post the dates of accident-free workdays for the public to examine.

ACHIEVEMENT REVIEW

The following questions refer to the six major steps in manufacturing concrete block. Write the correct answer to each.

1. How do raw materials feed into the batcher from the storage area on the roof of the plant?
2. The materials for concrete block mix are measured by what device?
3. By what two principles is concrete mix packed into the machine molds?
4. What are the two main methods by which concrete block is cured?
5. Which of the two methods of curing is used most of the time in the concrete manufacturing business?
6. What is the advantage in having a forklift device mounted on trucks delivering concrete blocks?

SUMMARY, SECTION 1

- The biggest change in the manufacture of bricks is the burning process – from the older method of baking in the sun to burning in a kiln under a controlled temperature.
- Clay and shale are the principal materials in fired bricks.
- Modern modular bricks are constructed so that they may be used in the modular grid system of building (based on the 4″ module).
- The entrance of computers and highly trained workers into the masonry industry has resulted in a great increase in the manufacture of bricks.
- The method most often used to prepare bricks for burning in the kiln is known as the stiff-mud process.
- The tunnel kiln has replaced most of the older types of kilns, as it is highly efficient and therefore allows mass production of brick units.
- The properties of bricks are directly affected by the raw materials contained and the manufacturing process to which the bricks are subjected.
- Bricks can be given texture during the manufacturing process by special machine processes and, as they are laid in the wall, by treatment of mortar joints.
- The burning of brick units in the kiln will cause some differences in size to occur, but all bricks must conform to measurements estabished by the masonry industry and government.

- The way in which bricks are laid in mortar is critical. Properly done, it results in a strong bond and efficient work.
- Outstanding characteristics and properties of bricks as a building product include color, durability, fireproofing quality and flexibility.
- The increased production of concrete block is credited primarily to the invention and use of the vibration machine.
- Concrete masonry units are made according to the modular grid system.
- The success of the concrete masonry unit is due largely to the fact that the larger units allow the structure to be built more quickly and economically.
- Concrete block is made primarily of portland cement, aggregates, and certain additives.
- The development of lightweight aggregates with no strength loss has greatly speeded the process of laying units.

SUMMARY ACHIEVEMENT REVIEW, SECTION 1

Complete each of the following statements referring to material found in Section 1.

1. Many of the bricks made in ancient times have a recessed panel in which seals and inscriptions were imprinted. This recess is called a ______________ .
2. The basic raw materials used in the manufacture of bricks are ______________ .
3. The three major methods of preparing bricks for burning in the kiln are the ______________, ______________, and ______________.
4. The most often used of the three methods mentioned above is the ____________.
5. Reducing the amount of oxygen in the kiln to cause coloring of the bricks is called ______________.
6. The color of the finished bricks is due directly to a change in the ______________ of natural clay and minerals.
7. Size variation in brick is primarily due to ______________.
8. The production of concrete block increased tremendously, and a block with a more uniform texture was achieved by the invention and use of the ______________.
9. Francis J. Straub, a Pennsylvania mason, invented a block in 1913 that was lightweight and could receive nails without breaking. This invention, which greatly accelerated the concrete block industry, was called ______________.
10. The size of concrete block, including the mortar joints, is sometimes expressed as 8" x 8" x 16". This is known as the ______________ size of the unit.
11. During the first hour of curing concrete block in a steam kiln, the temperature remains at 72°F. This process is known as ______________ .
12. The process of stacking concrete masonry units on wooden pallets, each containing the same number of blocks, in preparation for delivery to the job is known as ______________.

Section 2
Tools and Equipment

Unit 6
Basic Tools of the Trade

OBJECTIVES

After studying this unit, the student will be able to

- describe the basic tools used in masonry.
- select good-quality tools.
- list safety measures to be practiced while handling tools.

The quality of the work masons do depends to a great extent on the condition and quality of their tools. Poorly made, dirty tools may be a factor in poor work. With this in mind, masons should select their tools with care.

The standard set of tools used by masons is fairly small compared to that of some other trades. A modest sum will buy a good-quality set of standard tools. Masons should purchase tools which are made by a well-known company when possible.

TROWELS

The brick trowel, used to cut, spread and handle mortar, is the most important tool of the mason. The ability to use the trowel correctly and efficiently demands time and practice. The two major styles of trowels made for brick masons are the London pattern and the Philadelphia pattern. Trowels may vary in width and length, but they still are considered to be one of these two patterns.

Trowels are made from highly tempered select steel. There are 6 parts of a trowel, Figure 6-1. Handles may be wood, plastic, or leather, but wood is preferred. Notice that the wooden handle in the illustration has a metal band at the point where it attaches to the shank. The band, known as a *ferrule,* prevents the handle from splitting. The handle tapers where it attaches to the shank to hold it securely in place.

The *London pattern* has either a wide or narrow heel, called a *diamond* heel. Many masons prefer the narrow London pattern for laying brick and the wide London pattern for laying concrete block or handling

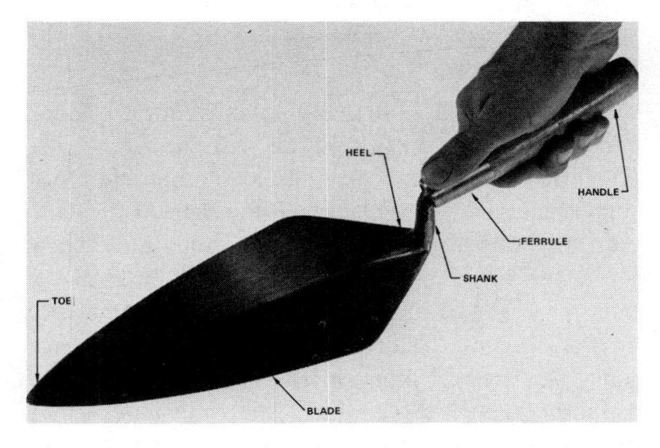

Fig. 6-1 Trowel in the narrow London pattern. On a well-made trowel, the weld between the blade and the shank is hardly visible. The diamond-shaped heel is a feature of a trowel in the London pattern.

mortar for stonework. However, either pattern can be used depending on the preference of the individual bricklayer. The *Philadelphia pattern* has a square-shaped heel and will hold more mortar than the London patterns. Figure 6-2 shows the two different patterns of trowels discussed here.

The *pointing trowel* is a miniature model of the standard mason's trowel. It must be of the same quality and constructed of the same materials to withstand hard use. It is used in places where a large trowel cannot fit or be easily maneuvered.

Selecting a Trowel

A good trowel should possess certain characteristics. Only high-quality materials should be used in the construction of this important tool. If the steel used is of good quality, the trowel will ring when tapped against a hard object. The best-quality steel will make the longest ring. Flexibility of the blade is very important. The bending of the blade during the process of spreading mortar and cutting brick reduces tension on the mason's wrist.

A good trowel should be lightweight and well balanced to be handled easily. The angle at which the handle is set in the trowel, known as the *set* of the trowel, is very important. If the set is too low or too high, the mason's wrist will be strained. The set of the trowel also aids in keeping the mason's hand out of the mortar. The length of a trowel depends on the individual's preference and how much weight the mason can comfortably lift with a trowel. The standard length is 11" to 11 1/2". If possible, masons should handle the trowel in the store before purchasing it to be certain that it is the proper one for them.

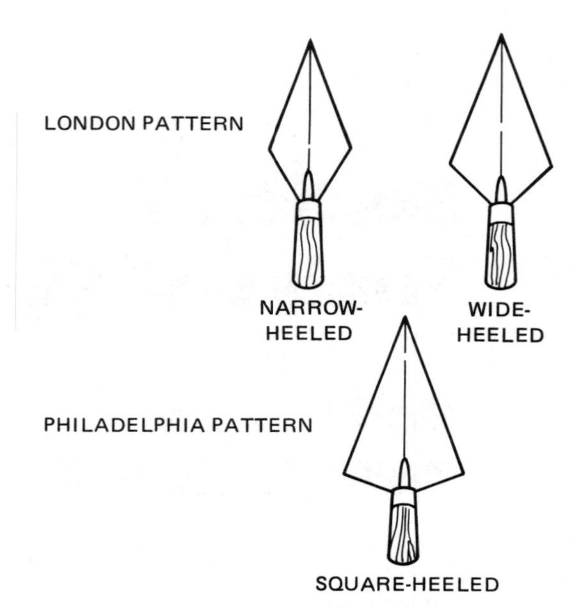

Fig. 6-2 Three types of brick trowels

HAMMERS

Hammers used in bricklaying and concrete block construction are usually of the same design, Figure 6-3. However, hammers used to cut concrete block sometimes have wider blades. The weight of the hammer may vary according to the mason's preference. Since the hammer is used to cut hard materials, it must be made of high-quality steel. The square end on the head of the hammer is used for cutting masonry materials, striking chisels, hammering nails, and tooling certain mortar joints. The other end of the hammer is wedge shaped. It forms a chisel peen for cutting and dressing masonry materials. The handle may be constructed of wood, fiberglass, or steel. Most masons prefer the wooden handle.

Fig. 6-3 Standard mason's hammer

> **Caution:** If the wooden handle of a hammer becomes loose, replace the worn handle immediately. The hammer should be sharpened by a blacksmith since the temper may be lost if it is ground on a wheel.

When selecting a hammer, masons should choose the style and weight which suits them best. A good hammer of a medium weight (about 18 oz to 24 oz) is suitable for most masonry work.

The Tile Hammer

The tile hammer, Figure 6-4, resembles the brick hammer except that it is much smaller and lighter (approximately 9 oz). It is used to cut tile or to make

Fig. 6-4 Tile hammer

a thin cut on bricks or block. The mason should not drive nails, strike chisels, or do other heavy work with a tile hammer.

THE PLUMB RULE OR LEVEL

The plumb rule or level is used to establish a *plumb* line (aligned vertically with the surface of the earth) and *level* line (aligned horizontally with the surface of the earth), Figure 6-5. Plumb rules (known in the trade as *plumbrule)* may be constructed of seasoned hardwood, various metals, or a combination of both. They are made as lightweight as possible without sacrificing strength, since the plumb rule must be able to withstand fairly rough treatment.

Plumb rules have a shape similar to a ruler and are equipped with vials enclosed in glass, Figure 6-6. Inside each vial is a bubble of air suspended in either alcohol or oil. When a bubble is located exactly between two marks, indicated on the vial, the object is either level or plumb. Whether an object is level or plumb depends on which position the mason is using the level. Alcohol is a more accurate substance to be used in a level. This is because heat and cold do not affect alcohol as much as they affect oil. (The term *spirit level* indicates that alcohol is used in the vial.) The vial is usually imbedded in plaster or plastic so that it remains secure and true.

Short levels are made for jobs where the longer plumb rule will not fit. The most popular of these are 24″ and 18″ long.

Plumb rules may be equipped with single vials or double vials. Rules with double vials are preferred since they can be used either horizonally or vertically, regardless of which way the mason has picked up the tool.

Selecting a Plumb Rule

Plumb rules are tested in the factory, but many times are broken during shipping. Therefore, levels should always be checked before purchasing them.

Fig. 6-5 Plumbing a corner with a plumb rule

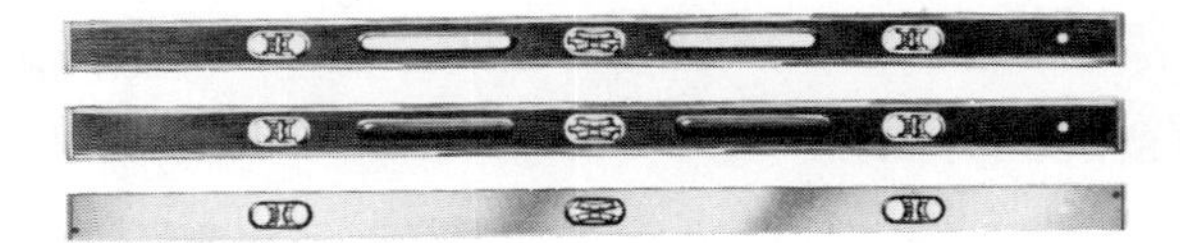

Fig. 6-6 Plumb rules

Hold the plumb rule against another in different positions to see if the bubbles match. Reverse one rule and check the bubbles again. The plumb rule is the most expensive of the basic tools and should be treated as such. Handling the plumb rule with care will give many years of service.

FOLDING POCKET MASON RULES

Masons use two types of folding rules to evenly space and divide courses of masonry units, including the mortar joints. These are the *course counter spacing rule* and the *modular spacing rule,* Figure 6-7.

The course counter spacing rule, Figure 6-7A, is used to lay out and space standard brick courses to heights that are not modular. For example, the rule could be used to lay out courses to the top of a door frame that is 1″ higher than modular spacing.

The modular rule, Figure 6-7B, is based on the module of 4″. Six different scales represent the different sizes of modular masonry units.

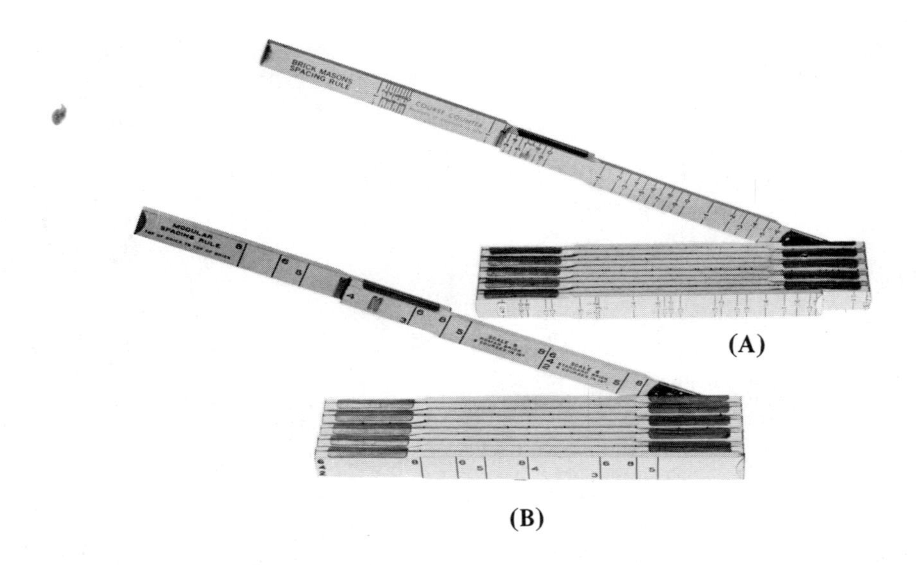

Fig. 6-7 The two rules often used by masons are the (A) course counter spacing rule and the (B) modular spacing rule. Notice the different scales on the two rules.

Unit 25 explains how the mason reads and uses both these rules. Masons should purchase both types of rules, since situations will arise when each rule is needed.

Selecting Rules

When buying a folding rule, purchase one which is heavy duty since it will last longer. Brass joints between the sections of the rule make the best joints. A brass tip should be located on each end of the rule to keep it from wearing. Most masons prefer wooden rules. Oil the joints of the rule occasionally with a light oil or paste wax and wipe off the excess. If it is not oiled, the rule may break due to dirt and mortar in the joints.

> **Caution:** Always open the rule with your fingers in the middle of the rule. Avoid opening with the hand on the end, which may pinch your fingers.

THE BRICK SET OR BLOCKING CHISEL

In everyday masonry work, bricks and blocks are cut with a hammer. However, when a very clean, sharp edge is required, a *brick set chisel* (also known as *blocking chisel* or *bolster chisel*) is used, Figure 6-8A.

The width of the blade is usually the same as the width of a standard-sized brick. Because of this, the mason is able to make a straight cut with one blow of the hammer on the chisel. Chisel weights vary to meet the needs of individual masons.

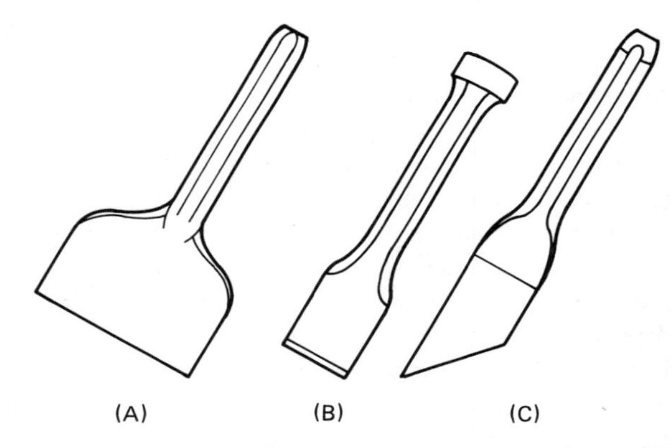

Fig. 6-8 Three chisels: (A) brick set chisel, (B) mason chisel, and (C) plugging or joint chisel

The blade of the brick set chisel has a beveled edge on one side and a straightedge on the other. When cutting with the brick set chisel, be sure that the straight edge of the blade faces the finished cut. The purpose of the beveled edge is to cut off protruding parts of the brick on an angle, thereby allowing the head joints to fit neatly together.

The cutting edge of the blocking chisel must be perfectly straight or the bricks may be chipped and ruined.

> **Caution:** The steel striking end of the brick set will become burred after constant use. Grind off the burred metal to avoid the hazard of flying steel.

STANDARD MASON'S CHISEL

The standard mason's cutting chisel, Figure 6-8B, is designed for cutting block, brick, and veined stones. It has a narrow, strong, tempered blade which gives a neat, clean cut. This chisel is often used to cut out masonry work and make repairs. The enlarged head of the chisel provides more striking area for the hammer.

PLUGGING OR JOINT CHISEL

The *plugging* or *joint chisel,* Figure 6-8C, has a tapered blade for cleaning mortar joints or for removing a brick or block from a wall.

LINES

A *mason's line* is used to lay out walls and masonry materials over 4′ in length since plumb rules are only 4′ long and therefore, cannot serve as a guide. The line is used as a guide for building straight, true walls.

For many years, cotton and linen line was used to construct masonry walls. However, the mason's line must be able to stretch without breaking, and cotton line has little stretch. Cotton line also deteriorates quickly if it is wet or exposed to weather for any period of time.

Most of these problems have been solved with the development of nylon line. Nylon line is long lasting and is not affected by moisture. It also resists the effects of mortar very well and tolerates a great deal of strain without breaking. Twisted nylon withstands well over 100 lb of pressure, while braided line may test as high as 170 lb. Braided line is slightly higher in cost than twisted line. However, braided line is smaller in diameter than twisted line and therefore, does not tend to sag as much when pulled tight. Pulling the line tight without breaking it may at first be difficult for the mason and will require practice.

FASTENING THE LINE

Pins and nails, line blocks, and *trigs* are several fasteners that attach the line to the wall, Figure 6-9.

Pins and Nails

Most masons use a line pin and nails to fasten line to the structure. The *line pin* is a steel pin about 4″ in length which tapers to a point. The better pins are made of tempered steel. First, the end of the nail is driven into the wall and the line wrapped around it. The line pin is driven into the wall on the opposite end at the top of the course of the material being installed. The line is pulled tight and wrapped around the pin.

The Line Block

Another method of fastening the line to the corner for use as a guide is with line blocks. *Line blocks* are L-shaped blocks made of wood, plastic, or metal which have a slit in the center of the block. Wooden blocks are preferred because they grip the corner securely. The line is drawn tight through the slit in the block and held by tension against the finished corner. The block on the other end of the line is fastened in the same manner.

The advantage in using the line block is that the block can be moved up the corner for each succeeding course without measuring or tightening the line again. Also, line blocks do not leave holes in the wall.

> **Caution:** Since the line block projects slightly from the corner of the wall, exercise extra care when working with it. Dislodging the block and line can cause severe injury to anyone in its path. Line pins are considered safer than line blocks since they are driven more securely into the wall.

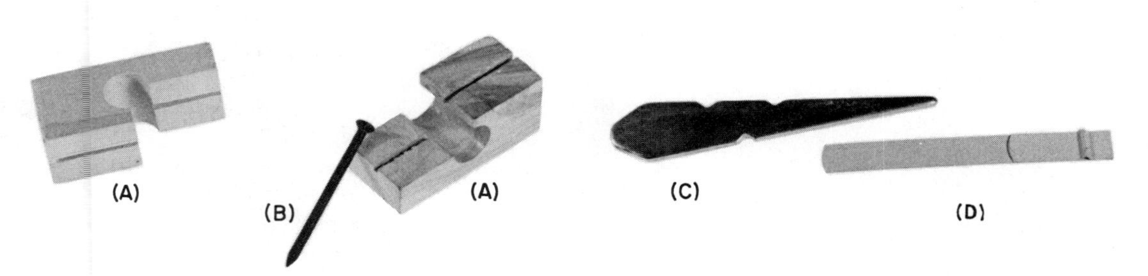

Fig. 6-9 Various fasteners used on line by the mason include (A) wooden blocks, (B) the tempered nail, (C) the steel pin, and (D) the metal trig.

The Trig

Even though nylon line may be pulled tight on a long wall, the line will sag due to the weight of the line itself. To overcome this problem, a metal fastener known as a *trig* (or *twig*) holds the line in position. A trig can also be made from a small piece of line looped around the main line and fastened to the top of the brick. It may be necessary to use several trigs to control the line on a long wall.

There are other ways of fastening and controlling line, but the basic methods are pins and nails, blocks, and trigs. Normally, these fasteners are not purchased by the mason. Suppliers of masonry materials are usually glad to provide them free of charge.

JOINTERS

A *jointer* (also known as a *striking tool* or *striking iron*) is used to finish or tool vertical or horizontal mortar joints. Tooling is necessary to waterproof and beautify joints. Jointers are classified according to the specific joint which they are used to form, such as the concave joint or the V-joint, Figure 6-10. The type of joint which the mason forms depends mainly on the architectural design and specifications of the structure.

Convex Jointer

The most commonly used jointers in various parts of the country are the *convex jointer* and the *V-jointer.* The convex jointer is made of a long, rounded piece of steel which forms a depressed, rounded indentation (often called a *round* or *concave* joint) in mortar. The convex jointer can be purchased in a pocket size or in longer lengths with a wooden handle. This type, *sled runner convex jointer,* is preferred over the short one. The short and long convex jointers are available in different widths to accommodate the various thicknesses of mortar joints being formed, Figure 6-11B and C.

Note: Remember that a *convex* jointer forms a *concave* joint.

V-Jointer

The V-jointer is made from an V-shaped piece of angled steel, Figure 6-11A. When passed through mortar, the angled edge of the V-jointer forms a V-shaped groove. This jointer can be obtained in the short or long styles. As with the convex jointer, the mason will find that the longer style is more effective. A solid, straight piece of wood may be used to obtain the same results as the metal V-jointer, but would not retain its straight, sharp edge very long.

Grapevine Jointer

The grapevine jointer has a raised bead of steel in the center which causes an indented line to form in the joint. This jointer is used mainly on Early American brickwork. The grapevine jointer is available only in the shorter size.

Rake Out Jointer

Originally, joints were raked out by a nail which was driven into a piece of wood and dragged across the mortar. This very inaccurate tool was replaced by the *rake out jointer.* A fairly recent development is a rake out jointer with wheels. This type of jointer has a screw allowing the mason to set the tempered nail to the desired depth, Figure 6-12. The advantage of

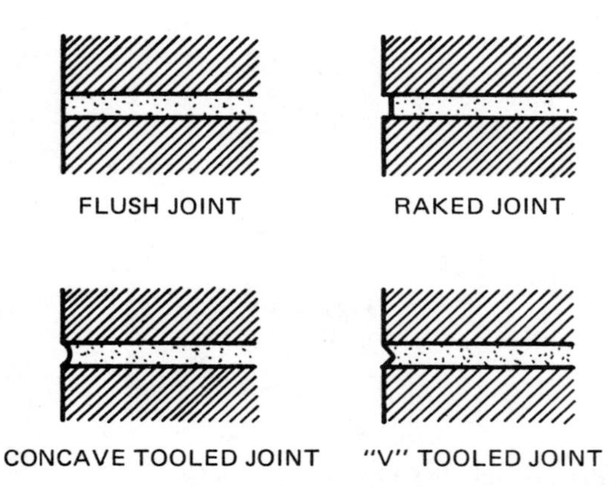

Fig. 6-10 Common mortar joints

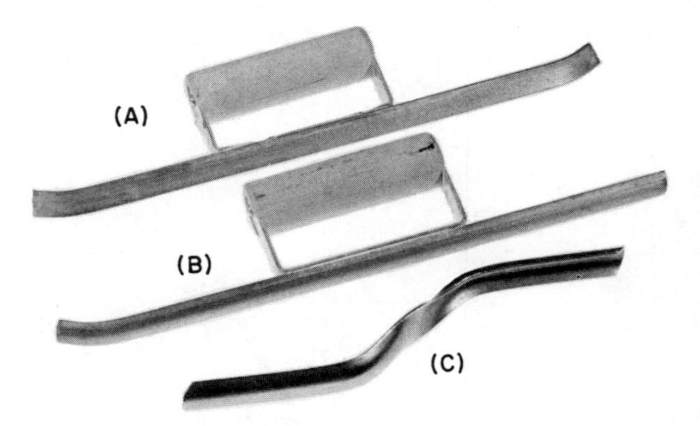

Fig. 6-11 Jointers commonly used by masons include (A) the V-jointer and (B and C) the long and short convex jointers.

Fig. 6-12 A rake out jointer equipped with wheels. The screw located between the wheels allows the mason to set the desired depth of the nail.

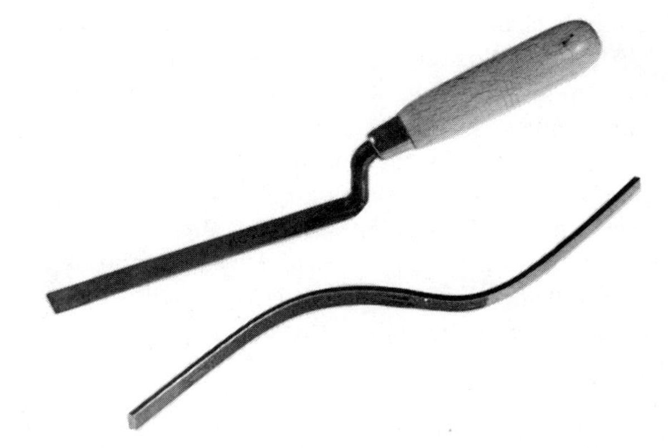

Fig. 6-13 Two styles of slickers. The slicker which does not have a handle may be used on two sizes of mortar joints.

the skate wheel rake out jointer is the speed with which it may be operated, and the very neat, straight, hole-free joint which it forms.

Slicker

A *slicker*, Figure 6-13, is a flat piece of steel used to smooth mortar and form flush, flat joints. It is often an S-shaped tool, either end of which may be used for striking. It is available in various sizes (usually with different sizes on each end) to form different mortar joints. The slicker is invaluable in working with stone and when paving. The slicker is also used as a pointing tool in tight places or in corners of buildings where a neat, clean joint is required. *Pointing* is another term for finishing joints.

MASON'S BRUSH

The finishing touch on masonry work includes brushing it with a good-quality, medium-stiff brush, Figure 6-14A. Most masons prefer the *stove* brush, which is equipped with a long handle. The long handle keeps the fingers away from the work. An old floor brush sawed in half works well, too.

In addition to brushing down work, the brush may also be used for brushing off footings and cleaning the work area, since a neat, clean work area is necessary to prevent accidents.

STEEL SQUARE

A *steel square* is a measuring tool which has at least two straight edges and one right angle, Figure 6-14B. A mason should carry a small square (about 12" long) so that he can square *jambs* (sides of door and window openings) and returns. Large, 2′ squares are also used many times by the mason. The square can also be used to measure smaller materials.

CHALK BOX

A *chalk box* consists of a metal or plastic case with a good cotton line (usually in 50′ lengths) inside on a spool, Figure 6-14C. There is a hole in one end of the chalk box case through which the line is pulled. As the line is pulled from the case, it passes through finely

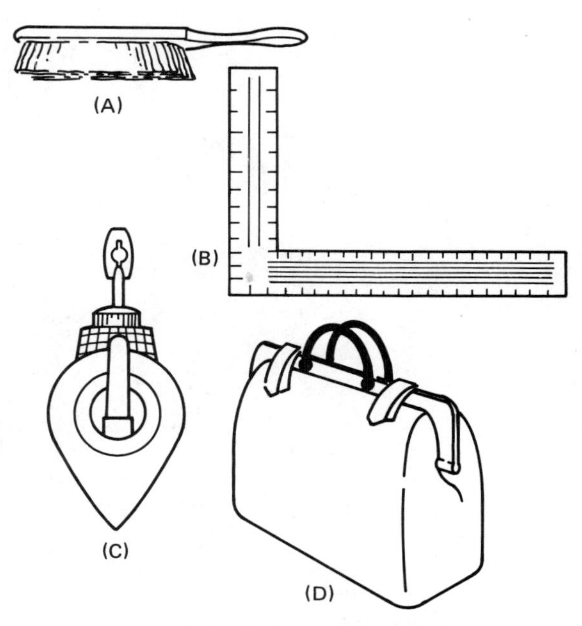

Fig. 6-14 Additional tools include (A) mason's brush, (B) steel square, (C) chalk box, and (D) mason's tool bag.

ground chalk, usually blue or red in color. The chalk box helps the mason establish straight, accurate lines in layout work. The chalk box should be a standard item in every mason's tool kit.

Be careful that the line never gets wet, since the line will deteriorate, the chalk will get soggy, and the metal box will rust. However, if the line does get wet, unwind it from its spool and dry it out thoroughly before using again. Replacements for worn line may be purchased in hardware stores.

TOOL BAG

Every mason should have a tool bag so that all tools are kept together and are always within reach. The average-sized tool bag measures 14" to 18" across and is equipped with an inside pocket in which small items are stored, Figure 6-14D. Some tool bags may be converted to shoulder bags so that the mason's hands are free to grip scaffolds and ladders when climbing. The plumb rule is pushed through the handle when the hands must be free. Metal tool boxes are sometimes used. However, they are not preferred since the boxes tend to rust when wet and may be bulky and difficult to handle.

Remember, take time and effort when selecting tools. Good-quality tools and equipment will pay off in the service you will receive from them.

ACHIEVEMENT REVIEW

A. Selecting a set of tools is an important task for the beginning mason. Quality and handling features are the main factors to be considered in selecting basic tools, Figure 6-15. Listed are some of the basic tools discussed in this unit. For each tool, list the outstanding features you would look for when purchasing a tool set.

1. Trowel
2. Brick hammer
3. Brick set
4. Plumb rule
5. Rule
6. Striking tool
7. Mason's line

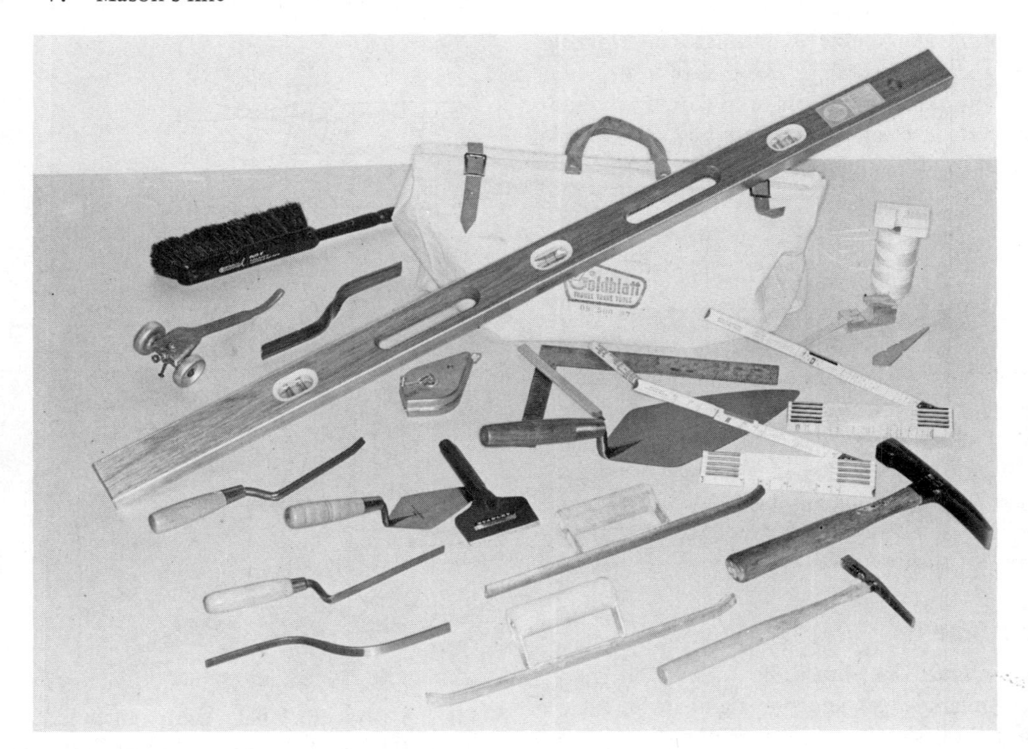

Fig. 6-15 Basic tools of the mason

B. The column on the left contains a statement associated with mason's tools. The column on the right lists terms. Select the correct term from the right-hand list and match it with the proper statement on the left.

1. Used to form a depressed, rounded indentation in a mortar joint
2. Angle of the handle on a trowel
3. Flat tool used to smooth and form flat mortar joints
4. Tool equipped with wheels
5. L-shaped block used to hold line tight
6. Measurement tool based on the 4" grid
7. Tool used for cutting brick accurately
8. Metal or plastic box filled with chalk and line; used for lines
9. Jointer with a raised bead of steel in center
10. Metal fastener used to keep line from sagging
11. Horizontal to the surface of the earth
12. Trowel pattern with a diamond-shaped heel
13. Vertical to the surface of the earth
14. Tool used to establish level and plumb points in masonry work

a. Philadelphia pattern
b. Plumb
c. Rake out jointer
d. Slicker
e. Modular rule
f. Grapevine jointer
g. Convex jointer
h. Chalk box
i. Trig
j. Level
k. Brick set chisel
l. London pattern
m. Line block
n. Set of the trowel
o. Plumb rule

Unit 7
Learning to Use the Basic Tools

OBJECTIVES

After studying this unit, the student will be able to

- describe how to use the basic tools of the masonry trade.
- demonstrate the correct methods of using the basic tools.
- explain why both hands should be used when laying bricks.
- demonstrate good safety practices when using the tools.

The importance of learning to use the basic masonry tools correctly cannot be overemphasized. To a large extent, ability depends on how well a mason handles and uses the basic tools. While the masonry student must perfect skills with all masonry tools, the trowel is used more than any other tool. Therefore, more skill and practice are required to use it correctly.

As students begin working with the masonry tools, they will make mistakes. When mistakes are made, a demonstration of the correct technique or procedure by the masonry instructor or another mason will give the students guidance. This guidance will enable the students to continue their practice until they have mastered each problem area.

Students should try to keep work neat while they are learning and should not try to develop a great deal of speed. Students will find that as they become more skillful in the use of tools, speed will increase. After a time, they will develop a rhythm or a gait in work movements. A *gait* is the speed at which the student is most productive. To attain this rhythm, masons must eliminate unnecessary movements as they work.

At the end of this unit, there is a progress checklist of the basic skills used in masonry work. The students should check their work with this list. Also, with the help of the instructor, the students should determine their strong points and weak points. Students should then work on the weak points until they have mastered the basic skills given in the list.

SAFETY NOTES

There are safe ways to use tools to prevent injury. Do not harm yourself and your fellow workers through carelessness.

- Always cut masonry materials away from a fellow worker's face.
- Always replace mortar cautiously in the mortar pan or on the mortar board. Do not throw it so that it splashes.
- Keep tools out of the paths of other people working on the job.
- Always hand tools to another person, never throw them.
- Use good tool habits and safety practices as tool skills are learned.

SPREADING MORTAR WITH THE MASON'S TROWEL

A great deal of the mason's working time is spent either spreading mortar or handling mortar with the trowel. There are several different ways of using the trowel to cut mortar free from the mortar pile. This unit will illustrate two accepted methods of doing this.

The Clock Method of Picking Up Mortar

The first method of picking up mortar is called the *clock method*. It is called the clock method be-

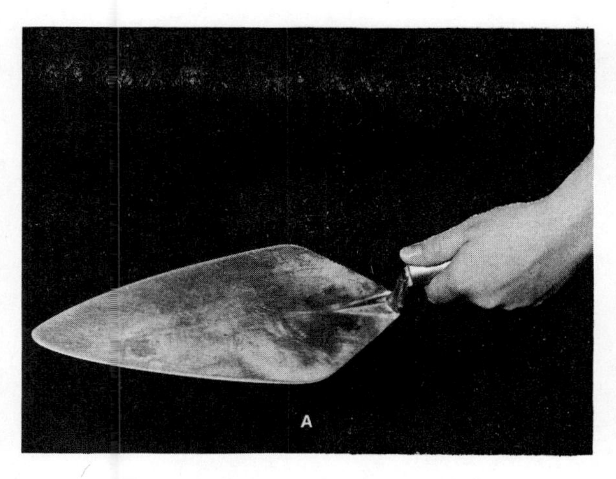

Fig. 7-1 Holding the trowel: (A) Correct way to hold the trowel (note that the thumb remains on the handle), and (B) incorrect way to hold the trowel

cause the movements of the mason are related to the location of the numbers on the face of a clock.

1. Using mortar mixed to the proper stiffness, fill the mortar pan. Mortar which is too stiff or too soft will not spread evenly. Place the mortar pan about 2′ from the wall to allow working space. It is recommended that the student use a mortar pan with the clock method. A mortar board can be used, but the pan will probably be easier to use.

2. Grasp the handle of the trowel between the thumb and first finger with the thumb resting well forward on the top of the handle. Figure 7-1A shows the correct way to hold the trowel. Note that the thumb should not drape over the handle and down the shank, Figure 7-1B. In this position, the thumb will be in the mortar. Prolonged contact with the mortar can cause irritation or cracking of the skin. The thumb stablizes the trowel and guides its movements from side to side. The thumb also helps support the wrist when lifting mortar from the board or pan. As shown in Figure 7-1A, the second, third, fourth, and fifth fingers are wrapped around the handle of the trowel. The muscles of the hand, wrist, and forearm must remain relaxed so that the trowel can be handled freely.

3. Figures 7-2A through 7-2D show the method of removing mortar from the pan. A clock face painted on a wooden frame is set over the pan so that the student can become used to the trowel positions. In Figure 7-2A, the trowel is in the starting position at 6:30 o'clock. The trowel is held at an angle so that as it is moved, it will cut the desired mortar away from the rest of the mortar.

4. The trowel is now moved upward to the 12:00 o'clock position, Figure 7-2B.

5. Next, the trowel is moved down to the 5:30 o'clock position, Figure 7-2C. Do not turn the trowel over and do not lift the trowel out of the mortar pan while making the final downward movement. If done correctly, the mortar is cut free and resembles a long church steeple. If the mortar looks like a wide piece of pie, it has not been cut at the currect angle. Therefore, it cannot be picked up properly.

6. Now, without removing the trowel from the mortar, slide the trowel over until it is face up. Then firmly push the trowel toward the 12:00 o'clock position. Finally, lift the trowel from the mortar at the end of the stroke. If the entire operation has been done correctly, the trowel will be fully loaded with a tapered section of mortar. The position of the mortar on the trowel is very important if the spreading is to be accomplished successfully. Figure 7-2D shows a fully loaded trowel and the mortar pan from which the mortar was removed.

To prevent the mortar from falling off the trowel, a slight snap of the wrist will set the mortar on the trowel. This is one of the most important steps to master in spreading mortar. Do not give too much of a snap or the mortar will become dislodged and it will

Fig. 7-2A Starting position at 6:30 o'clock

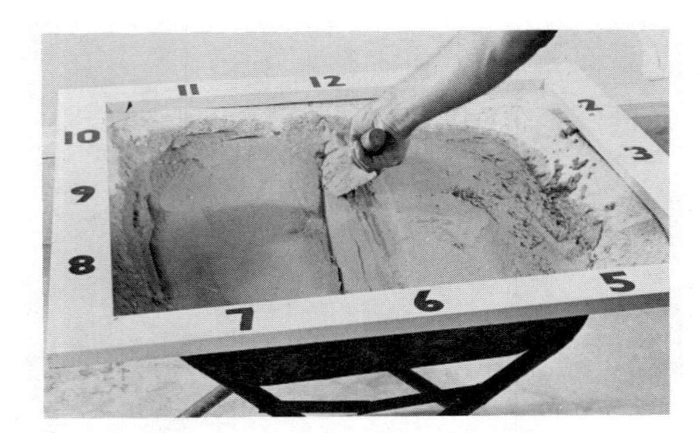

Fig. 7-2B Trowel moved upward at an angle to 12:00 o'clock position

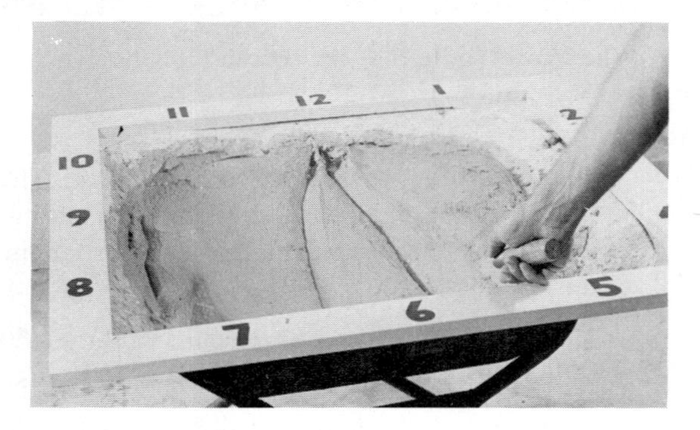

Fig. 7-2C Movement of trowel downward to 5:30 o'clock position

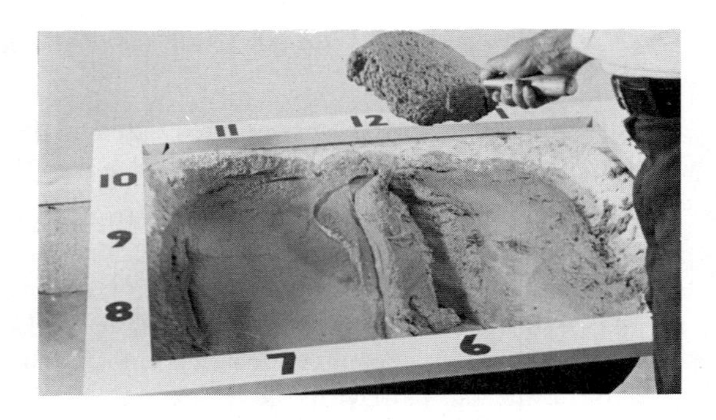

Fig. 7-2D Fully loaded trowel ready for spreading

be necessary to start over again. Experienced masons can set the mortar so smoothly that one has to observe very carefully to see them do it.

The process of applying the mortar to a wall is called *stringing* the mortar. Approach the wall (in this case a practice 2" x 4") and keep the elbow of the right arm slightly away from the body so that the movements of the trowel hand are not restricted.

Figure 7-3 shows the mortar ready to be spread. Use a sweeping motion to spread the mortar on the wall. Keep the trowel in the center of the wall for the length of the spread. In Figure 7-4, note the angle at which the trowel is held while the mortar is spread. Try to spread the mortar along the length of at least 2 bricks. Beginning students often try to obtain too long a spread. This results in an uneven bed of mortar on which it will be impossible to lay bricks without filling in the spaces. Students should repeat this spreading operation over and over until they have mastered the operation. This operation should be repeated until they can spread the correct amount of mortar with one motion of the trowel with little or no mortar thrown over the edge of the bricks. The ideal amount of mortar to be spread with one trowel motion equals the length of 3 to 4 bricks.

By holding the trowel face up at a slight angle, Figure 7-5, excess mortar hanging over the face of the wall can be cut off. The flat angle at which the trowel blade is held enables the mason to catch any mortar as it is cut off, thus preventing the excess mortar from smearing the wall. The excess mortar can be used to fill in any spaces in the bed joint or it may be returned to the mortar pan.

Caution: Mortar splashed in an eye is very painful. Immediately flush eyes with water if this happens. To avoid accidents, mortar should not be thrown in the pan. Replace it carefully.

The next step in spreading the mortar is to *furrow* it, Figure 7-6. Hold the trowel at a 45° angle. Make a shallow furrow in the center of the bed joint by tapping with the point of the trowel along the length of the mortar spread. A shallow furrow in the bed joint allows enough movement of the mortar to permit the mason to adjust the brick to its proper position. Avoid

Fig. 7-3 Trowel with load of mortar ready for spreading

deeply furrowed joints. A deep furrow may expose the building unit below and may eventually cause a leak in the wall.

After furrowing, the excess mortar again is cut off. This mortar may be returned to the mortar pan or it may be used for the head joint on the brick to be laid.

The Cupping Method of Picking Up Mortar

The second method of picking up mortar is called *cupping* the mortar. This method is recommended when a mortar board is used.

1. In the cupping method, the amount of mortar to be picked up with the trowel is separated from the main pile on the mortar board. The mason slices down with the trowel and then pulls the mortar away from the main pile, Figure 7-7A.

2. The smaller pile of mortar is pulled and rolled to the edge of the mortar board.

3. At the edge of the board, the trowel works the mortar until it is shaped into a long, tapered form for convenient pickup, Figure 7-7B.

4. The trowel is then slid under the smaller, shaped pile of mortar. The trowel is held flat against the mortar board and then is pushed forward under the mortar. At the same time a quick, upward motion of the trowel lifts the mortar free from the mortar board at the completion of the stroke, Figure 7-7C.

From this point on, the spreading operation is the same as that shown in the previous clock method.

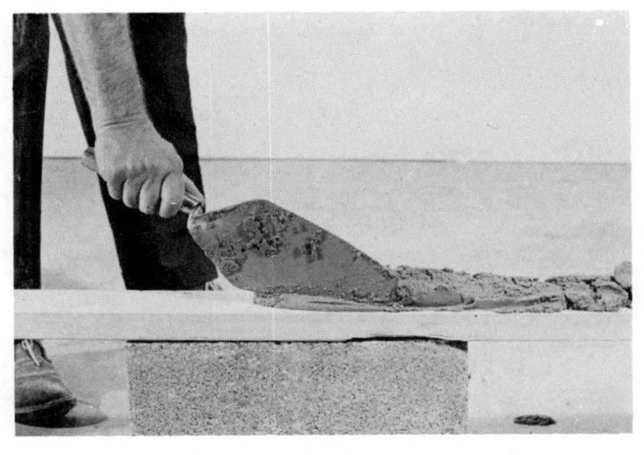

Fig. 7-4 Spreading the mortar on the wall. Note the location of the trowel in the center of the board and the flat back side of the trowel turned away from the mason.

Fig. 7-5 Using the trowel to cut off the excess mortar. Note the flat angle at which the trowel is held. The mason can catch the mortar so it does not fall on the floor.

Fig. 7-6 Furrowing the mortar with the point of the trowel. Note the angle at which the trowel is held.

Fig. 7-7A Cutting the mortar loose from the main pile with a downward slicing motion of the trowel. The smaller pile of mortar is then pushed to the edge of the mortar board with a rolling motion.

Fig. 7-7B The mortar is shaped into a long, tapered form ready for pickup on the trowel.

Regardless of which pickup method is selected, the speed with which students lay bricks depends on their ability to pick up and spread the mortar properly.

LAYING THE BRICKS

There is a correct way to hold a brick when laying it on the mortar bed. The mason's thumb should curl down over the top edge of the brick and be held slightly away from the face of the brick. With the thumb held in this manner, the line can fit between the thumb and the top edge of the brick. As a result, the mason does not cause the line to swing or bounce (called *crowding the line*) and other masons working on the same line do not have to wait for the line to stop swinging. Figures 7-8A and B show the correct thumb placement for laying the brick and cutting off the excess mortar.

Fig. 7-7C The trowel is moved under the mortar pile and is lifted to break the mortar free of the board. Note the tapered shape of the mortar on the trowel.

Using Both Hands for Efficiency

The speed of bricklaying can be improved and the physical strain decreased if the operation is done with as few motions as possible. A common mistake made by many masons is to use only one hand to pick up and spread the mortar and place the bricks. Thus, one hand does all the work while the other remains idle. Masonry students should practice using both hands, Figure 7-9. They will find that not only does the work go faster, but they will experience less fatigue.

For example, when working with *brick veneer* (brick that carries only its own weight and is used mainly as a facing material), the student should pick up bricks with one hand while the other hand picks up the mortar with the trowel, Figure 7-9. In addition, when working on a *solid wall* (brick backed by some other masonry material or another thickness of brick), a practice called *walling* the brick is recommended, Figure 7-10. In this procedure, both hands are used to pick up bricks and stack them on the finished back section of the wall. The bricks are stacked before any mortar is spread on the front wall. This eliminates a great deal of bending over to reach each brick on the brick pile.

Another time-saving and labor-saving technique is shown in Figure 7-11. After the bricks are stacked on the completed back section of the wall, the mason picks up 1 brick and lays it in the mortar bed. While one hand cuts off the excess mortar and places it on the head of the brick just laid (to form the head joint), the mason should pick up another brick with

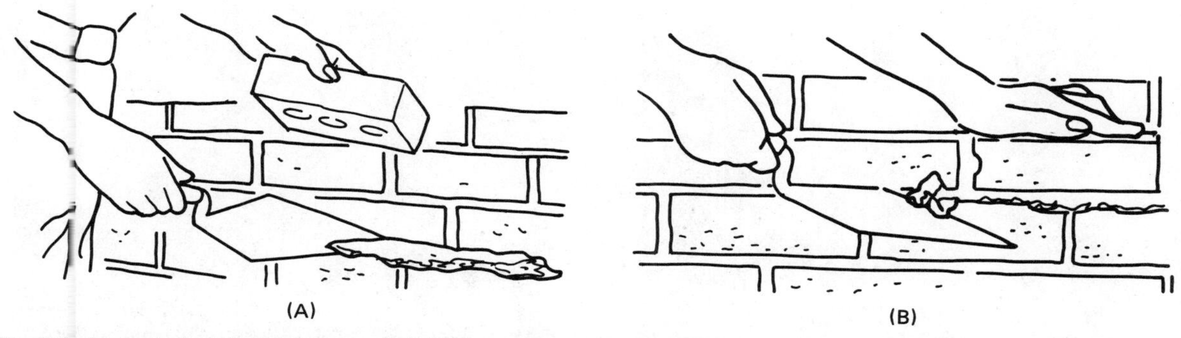

Fig. 7-8 Correct thumb placement: (A) Thumb in this position will not interfere with the trowel cutting off the mortar, and (B) thumb in this position will not interfere with the line as the brick is placed on the wall

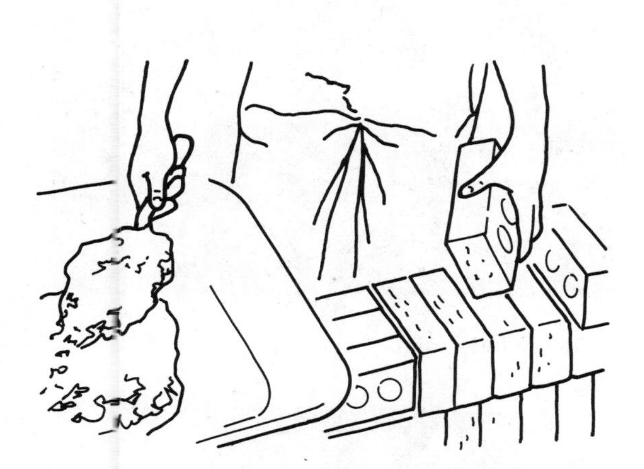

Fig. 7-9 Use both hands to spread mortar and lay brick.

the other hand to have it ready to lay as soon as the one hand completes the head joint.

Making the Head Joint

The proper way to form a head joint is shown in Figure 7-12. In Figure 7-12A, the first brick is placed on the bed joint. In Figure 7-12B, the brick is pressed into position on the mortar and the excess mortar is cut off with the trowel. The second brick is picked up and the mortar joint is applied to the end of the brick, Figure 7-12C. Finall, the second brick is pressed into position in Figure 7-12D. Note the way in which the mason's palm presses across the ends of both bricks. This step enables the mason to determine when the brick is in a level position. The mason then repeats the procedure and continues adding bricks to the course.

LEVELING AND PLUMBING THE COURSE

A course of bricks must be leveled and plumbed if a wall is to be constructed properly.

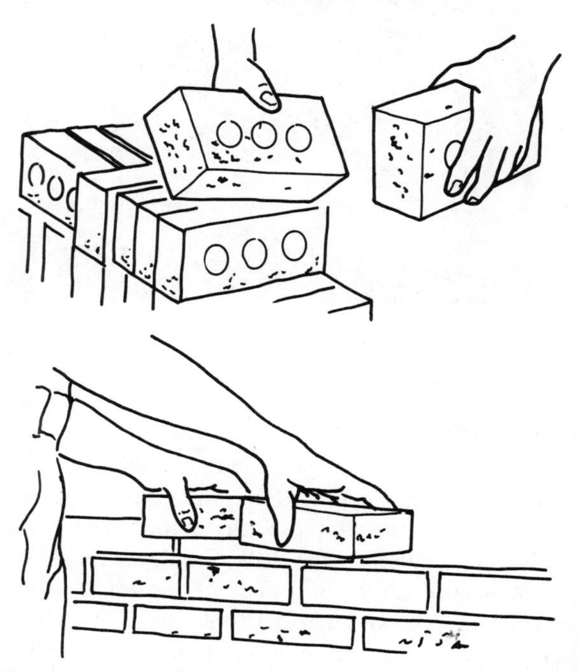

Fig. 7-10 To wall brick, use both hands to place bricks on the completed back section of a solid wall before beginning work on the front wall section.

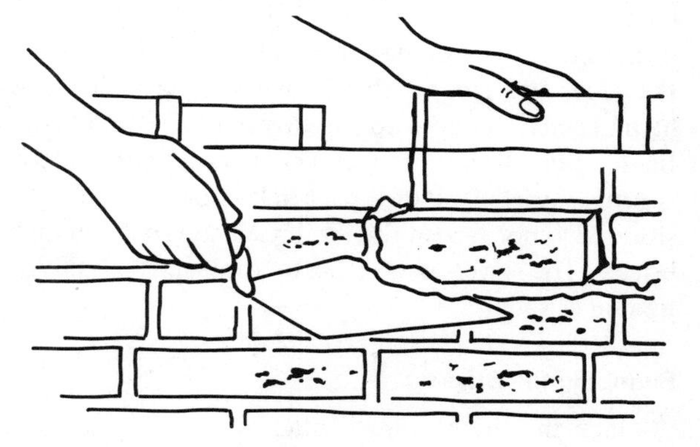

Fig. 7-11 Cutting off mortar with one hand while the other hand picks up the next brick to be laid.

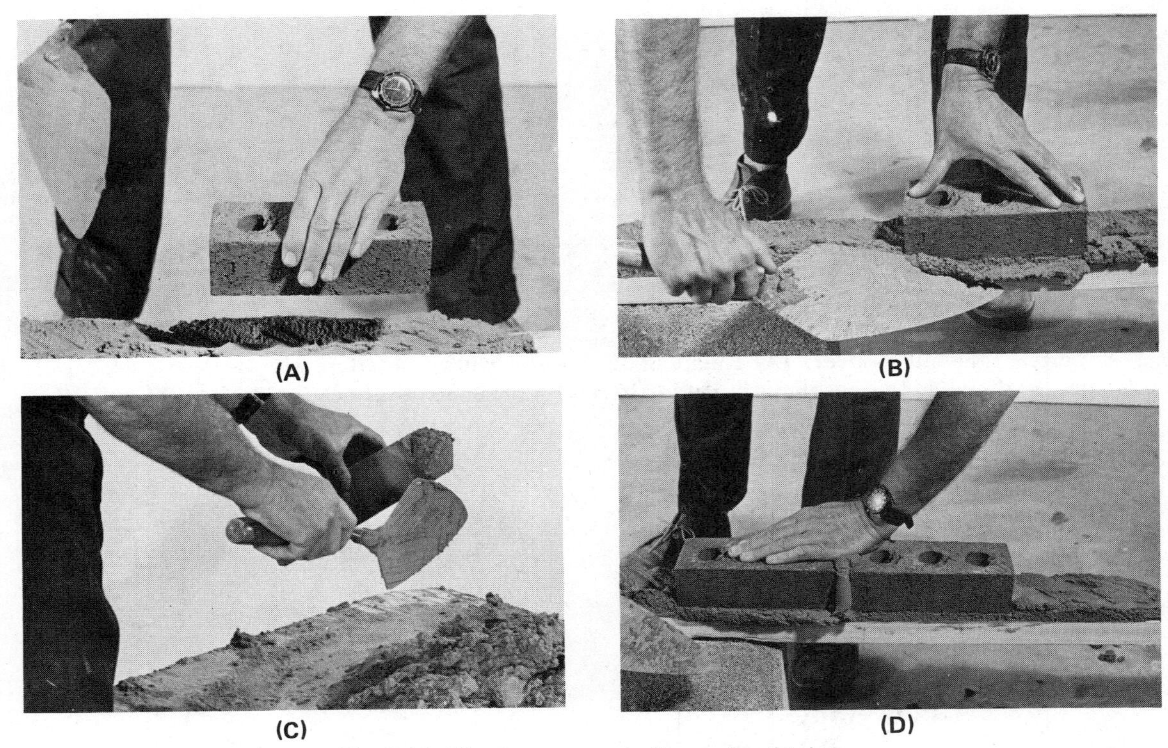

(A) (B) (C) (D)

Fig. 7-12 The proper way to form a head joint

Leveling Bricks

After 6 bricks have been laid they must be leveled using the following procedure. Remove any excess mortar on the top of the bricks before placing the level (plumb rule) on the wall. The level should be placed lengthwise in the center of the bricks to be checked. Tap down any bricks that are high with relation to the level, Figure 7-13. If the bricks are low with relation to the level, they must be picked up and relaid in a fresh bed of mortar until they are level.

When tapping down high bricks, keep the trowel blade parallel to the course of bricks and in the center of the bricks. By doing this, splattered mortar around the edges and chipped brick faces will be avoided. A hammer may be used to tap down a brick that is difficult to handle with the trowel. However, the trowel is usually sufficient to tap a brick into the proper position. Do not tap on the level itself to tap down high bricks. The level is a delicate tool and should be treated as such.

Plumbing the Bricks

Once the bricks are adjusted so they are all level, they must be plumbed. Hold the level in a vertical position against the end of the last brick laid, Figure 7-14. Tap the brick with the trowel as shown to adjust the brick face either in or out. Do this until the bubble in the glass vial of the level rests between the two lines. As the bricks are being plumbed, do not exert a great deal of pressure to force the level against the bricks. If too much force is used, the bricks may spring out of plumb when the level is removed. Next, move the level to the end of the first

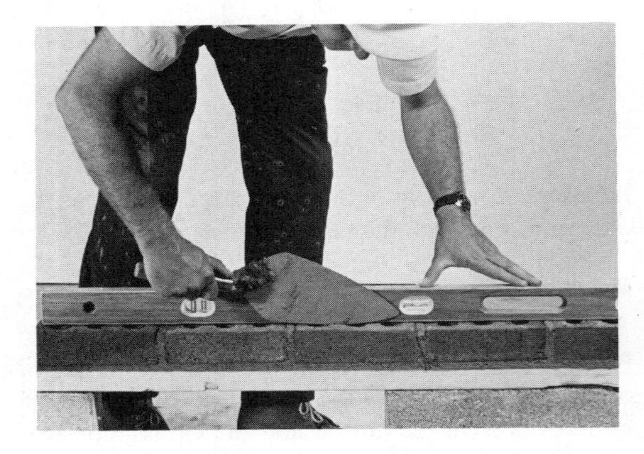

Fig. 7-13 Leveling bricks by tapping them down with the blade of the trowel. Note the correct position of the trowel, parallel with the course of bricks and back from the edge of the bricks.

Fig. 7-14 Plumbing the end bricks with the level

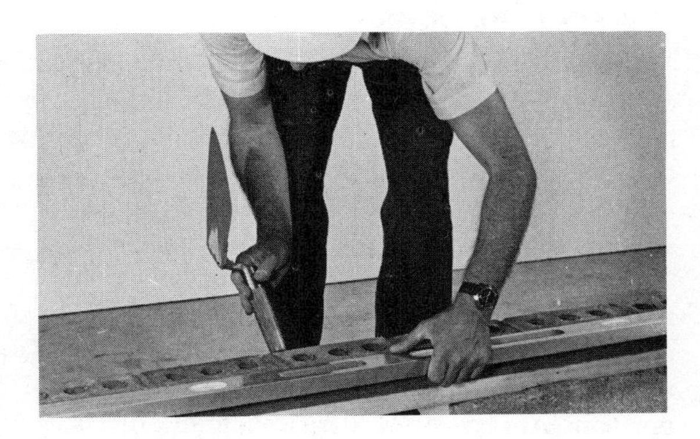

Fig. 7-15 To align the front faces of the remaining bricks, the mason places the level horizontally along the front upper edges of the bricks and taps the bricks until they are aligned. The level must be kept even with the top of each end brick while the mason adjusts the middle bricks.

brick laid and repeat the plumbing process. Now that the first and last bricks have been plumbed, hold the level in a horizontal position against the top outside edge of the bricks, Figure 7-15. Tap the bricks either forward or back until they are all aligned against the level. Be careful not to move or change the original plumb points while the remaining bricks are being aligned.

Because bricks often have an uneven or curled top edge, the level must be held at this edge of the brick to assure that the edge is aligned. True alignment must be achieved at the top edge to avoid uneven humps in the finished wall.

After the plumbing process is complete, place the level out of the way. Hang the level on a nail using the small hole in the top section of the level, or place it upright in one of the holes of a concrete block. Never place the level on the scaffold or on the ground near the work area as it will surely be broken.

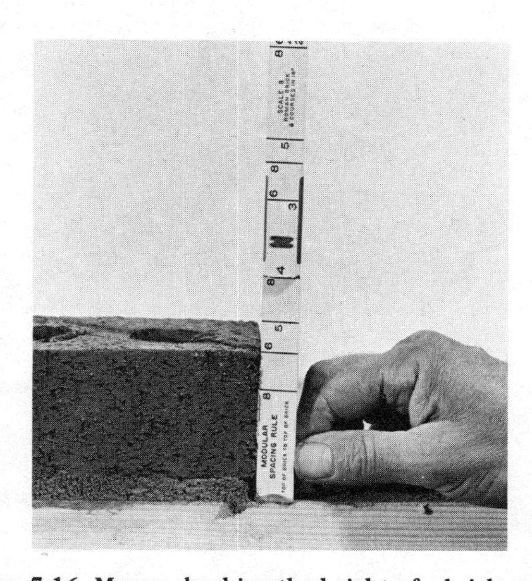

Fig. 7-16 Mason checking the height of a brick course with a modular spacing rule. Note that the top of the brick is even with number 6 of the rule. This is the correct value for standard brick spacing.

CHECKING THE HEIGHT WITH THE MODULAR RULE

Of the two types of mason's rules listed in Unit 6, the modular rule and the course counter spacing rule, only the modular rule will be covered at this point. The correct spacing height for 1 course of standard brickwork equals the number 6 on the modular rule, indicating that there are 6 courses to every 16" of vertical height. Since this value includes the mortar joints, the mason does not have to allow extra height for the mortar.

After the bricks are laid in the wall, unfold the modular rule and place it on the base used for the mortar and bricks. Hold the rule vertically, Figure 7-16. Be sure the end of the rule is flat on the base so that the reading is accurate. The top of the course of bricks should be even with the number 6 on the rule.

Note: If more than 1 course is being laid, always set the modular rule on the top of the first course to check the height as the base may have been irregular and a large joint may have been used to level the first course.

STRIKING THE WORK

To determine when the mortar is ready for striking or tooling, the student should press a thumb firmly into the mortar joint. If an impression is made and the mortar does not stick to the thumb, the mortar is ready to strike, Figure 7-17. A sled runner convex striking tool may be used to tool the joint, Figure 7-18. It is recommended that masonry students use the concave joint in their work. This is the joint used most often today. The tool should be slightly larger than the mortar joint to achieve the correct tool impression. Always strike the head joints first and then the bed joints. This order is followed because the bed joint should be straight and unbroken from one end of the wall to the other. If the head joints are struck last, there will be mortar ridges evenly spaced on the bed joints. These ridges will detract from the appearance of the joints. Masonry students should practice striking first the head joint and then the bed joints of their work. Most contractors require construction projects to show the smooth appearance resulting when the bed joint is struck last.

If a joint is struck while the mortar is too soft, the joint will be wet and runny. Also, the imprint of the striking tool will not be as sharp and neat as it should be. However, if the joint is struck when the mortar is too hard, a black mark will appear on the surface of the joint. This is called *burning the joint.* Striking the joints when the mortar is too hard will also result in unnecessary fatigue for masons. Masons must test the joints repeatedly for the correct hardness so that they can strike them at the proper time.

Fig. 7-17 Testing the mortar to determine if it is ready for striking. If the thumb impression remains, the mortar is ready.

Fig. 7-18 Striking the mortar joint with a sled runner tool. Note that the thumb is placed on the top outside edge of the handle so it does not scrape on the bricks.

BRUSHING THE WALL

Brushing is the final step in any masonry work and should be done before the mortar sets. A medium-soft brush should be used to brush the wall after the mortar is struck. In general, if the striking is done at the correct time, the brushing of the wall can be done immediately after. The masonry student should first brush all of the head joints vertically to remove any excess mortar. The brushing is completed by brushing lightly across the bed joints, Figure 7-19. The mortar joint may be restruck after brushing to obtain a sharp, neat joint without imperfections.

CUTTING BRICKS

Masons often must cut bricks to size to fit the construction job. Bricks can be cut with the brick hammer, brick set chisel, or trowel.

Fig. 7-19 Brushing the wall

Caution: Students should not wrap their fingers around the brick set near the brick to be cut. When the cut is made, the downward pressure can force their fingers into the brick, causing abrasions. Whenever bricks must be cut, wear safety glasses or other eye protection.

The Brick Hammer

The brick hammer is the most frequently used cutting tool. To cut the brick with the brick hammer, hold the brick in one hand. Students must remember not to place their fingers and thumb on the face side of the brick where the cut will be made, Figure 7-20. Mark the desired cut on the face side of the brick either with a pencil or by scoring the brick with the hammer. Next, strike the brick with light blows of the hammer blade along the line marked to score the brick. Turn the brick on its side and strike with the hammer in line with the scoring just completed on the front of the brick. Repeat this procedure until the cutting line is marked on all 4 sides of the brick. Then strike the face of the brick with a hard, quick blow of the hammer blade. The brick should break cleanly along the lines scored. If the brick does not break the first time, restrike it gently until it has weakened enough to break. Scoring the brick causes it to break more evenly. Striking an unscored brick usually results in an unacceptable break or shatters the brick. The cut made with the brick hammer is not a perfect cut. However, it will be accurate enough for use in the average construction job.

The Brick Set Chisel

When an accurate, straight cut is required, the brick set chisel is used. The mason can save time on the job by marking and cutting to size all of the bricks required at the edges of windows, around doors, or for window sills.

First, mark the cutting line on the face side of each brick using a pencil and a straightedge. Then place the brick, marked face up, on a wooden plank or on soft dirt. The bricks must never be placed on a hard surface such as concrete. This type of surface cannot give when the brick is struck with the brick set chisel. As a result, the brick may not cut true or it may shatter. The chisel of the brick set must be held vertically with the flat side of the blade facing the direction of the finished cut, Figure 7-21. The blade is beveled, so when the chisel is struck with the hammer, the blade travels downward at a slight angle and cuts off any particles of brick that may be protruding. As a result, the mason can place the cut brick in the wall immediately and does not have to do more chipping or dressing.

Cutting with the Trowel

The brick trowel is not recommended for most cutting, but may be used on medium-hard bricks when the hammer is not readily available. Masonry students

Fig. 7-20 Cutting brick with the brick hammer. Note that the mason's fingers and thumb are not placed on the side of the brick being scored.

Fig. 7-21 Cutting brick with the brick set. The flat side of the chisel must face the finished cut being made. Keep fingers above the cutting edge of the chisel to avoid injuries.

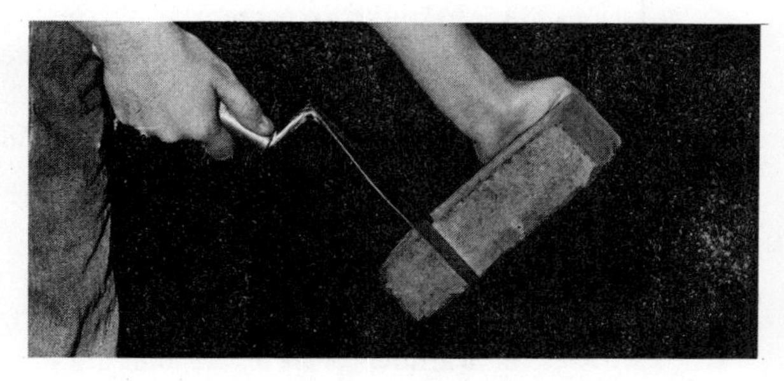

Fig. 7-22 Cutting brick with the trowel. Keep the fingers of the hand holding the brick well under the brick to avoid injuries from the blade.

should use the hammer and the brick set chisel as much as possible for cutting bricks until they have mastered cutting with these tools. Then they can learn how to cut with the trowel, which is a common practice on the job.

As shown in Figure 7-22, the brick to be cut is held in one hand. Keep the fingers and thumb away from the side of the brick to be cut. Strike the brick with a quick, sharp blow at the point where the break is desired. To cut the brick with the trowel, a sharp snap of the wrist is required. A slow, swinging blow will not cut a brick. If the brick does not cut easily, use a brick hammer.

Caution: Always hold the trowel downward when cutting bricks. Never cut bricks over the mortar pan or near other workers. When the weather is extremely cold, bricks should not be cut with the trowel as the blade of the trowel may break.

PROGRESS CHECK

At this point, masonry students should have acquired certain skills. It is important to practice these skills. Students should demonstrate for their instructor each of the following skills. Any skills that are not mastered should be practiced until they can be demonstrated correctly.

- Spread mortar with the trowel for a length of at least 2 bricks.
- Apply the head joints on a brick.
- Level a course of bricks.
- Plumb a course of bricks.
- Check the brickwork for height with a modular rule.
- Strike the mortar joints with a concave tool.
- Brush the wall and restrike the joint.
- Cut bricks using the hammer, the brick set, and the trowel.

ACHIEVEMENT REVIEW

Select the best answer to complete each statement from the choices offered. List your choice by letter identification.

1. The mortar pan is placed away from the wall to allow for working space. The correct distance is
   a. 3′. c. 1 1/2′.
   b. 2′. d. 5′.
2. The most efficient method of cutting mortar from the mortar pan with the trowel is called the
   a. cupping method. c. clock method.
   b. rolling method. d. scooping method.
3. If the mortar in the mortar pan is cut correctly with the trowel, it resembles
   a. a slice of pie. c. a church steeple.
   b. a circle. d. a square.
4. One of the most important steps in spreading mortar is the slight snap of the wrist after picking the mortar from the pan. The purpose of this movement is to
   a. shake the excess mortar off the trowel.
   b. distribute the mortar evenly on the trowel.
   c. set the mortar on the trowel to prevent it from falling off.
   d. shape the mortar for easier spreading.
5. The ideal amount of mortar spread with the trowel should equal the length of
   a. 1 to 2 bricks. c. 5 to 6 bricks.
   b. 3 to 4 bricks. d. 7 to 8 bricks.
6. Masons can improve their efficiency by
   a. learning to use both hands at once.
   b. using a larger trowel.
   c. spreading the mortar across the entire wall before laying any bricks.

7. The term *furrowed* is used when spreading mortar with the trowel. Which of the following best describes the term?
   a. Furrowing is the process of cutting mortar from the pan.
   b. A furrow is a slight indentation made in the center of the mortar joint with the point of the trowel.
   c. Furrowing is the spreading of the mortar on the wall.
   d. The trimming of the excess mortar from the wall after it is spread with the trowel is called furrowing.

8. The mason can determine the correct time to strike the mortar joint by using a simple test. Which of the following best describes the test?
   a. Measuring the rate of drying with a watch
   b. Pressing the thumb in the mortar of a bed joint
   c. Checking the mortar with the point of the trowel
   d. Light tapping on the face of the wall to see if the bricks have set correctly

9. A common expression in masonry work is *burning the joint.* This is defined as
   a. drying the mortar joint with heat.
   b. striking the mortar joint when it is too soft.
   c. striking the mortar joint when it is too hard, causing black marks on the joint.
   d. special effects created by raking out the joint.

10. The height of a standard course of bricks is indicated on a modular rule by the number
    a. 2.
    b. 5.
    c. 4.
    d. 6.

11. After the course of bricks is laid in mortar, leveled, and plumbed, it must be aligned horizontally with the level. This is done by
    a. holding the level across the top outside edges of the bricks and adjusting the bricks against the level.
    b. holding the level across the center of the course of bricks and adjusting the bricks until they are in line with the level.
    c. holding the level across the bottom of the course and adjusting the bricks to the level.

12. When bricks must be cut accurately, the best cutting tool to use is the
    a. brick hammer.
    b. brick set chisel.
    c. trowel.
    d. mortar hoe.

PROJECT 1: SPREADING MORTAR

OBJECTIVE

- The masonry student will be able to pick up and spread mortar for a length of from 1 to 3 bricks with the brick trowel.

EQUIPMENT, TOOLS, AND SUPPLIES

Mortar board or mortar pan and stand
Standard dirt shovel
Brick trowel (not to exceed 11" in length)
2" x 4" lumber, at least 5′ long
2, 4" concrete blocks
Mortar hoe
Bucket of clean water for tempering
Mortar

SUGGESTIONS

- Have your instructor check that you are holding the trowel correctly.
- Spread the mortar a short distance in the beginning. As you become more skillful, you will be able to spread the mortar for a length of 3 or 4 bricks.
- Do not furrow too deeply with the point of the trowel. Do not grip the trowel too too tightly.
- Practice spreading mortar until it can be done smoothly and with ease.

PROCEDURE

Note: Mixed mortar will be provided by the instructor for this project.

1. Place the mortar pan or board on a mortar stand. (The stand may be metal or may consist of several concrete blocks placed under the mortar pan or board.)
2. Place the 2, 4″ blocks in the work area and place the 2″ x 4″ piece of lumber on top of the blocks. The 2″ x 4″ should be at least 2′ away from the mortar pan to allow working room.
3. Wet the mortar pan or board with some water before placing mortar in it. This will prevent the mortar from drying out too quickly.
4. Fill the mortar pan with mortar provided by your instructor.
5. Cut the mortar from the pan with the trowel as described in the unit and spread the mortar on the 2″ x 4″.
6. Furrow the mortar with the point of the trowel. Cut off the excess mortar and return the excess to the pan.
7. Temper the mortar with water as needed to keep it in a workable condition. If the mortar is kept rounded in the pan at all times, it requires less tempering.
8. Clean all tools with water immediately after use.

PROJECT 2: LAYING 6 BRICKS ON THE 2″ x 4″

OBJECTIVE

- The student will be able to use the basic tools in laying bricks without a line as a guide.

EQUIPMENT, TOOLS, AND SUPPLIES

Bucket of clean water
Mortar (from instructor)
6 bricks
Mortar pan or board
Standard dirt shovel
2″ x 4″ lumber, at least 6′ long
Mortar hoe
2, 4″ blocks
1, 8″ block (use as level holder)
Brick trowel
Plumb rule (level)
Mason's modular rule
Convex jointer or V-jointer
Brick hammer
Brush

SUGGESTIONS

- Stand close to the 2" x 4" and sight along the front edge of the bricks to determine if the wall is plumb.
- Press the bricks into position on the bed joint firmly. Do not use too much side motion. Do not tap the brick into position.
- When applying the head joint, do not smear the face of the brick.
- When laying the bricks, rest the palm of one hand half on the brick just laid and half on the brick being laid. This procedure helps the mason to develop the feel of level bricks.
- Keep the plumb rule in the center of the wall when leveling.
- In the plumbing operation, hold the plumb rule at the bottom of the brick and tap the brick true to the plumb rule.
- Never tap the plumb rule itself. When not in use, always place the plumb rule upright in one of the holes of a block.
- Never use a 4" block as a plumbrule holder. This block is not wide enough and the level will fall out easily. Any of the larger size blocks will work well.
- If a 2" x 4" is not available, a chalk line can be struck on the floor and the same procedure followed. A 2" x 4" is preferred because it is the same width as a standard brick wall.

PROCEDURE

1. Wet the mortar pan or board and fill it with mortar.
2. Set up the 2" x 4" as in Project 1.
3. Spread enough mortar on the 2" x 4" for 6 bricks.
4. Lay the 6 bricks on the mortar bed.
5. Level the bricks with the plumb rule.
6. Plumb the ends of the course and then straightedge the bricks with the level.
7. Check the height of the brickwork with the modular rule.
8. When the mortar is thumbprint hard, strike the joints with the convex jointer.
9. Brush the brickwork, removing all excess dirt and mortar.
10. Clean all tools after use.

PROJECT 3: CUTTING BRICKS WITH THE HAMMER, BRICK SET CHISEL, AND TROWEL

OBJECTIVE

- The student will be able to cut bricks with the hammer, the brick set chisel, and the trowel. With practice, the quality of the cut will be such that the cut brick can be used in the finished wall.

EQUIPMENT, TOOLS, AND SUPPLIES

6 bricks
Pencil
Brick hammer
Brick set chisel
Trowel
A piece of lumber at least 2′ long
Safety glasses or goggles

SUGGESTIONS

- Wear safety glasses or goggles to protect eyes when cutting.
- Keep fingers off the top edge of the brick where the cut is to be made to avoid injury.
- Have your instructor check the various cuts to see if they are of sufficient quality to use in the wall.

PROCEDURE

A. Cutting with the Hammer

1. Mark the brick where it is to be cut.
2. Score the brick on all 4 sides with the hammer.
3. Strike the brick on the face side to complete the cut.

B. Cutting with the Brick Set Chisel

1. Place the brick on the piece of lumber.
2. Mark the desired cut with the pencil.
3. Cut the brick on the pencil mark with the brick set chisel as explained in this unit.

C. Cutting with the Trowel

1. Mark another brick with a pencil or scratch a mark on the brick with the trowel.
2. Hold the brick down and away from the body. Strike the brick with a quick, sharp blow of the trowel. A second blow may be needed if the brick does not fracture the first time.

Unit 8
Related Equipment

OBJECTIVES

After studying this unit, the student will be able to

- describe important hand-powered equipment used in masonry work.
- describe important power-driven equipment used in masonry work.
- maintain power equipment and use it safely.

For masonry work to be done efficiently, both manual and power equipment are necessary. Most of the extremely heavy work in masonry is now accomplished by using power-driven machinery. Work which does not require as much strength is done with manual tools.

MANUAL TOOLS

Brick Tongs

Brick tongs (or *brick carriers*), Figure 8-1, are used to manually move or place bricks near the mason. The ends of the tongs are equipped with a special flange so that the bricks are not damaged when they are picked up. They are adjustable to fit different-sized bricks. As a rule, 10 bricks may be picked up at one time. The tongs in the figure have a curved rubber handle which enables the mason to use the tongs with the least amount of discomfort. There are several types of brick tongs in use today. They are all built to withstand heavy use.

Wheelbarrows

There are two types of hand-pushed wheelbarrows used in masonry construction. One type, known as a *contractor's wheelbarrow,* is made of heavy-duty steel and is equipped with wooden handles and an air-filled rubber tire, Figure 8-2. Mortar or masonry materials can be transported directly to the work area with this wheelbarrow. The contractor's wheelbarrow is used for this reason instead of other types of wheelbarrows because it has a deeper bed.

The second type of wheelbarrow, equipped with a wooden body, is used for transporting bricks, blocks, or other materials to the work area, Figure 8-3. This

Fig. 8-1 Moving units with the brick tongs. The weight of the tongs shown is about 4 pounds.

Fig. 8-2 Contractor's wheelbarrow

Fig. 8-3 Wooden body wheelbarrow

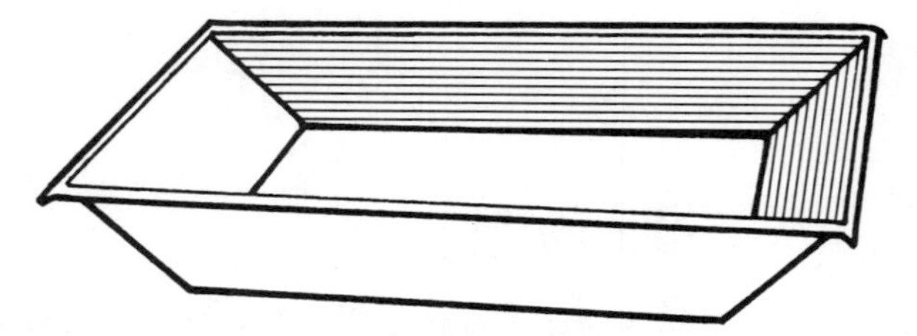

Fig. 8-4 Steel mortar mixing box

wheelbarrow differs from the contractor's wheelbarrow in that it has no sides so that the material can be unloaded from any direction.

Mortar Box

Mortar is mixed in a *steel mortar box.* The box is leakproof and has smooth, welded edges for easy mixing, Figure 8-4. Mortar boxes are available in different lengths, widths, and depths. An average-sized box measures about 32" x 60" and weighs approximately 57 lb.

Mixing Tools

Several tools are used to mix mortar, Figure 8-5. Two different types of shovels are used in the mixing process. To proportion the ingredients in the mortar box, a standard dirt shovel is usually used. After the mortar has been mixed, it is transferred from the box to the wheelbarrow with a square-bladed shovel. The square blade allows the mason to scrape all corners of the mortar box, thereby picking up more mortar and cutting down handling time. Both shovels should be of a good-quality forged steel so that they can withstand hard use.

A large hoe is used to mix the mortar. The mortar hoe has 2 holes in the blade, centered on each side of the point at which the wooden handle is attached. The holes reduce the amount of energy exerted to pull and push the hoe through the mix and help break up ingredients in the mix. After mixing, all mortar should be cleaned immediately from mixing tools with water.

The Mortar Pan, Stand, and Board

The *mortar pan* and *mortar board* are used to hold the mortar after is has been mixed.

Fig. 8-5 Mixing tools shown (left to right) are the standard dirt shovel, the mortar hoe, and the square-bladed shovel.

The metal mortar pan, Figure 8-6, is approximately 30" square at the top, tapering to approximately 16" x 16" at the bottom. It is about 7" deep. Many masons and contractors prefer the mortar pan to the board for two reasons. It holds more mortar than the board and does not deteriorate as quickly as the mortar board since it is made of steel or rigid plastic.

The mortar pan also fits on the mortar stand more securely. The mortar stand holds the pan at a convenient height, reducing excess strain and fatigue. The mortar stand consists of a folding tubular steel frame held together by a set of steel hinges, Figure 8-6.

Mortar boards, Figure 8-7, should measure about 30" square. They can be made from any small lengths of lumber found around the job. The best selection of material for a mortar board, however, is a piece of 3/4" exterior plywood with 2" x 4" wooden runners on the bottom.

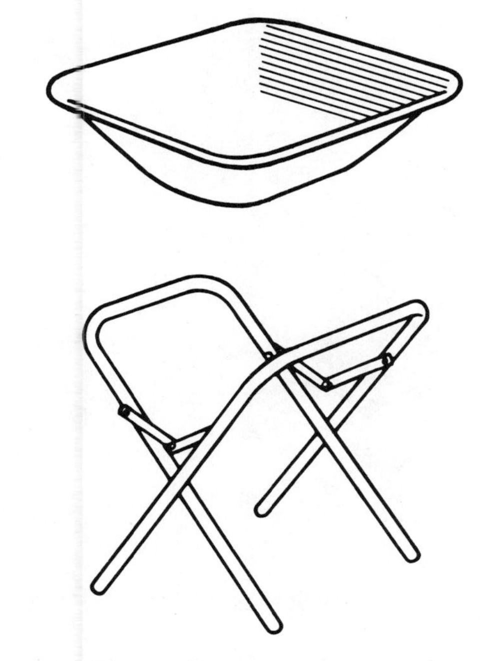

Fig. 8-6 Mortar pan and stand

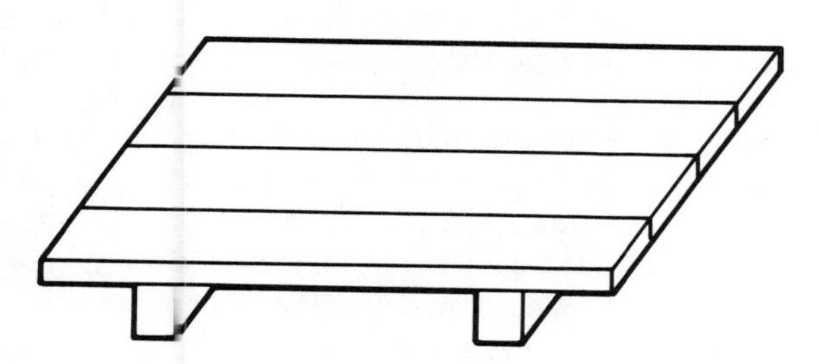

Fig. 8-7 Wood mortar board

Mortar boards are limited in that they are not extremely strong and, since the mortar board has no sides, only a limited amount of mortar can be placed on the board. The dry wood also rapidly absorbs moisture from the mortar, making it necessary to temper the mortar with water more often than mortar which is mixed in a steel pan.

The Rope and Pulley

A heavy, reinforced steel pulley wheel and a strong rope, Figure 8-8, are used to hoist materials onto the scaffold. The pulley is hooked to a steel brace on the scaffolding The mason pulls the rope over the wheel, raising the materials to the working level. The rope and pulley are usually used on small jobs or in construction where costs must be kept to a minimum.

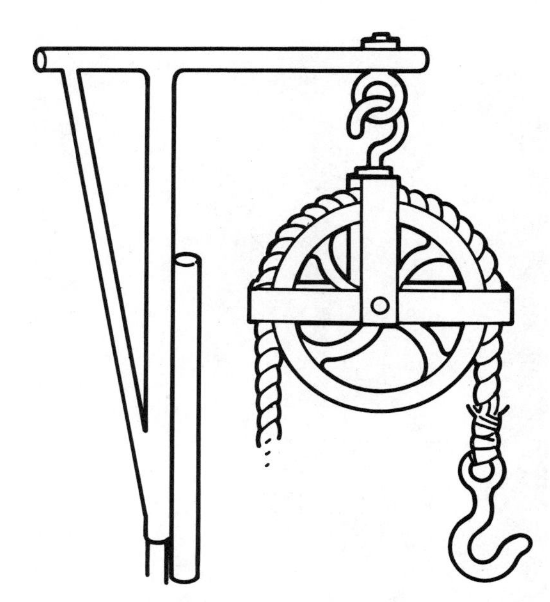

Fig. 8-8 Boom, rope, and pulley wheel

Corner Pole

A *corner pole* (or *masonry guide*), is used when the mason wishes to lay bricks in a wall without building corners, Figure 8-9. This means great savings and faster construction to the contractor. Most commercial corner poles are made of metal and have line blocks or line carriers that can be attached to the pole. The corner pole has masonry units marked on it. In this way, the line can be moved to the desired position.

Corner poles may be erected differently, according to the type of pole used. All corner poles must be braced and plumbed. They must also be located at a predetermined distance from the finished masonry work to enable the mason to work freely. The corner pole should be checked constantly to be certain that it remains plumb and tight in its bracing.

The mason can usually accomplish jobs much faster by using corner poles. It has been estimated that on the average, masons lay at least 5 bricks in the time it takes them to lay, level, and plumb 1 brick on a corner. With a little experience, it should not take over 10 to 15 minutes to erect a corner pole. Except for occasionally checking a doorway or window jamb, the mason can lay bricks without using the plumb rule.

There are many places on a job, however, where the corner pole is not practical to use. It is not recommended for very small sections of wall even though it may be used to build the longer sections of that same wall.

Fig. 8-9 Corner pole used in construction

POWER-DRIVEN EQUIPMENT

Mortar Mixer

Mortar mixers may be powered by gasoline engines or electric motors, Figure 8-10. The gasoline-powered mixer is preferred since electricity is not always available on the job. The mortar is mixed by a turning shaft with attached blades which revolve through the mix. The mixer has a large dumping handle which allows the mason to empty the drum at the completion of the mixing process. Mixers are available in several different sizes. A typical mixer will mix about 4 cubic feet (cu ft) of mortar at one time.

The mixer should be washed immediately after use so that the mortar does not accumulate and ruin the mixing drum. Never beat the mixer drum with a hammer or other heavy equipment.

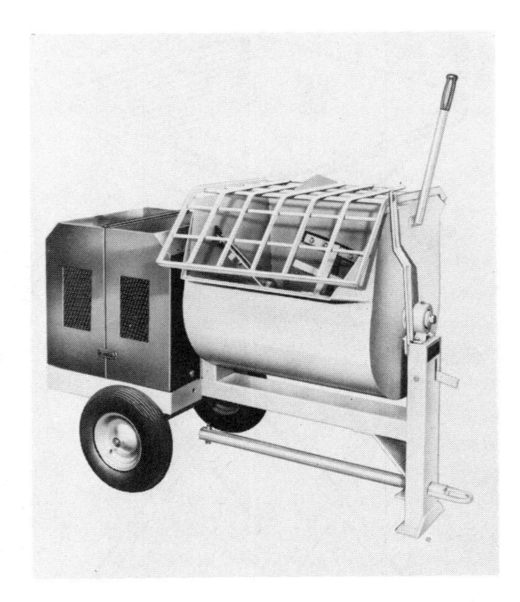

Fig. 8-10 Gasoline-powered mortar mixer

Masonry Saw

When a very precise cut is necessary or when material cannot be cut with a hand tool, the *masonry saw* is used, Figure 8-11. Many architects specify that all masonry material on a job be cut with the masonry saw. This is because the saw does not weaken or fracture the material, with a straighter cut being the result.

The saw can be operated dry or with water. If it is used dry, a blade made of silicon carbide is usually used. If operated with water, an industrial diamond-chip blade is used. Diamond blades should never be used for dry cutting or they will burn and be ruined.

> **Caution**: Safety goggles are required when cutting with a wet or dry saw, since serious eye damage could result if a chip flew into the operator's eye. The operator should also have adequate ventilation and a dust mask or respirator if the material is cut dry.

Hydraulic Masonry Splitter

Hydraulic splitters are available in a variety of sizes and models. Hydraulic force exerted by a pump presses the masonry unit against a cutting blade. It is often practical to use one of these cutters instead of a masonry saw.

Figure 8-12 shows a portable splitter used for stone, splitblock, face brick, and firebrick. It is especially popular with contractors because it is both portable and simple to operate.

Fig. 8-11 Cutting concrete with the masonry saw

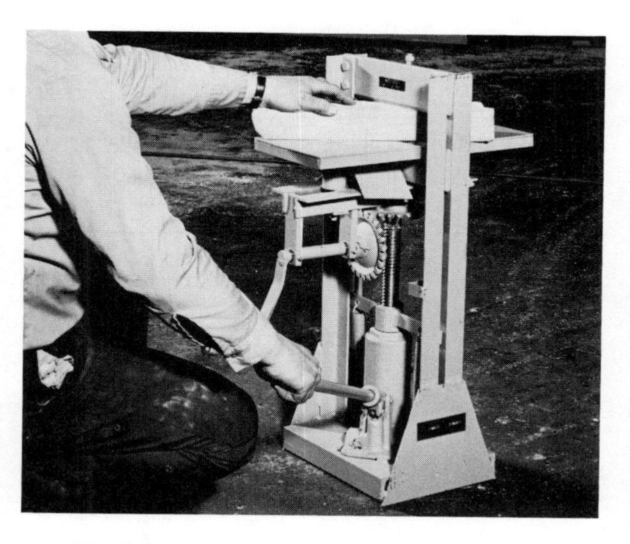

Fig. 8-12 Portable hydraulic masonry splitter

Forklift Tractor

The *forklift tractor,* Figure 8-13, enables the mason to move and position materials mechanically. Mortar can be hoisted in a special steel cart or mortar box and placed automatically in the immediate work area.

Fig. 8-13 Forklift hoisting mortar box onto scaffold

Elevator

Elevators are used to carry the material to the correct level so that it can be delivered to the mason, Figure 8-14. To protect the worker from injury, safety rails and a heavy steel mesh screen are required around the elevator platform. The elevator is operated by the person who is loading or unloading from either the top or ground level by the use of weights. This particular type of elevator is preferred by many contractors since it does not require a full-time operator. These elevators are meant to carry materials only. Construction workers should never attempt to ride in them.

Fig. 8-14 Material elevators are used to deliver supplies to the mason.

Electric Tuckpointer's Grinder

Tuckpointing is the replacing of old mortar joints with new mortar to repair the joint. In this way, the mason restores old masonry work. The term *tuck* is used since the mortar is tucked into the joints in the process with a steel tool such as a slicker. The *tuckpointer's grinder* is a hand-operated machine designed specifically for grinding out joints in the mortar bed and head joints of a masonry wall, Figure 8-15. It is equipped with shatterproof blades and a safety guard on that part of the machine which faces the mason.

> **Caution**: Always wear safety goggles while operating the grinder to protect your eyes from flying debris.

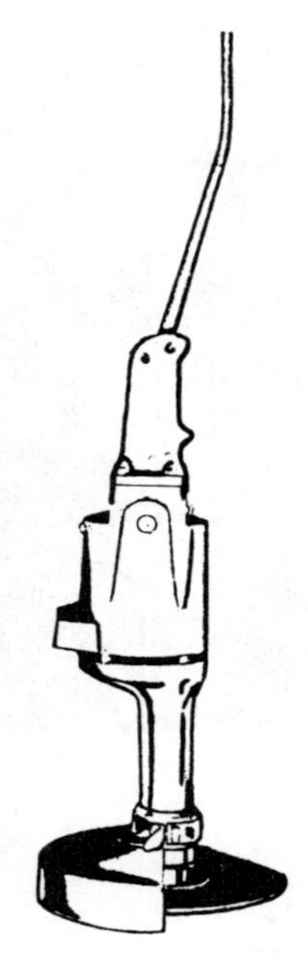

Fig. 8-15 Truckpointer's grinder

ACHIEVEMENT REVIEW

Select the best answer from the choices offered to complete the statement or answer the question. List your answer by letter identification.

1. Brick tongs are used for
   a. cutting bricks.
   b. tooling mortar joints.
   c. carrying bricks.
   d. measuring materials.
2. The contractor's wheelbarrow is usually used to transport mortar instead of other types of wheelbarrows because
   a. it has a deeper bed.
   b. it is rustproof.
   c. it is less expensive.
3. The mortar box has smooth, welded joints and edges because
   a. they prevent the box from leaking.
   b. extra strength is provided.
   c. the mixing tools do not catch on the edge.
4. Most commercial corner poles are made of
   a. plastic.
   b. metal.
   c. wood.
5. The mortar hoe has 2 holes in the blade. Which of the following is *not* a valid reason for the holes being in the hoe?
   a. The holes greatly reduce the effort necessary to push and pull the hoe through the mix.
   b. The holes force air into the mix.
   c. The holes break up lumps in the material.

6. The mortar stand is used for
   a. supporting the mortar pan at the correct working height.
   b. transporting the mortar to the working area.
   c. mixing the mortar.
   d. storing the dry mortar before mixing.
7. The equipment used to raise material onto the scaffold is the
   a. wheelbarrow.
   b. brick tongs.
   c. corner pole.
   d. rope and pulley.
8. When masonry material is cut with a wet masonry saw, the best blade to use is the
   a. diamond blade.
   b. silicon carbide blade.
   c. steel blade.
   d. fibergalss blade.
9. Measuring ingredients for mortar should be done with a
   a. square shovel with a V-shaped handle.
   b. regular, standard dirt shovel.
   c. 5-gal bucket.

SUMMARY, SECTION 2

- Many hours of practice are required to learn to use the trowel properly. The trowel is the mason's most frequently used tool.
- The beginning mason should become familiar with the basic set of tools and acquire these first.
- Always select quality tools.
- The trainee should be able to use either of two methods for picking up mortar from the pan.
- It is essential to use both hands when laying bricks. This can be accomplished by continuous practice.
- Always observe good safety practices when using tools. Masons have a responsibility to protect fellow workers by not being careless with tools and by not taking shortcuts.
- There are various types of manual and power-driven machinery used in masonry work. The mason must be familiar with each tool and know related safety practices. Power-driven and automatic machinery have been developed to help masons, not to replace them.

SUMMARY ACHIEVEMENT REVIEW, SECTION 2

Complete each of the following statements referring to material in Section 2.

1. The tool used in masonry more often than any other is the ________________ .
2. Attempting to spread too much mortar at one time is a common mistake of the trainee. The result of this practice is ________________ .
3. Deeply furrowed bed joints are not acceptable because they may cause ________________ .
4. Stacking bricks on the finished back section of a wall before laying them in the front section of the wall is called ________________ .

5. The thumb is curled over the top edge of the brick as it is laid in the wall. This is to prevent the mason from ________________ .

6. When tapping down bricks with the trowel in the leveling process, the trainee should be particularly careful not to chip the face of the brick or splatter mortar on the front of the wall. This can best be done by being sure that the trowel is in the ________________ .

7. After the course has been plumbed, the plumb rule is used as a straightedge to form a straight line of bricks. When doing this, the plumb rule is always held ________________ .

8. Power-driven mortar mixers can be operated by a gasoline engine or an electric motor. Gasoline engines are more popular because ________________ .

9. When using a masonry saw, the mason must always wear ________________ for protection.

Section 3
Mortar

Unit 9
The Development of Mortar

OBJECTIVES

After studying this unit, the student will be able to

- explain the development of mortar from early times to the present.
- list the ingredients of mortar.
- describe the relation of lime to mortar.

A masonry wall is no better than the mortar that bonds its units together. Mortar must be capable of keeping the wall intact and it must create a weather-resistant barrier. Units 9 and 10 deal with the development of mortar and with different types of mortar. Masons should know how to recognize quality mortar and how to mix it.

Depending on the ingredients used, mortars have different compressive strengths, flexibility, and shrinkage rates. Therefore, masons should try to match the correct mortar to the needs of a particular job. The proper balance between the mortar and the structural requirements of a job is often the difference between a profitable job and an unsuccessful one.

HISTORY OF MORTAR

Clay was the first material to be used for mortar. It has been used through history in masonry walls of unburned brick, but the lack of a hard binding agent makes clay impractical in humid climates. In 2690 BC the Great Pyramid of Giza was built in Egypt. The huge blocks of this structure were cemented together with mortar made from burned gypsum and sand. Many years later, the Greeks and Romans developed mortar from volcanic waste and sand. Structures built with this mortar still stand.

Mortar made from lime-sand was commonly used until the late nineteenth century. However, in 1824, portland cement was developed, an occurrence that marked the beginning of modern-day cement.

Portland cement was a much stronger material than had been used before, whether it was applied alone or combined with lime. Mortar used today is a combination of portland cement, hydrated lime, and masonry sand. More recently, masonry cements have been developed which require that only sand and water be added for the formation of mortar.

The formula or percentage of ingredients are usually not printed on bags of mortar. However, the contents of the bag of mortar must pass specifications for a mortar of medium strength. This mortar may be sold under various brand names, but portland cement is always specified by type.

Besides binding the masonry materials into a permanent structure, mortar seals the joints against penetration of air and moisture. Mortar acts as a bond for various parts of the structure such as reforcement rods, anchor bolts, and metal ties so they may become an integral part of the wall. There are cases of leaks and cracks in masonry walls which can be attributed directly to the use of defective mortar.

INGREDIENTS OF MORTAR

To better understand how mortar reacts to different temperatures, stress, and prolonged use, the main ingredients of mortar must be examined. These main ingredients are portland cement, hydrated lime, sand, and water.

Portland Cement

Portland cement was discovered in 1824 by Joseph Aspdin, an English stonemason. He was attempting to produce mortar which would harden when water was added. The name *portland* was given to the new cement because Aspdin thought it resembled natural stone which was quarried on the Isle of Portland.

By the middle of the nineteeth century, portland cement was in wide use throughout England. It was manufactured in the United States from 1871 on and gradually replaced natural and lime-sand mortars. It became a major ingredient in mortar after 1880.

There are many lengthy procedures in the production of portland cement. The following are the most important operations:

- The quarrying of raw materials
- Raw grinding and mixing
- Initial blending of the raw materials
- Fine grinding in preparation for the kiln
- Burning in the kiln for the formation of clinkers
- Grinding of the clinkers
- Bagging and shipping

The end product is a fine, grayish powder made up of certain chemical compounds. The most common compounds include carefully blended limestone, clay, or shale. Although about 99% of all cement used in construction contains portland cement, it is always mixed with other materials. It is a common mistake for people to speak of a *cement sidewalk* or *cement highway*. There are always other ingredients present in the mixture.

Portland cement combined with water forms a paste. As the paste hardens, it binds the various materials in the mixture. The paste hardens more and more as it ages and eventually can become as hard as the rock or other aggregates mixed with it This is one of the outstanding features of portland cement. However, portland cement must be kept dry in storage. If it becomes wet, lumps form and the cement cannot be used.

Some designers falsely assume that what is good practice for concrete is also acceptable for mortar. Actually, mortar differs from concrete in consistency, methods of application, and use. Mortar binds masonry units into a single mass, Concrete is usually a structural element in itself. Most of the important physical properties of portland cement concrete concern its compressive strength. For these reasons, the requirements of masonry mortar are vastly different from those of portland cement concrete.

Note: All types of mortar have a percentage of portland cement in the mix.

Types of Portland Cement

The specifications of the ASTM list five types of portland cement which are used in masonry and the physical properties and chemical composition of each type. Any one of these types can be used for masonry mortar, but Types I and II are generally used in mortar. The other types are usually used for mixing concrete.

Type I. This is a general-purpose cement and is the one masons most often use. It may be used in pavements, sidewalks, reinforced concrete bridge culverts, and masonry mortar.

Type II (modified portland cement). This cement hydrates at a lower heat than Type I and generates heat at a lower rate. It does, however, have better resistance to sulfate than Type I. It is usually specified for use in such places as large piers, heavy abutments, and heavy retaining walls.

Type III (high-early-strength portland cement). Although this cement requires as long to set as Type I, it achieves its full strength much sooner. Generally, when high strength is required in 1 to 3 days, this cement is recommended. In cold weather when protection from freezing weather is important, Type III often is specified.

Type IV (low-heat portland cement). This is a special cement for use where the amount and rate of heat generated must be kept to a minimum. It is critical to hold the temperature down to ensure that the concrete cures properly. Since the concrete does cure slowly, strength also develops at a slower rate. Too much heat in the hardening process causes a defective or weak concrete. Low-heat portland cement is used in areas where there are huge masses of concrete, such as dams or large bridges.

Type V (sulfate-resistant portland cement). This is also a special portland cement intended for use only in construction which is exposed to severe sulfate actions. It also gains strength at a slower rate than normal portland cement.

There are special air-entraining cements which contain small amounts of a chemical which enables them to hold air bubbles in the mix. The bubbles improve the workability and increase the resistance of the cement to freezing and thawing. These cements are used on jobs with special conditions. Air-entraining portland cements, covered by ASTM number C150, include Types IA, IIA, and IIIA.

Lime

The production of quicklime is the first step in the manufacture of lime. *Quicklime* is formed by calcining (or burning) pieces of limestone under controlled heat. The quicklime is then crushed and *slaked,* or mixed with water, in a hydrator where hydrated lime is produced. Figure 9-1 shows a worker bagging hydrated lime.

Prior to the twentieth century, quicklime was always used in the production of mortar. Today, however, hydrated lime is usually used in the production of mortar since it can be added directly to the mixer in dry form.

Hydrated lime is essential to an all-purpose mortar because of the particular effects it has on the mortar. The following are the effects hydrated lime has on mortar.

Bond Strength. Bond strength is probably the most important property of mortar. The addition of lime to mortar helps it fill voids and adhere to the unit. Figure 9-2 shows an example of good bond strength. Other factors in bond strength are water used in mixing (maximum should be used with retempering as needed), air content, and absorption and degree of roughness of the unit.

Workability. Lime helps mortar to be very workable, enabling the mason to fill joints completely without great effort. The result of this is better workmanship and more economical construction. Mortar with the correct lime content spreads easily with the trowel. Cement which is not mixed with lime is more difficult to spread. Sand graduation and air and water content also greatly aid workability.

Water Retention. Lime-based mortar is high in *water retention.* This means that the mortar resists drying effects when combined with the masonry units to build a wall. Mortar with a low lime content soon

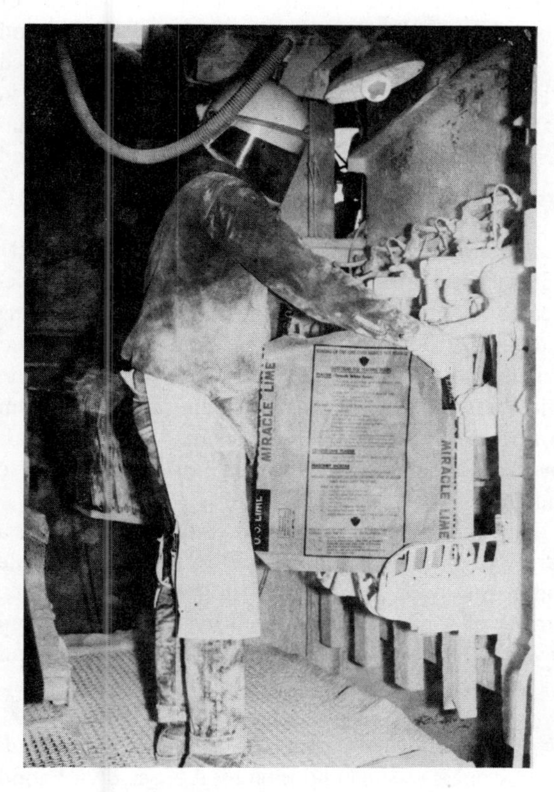

Fig. 9-1 Bagging hydrated lime in the plant

Fig. 9-2 Good bond strength. Mortar adheres tightly to the brick even after the brick is picked up.

loses its moisture and sets prematurely when spread on the wall. When the mason does not have enough time to lay the material on the mortar bed before it drys out, a poor bond results, causing a weak structure. High water retention also minimizes the need for constantly retempering mortar.

Tensile Strength. Good mortar possesses adequate compressive and tensile strength with a substantial safety factor. Mortar with an average strength tests 750 psi when cured for 28 days, which is the time factor used by engineers when checking mortar strengths. Mortar which tests stronger than 750 psi becomes progressively more rigid and brittle. Prepackaged masonry cement which requires that only sand and water be added to it should test within this range. Mortar should set with reasonable speed to enable construction to move forward without great delays. The addition of portland cement to a lime-based mortar helps to increase compressive and tensile strength.

Flexibility. For mortar to withstand stress by strong winds, lateral pressure, and hard jolts, it must have enough flexibility to prevent cracking. A tall masonry structure, such as a high chimney, may sway as much as 12". Lime gives mortar plasticity while portland cement adds strength.

Minimal Change in Volume. Very hard, high-strength mortar tends to shrink after hardening and can produce cracks between the mortar and units. This could result in a loss of bond strength. Of all the materials present in cement, lime undergoes the least change in volume.

Autogenous Healing. The ability of mortar to reknit or reseal itself if hairline cracks occur is called *autogenous healing.* Rainwater and atmospheric carbon dioxide react with the mortar to provide this feature. The hydrated lime dissolves and is then recarbonated by the carbon dioxide, which reseals hairline cracks in the mortar joint. A high-lime mortar shows little cracking, but it also has a lower compressive strength. A high-lime mortar is good when a tight sealing mortar joint, rather than a high compression mortar joint, is desired.

Resistance to Weather. Mortar should be able to resist strong winds, freezing temperatures, and alternate wet and dry weather. Wetting and drying cycles are actually beneficial to lime-based mortar and increase the overall strength of the mortar as it ages. The tightly knit mortar joints caused by the addition of lime to the mix help prevent water from entering the wall.

Sand

Sand, an aggregate of mortar, is the result of the deterioration of rock or stone. Good-quality sand is available in two types, *natural* and *manufactured,* and may be purchased almost anywhere.

Sand is sold by weight and delivered to the job by trucks. On an average job, 1/2 to 1 ton of sand will be lost and cannot be used. This should be considered when sand is purchased. Sand is one of the more inexpensive building materials and may be used on many different projects around the construction site.

Types of Sand. Manufactured sand is obtained by crushing stone, gravel, or air-cooled, blast furnace slag. Manufactured sand, with its sharp, angular grain shape, produces mortar with different properties than natural sand. Natural sand is used in most mortars. Manufactured sand is reserved for special purposes in most of the United States.

Natural sand is rounder and smoother than manufactured sand. It is usually found on lake bottoms or around riverbeds, pits, and riverbanks. Seashore sand is very high in salt content and is not recommended for use in masonry mortar. Most of the sand used in masonry work in the eastern United States is found near riverbanks or riverbeds and is mined by sand and gravel companies.

Characteristics of Sand. Sand acts as a filler in mortar which contributes to the strength of the mix. It decreases the shrinkage of mortar which occurs in setting and drying, therefore minimizing cracking. Water and cementitious materials form a paste which fills the voids between the sand particles and lubricates them to form a workable mix. The volume of voids or air spaces is determined by the range of the particle size and grade of sand being used.

It is important to use a good grade of sand. Over a period of time, it will be consistently more workable and form more satisfactory finished mortar joints. Grades of sand are determined by a standard screen sieve test. The Brick Institute of America has found that well-graded sand helps the workability of mortar.

The Importance of Choosing Good-Quality Sand. When sand is washed and treated, it is usually of good quality. The mason should never attempt to cut costs by using sand which has not been treated and washed.

It is important to understand why foreign substances are damaging to mortar. Silt causes mortar to stick to the trowel. This alone can cause a problem, as the mason's productivity is cut considerably. Silt also weakens mortar since it prevents the cementitious material from bonding to the sand particles. When tooling the mortar joints, silt causes mud pits and small holes on the surface of the joints. The mortar will possess a brownish, unnatural color. The color will vary from batch to batch, since the amount of impurities will not be the same in every batch mixed. Good-quality sand is worth the difference in price since it increases productivity and provides better-quality mortar.

The Siltation Test. If impurities (silt, clay, or organic matter) are suspected in the sand, a simple siltation test may be performed to determine if the sand should be used for masonry mortar.

Note: Many of these undesirable materials may be generally classified as loam.

1. To perform the silation test, fill a glass jar (preferably quart-sized) half full of sand. Add about 3″ of water. Shake the jar vigorously and let it set overnight.
2. The sand will settle to the bottom of the jar and the foreign matter will rise to the top of the sand.
3. If the accumulation of silt or organic matter is more than 1/8″, the sand should not be used for masonry mortar. Figure 9-3 shows one example of a siltation test.

Water

Water used in mortar should be clean and as free as possible from alkalis, salts, acids, and organic matter.

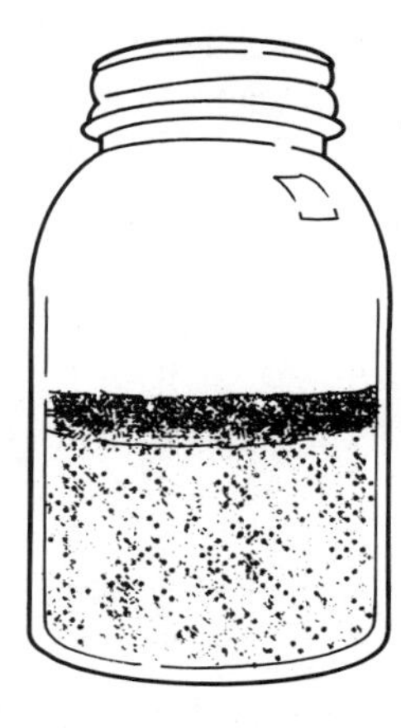

Fig. 9-3 The siltation test detects the presence of silt or loam. Notice the silt deposit on the top of the sand. This is in excess of the allowable 1/8-inch limit.

Water for mixing mortar should never be taken from a mud hole on the job. This is a very common practice on many construction jobs. Muddy water will cause the same reaction in mortar as sand with a high silt content. The amount of chemical deposits in the water should be determined by laboratory analysis if a problem is suspected. As a rule, water which is fit to drink is acceptable for use in mortar. Generally, purifying chemicals found in city water supplies have no adverse effect on mortar.

SUMMARY

The primary function of mortar is to bond masonry units into one integrated mass. The main components of mortars used today are portland cement, hydrated lime, sand, and water. No one combination of these ingredients yields the perfect mortar, however, since each particular job requires different properties in the mortar. Masons should be able to recognize good-quality mortar and mix ingredients in the proper proportions to obtain the best mortar for the particular job.

ACHIEVEMENT REVIEW

Select the best answer from the choices offered to complete each statement. List your choice by letter identification.

1. The early Greeks and Romans used a mortar made from
   a. lime and clay.
   b. volcanic waste and sand.
   c. portland cement and sand.
2. The addition of portland cement to mortar greatly increased mortar's
   a. strength.
   b. plasticity.
   c. workability.
   d. moisture resistance.

3. Quicklime may be defined as
   a. lime that has undergone a chemical change by treating it with water.
   b. crushed limestone.
   c. the fine, white powder resulting from burning in a kiln.
4. After lime has been slaked with water, it is called
   a. quicklime.
   b. hydrated lime.
   c. ground, burnt lime.
5. The term *bond strength* refers to
   a. compressive weight that the bond of mortar will tolerate before failing.
   b. tensile stress the bond of mortar will tolerate before failing.
   c. how well the mortar adheres to the masonry unit.
   d. the testing strength of the masonry unit.
6. Very high-strength types of mortar have one drawback. After hardening, they have a tendency to
   a. expand.
   b. shrink.
   c. absorb moisture.
   d. form pits.
7. One of the biggest advantages a lime-based mortar has is its autogenous healing. Briefly, this is
   a. the ability to resist excessive weight imposed on the mortar joint.
   b. the ability to withstand acids and alkalis from the atmosphere.
   c. the ability to reseal cracks in the joint by the process of recarbonation.
8. A good test used to check sand for foreign matter is the
   a. sieve test.
   b. siltation test.
   c. grading test.
   d. meter test.
9. Dirt in sand tends to make mortar difficult to handle with a trowel. The phrase best describing the condition of the mortar is
   a. sticky and gummy.
   b. gritty with a tendency to fall off the trowel.
   c. very liquid in form.
   d. rapidly setting and requiring excessive tempering.

Unit 10
Types of Mortar and Their Characteristics

OBJECTIVES

After studying this unit, the student will be able to

- identify the various types of mortar used in masonry work.
- describe admixtures and how they are used in mortar.
- explain the importance of the proper water content in mortar.
- describe the causes of efflorescence and how to prevent its development in masonry work.

CLASSIFICATIONS OF MORTARS

Two main classifications of mortars students should become familiar with are portland cement-hydrated lime mortars and masonry cement mortars. Both the ASTM and the BIA have established specifications these mortars must meet.

Portland cement-hydrated lime mortars are a combination of portland cement, hydrated lime, sand, and water, Figure 10-1. Depending on the proportions of the materials in the mix, the mortars will have different strengths and properties.

Masonry cement mortars are popular in masonry construction today because they come prepackaged. The mason needs to add only sand and water on the job, Figure 10-2. Correct proportions of sand and water are essential if masonry cement mortar is to meet standard specification ASTM Designation C270.

Masonry cement mortar has certain additives to provide workability, flexibility, and water-retention properties. One major complaint of architects and engineers is that the proportions of these additives are not printed on the bag. Portland cement-lime mortars can be mixed in exact proportions to meet specifications and, therefore, be the correct mortar for a particular job. One mortar is not necessarily better than the other. However, the mason should be aware of the two main classifications of mortars used in masonry construction.

PORTLAND CEMENT-HYDRATED LIME MORTARS

The Brick Institute of America, the leading authority on brick masonry in the United States, has developed the following specifications for mortars used in the construction of brick masonry work. Four types of portland cement-lime mortars are covered

Fig. 10-1 Hydrated lime and portland cement combine with sand and water to make mortar.

Fig. 10-2 A bag of masonry cement, mixed with sand and water, makes mortar.

under BIA Designation M1-72. The following is a description of the four types and their uses.

RECOMMENDED USES OF MORTAR

Type N

Type N is the most often used mortar. Type N is a medium-strength mortar suitable for general use in exposed masonry above grade and where high compressive or lateral masonry strengths are required. It is specifically used for the following:

- *Parapet walls* (that portion of a wall which extends above the roof line, usually used as a fire wall)
- Chimneys
- Exterior walls which are exposed to severe weather

Type N mortar has a compressive strength of at least 750 psi after being cured for 28 days, Figure 10-3.

When mixed, prebagged cement mortars sold under various brand names should conform to a mortar, like Type N, of at least medium strength.

Type M

Type M mortar has a higher compressive strength (at least 2500 psi in 28 days) and somewhat greater durability than some of the other types. It is especially recommended for masonry which is below grade and in contact with the earth, such as foundations, retaining walls, walks, sewers, and manholes. It will also withstand severe frost action and high-lateral loads imposed by pressure from the earth.

Type S

Type S mortar is recommended for use in reinforced masonry and for standard masonry where maximum flexural strength is required. It is also used when mortar is the sole bonding agent between facing and backing units. Type S mortar has a fairly high compressive strength of at least 1800 psi in 28 days.

Type O

Type O mortar is a low-strength mortar suitable for general interior use in walls that do not carry a great load. It may be used for a load-bearing wall of solid masonry. However, it may only be used in those walls in which the axial compressive stresses developed

Mortar Type	Compressive Strength*				Minimum Water Retention*** %	Maximum Air Content %	Efflorescence
	Minimum 7 days**		Minimum 28 days				
	psi	(kgf/cm^2)	psi	(kgf/cm^2)			
M	1600	(112)	2500	(175)			
S	1100	(77)	1800	(126)			
N	450	(32)	750	(53)	70	12	None
O	200	(14)	350	(25)			

*Average of three 2-9n. cubes.

**If the mortar fails to meet the 7-day compressive strength requirement, but meets the 28-day compressive strength requirement, it shall be acceptable.

***Flow after suction, percent of original flow.

Fig. 10-3 Physical requirements of laboratory mixed mortar

do not exceed 100 psi, and which will not be subjected to weathering or to freezing temperatures. The compressive strength of Type O mortar should be at least 350 psi.

PROPORTIONS

Each type of portland cement-lime mortar should be mixed in proportions shown in Figure 10-4.

ADMIXTURES IN MORTAR

Mortar *admixtures* are materials which are added to mortar mix, usually in small amounts, to achieve a particular result. The results may include the following:

- Increased workability
- Added color
- A stronger bond between the mortar and masonry units
- An acceleration in setting time

Little data has been published regarding the effect of admixtures on mortar bond or strength. However, experience on various jobs has indicated undesirable results may occur in some instances. Air entrainment, for example, has a definite detrimental effect on the bond between mortar and the unit. Admixtures should never be used unless they are definitely specified in the contract and the manufacturer's mixing directions are followed exactly.

Color

Colored mortar is produced by adding colored aggregates or mortar pigments to the mixture. The pigments have a fine consistency to assure thorough mixing and even color in the mortar. Only the minimum amount of pigment should be used, as the strength of the mix could be affected. Color pigments should be of a metallic oxide composition in amounts not exceeding 15% of the portland cement content. Carbon black may be used as a coloring agent if it does not exceed 3% of the cement weight of the mix. Coloring agents that have not been tested commercially or by previous experience should not be used.

Precolored masonry cement which will blend with certain colors of brick or help create unusual effects in masonry walls may be purchased.

Air-Entraining Agents

Air-entraining agents are used in masonry cement mortar to increase its resistance to freezing and thawing, thereby increasing the life and durability of the mortar. The air-entraining admixture traps microscopic air bubbles in the mix. These air bubbles allow the mortar to contract and expand. When freezing and thawing occur, the mortar is less likely to crack and break. Mortar also has a tendency to hold water longer when an air-entraining additive is used, thereby increasing workability of the mix.

Mortar with air content over 12% by volume will weaken the bond strength of the mortar mix, according to tests by the BIA. Manufacturers must, therefore, carefully control the amount of air-entraining agents in a mix.

Waterproofing Agents

Various types of waterproofing additives are added to mortar mixes to prevent moisture from penetrating the masonry wall. Bags of masonry cement or portland cement that contain waterproofing agents are marked to show this. Such cements are also recommended as a waterproofing coating for basement walls.

EFFLORESCENCE

Origins

Efflorescence is a major problem confronting architects, engineers, and masonry contractors. Efflo-

Mortar Type	Parts by Volume of Portland Cement	Parts by Volume of Hydrated Lime	Sand, Measured in a Damp, Loose Condition
M	1	1/4	Not less than 2 1/4 and not more than 3 times the sum of the volumes of cement and lime used.
S	1	1/2	
N	1	1	
O	1	2	

Fig. 10-4 Mortar proportions by volume

Fig. 10-5 Efflorescence on brick wall

Principal Efflorescing Salt		Most Probable Source
Calcim sulfate	$CaSO_4 \bullet 2H_2O$	Brick
Sodium sulfate	$Na_2SO_4 \bullet 10H_2O$	Cement-brick reactions
Potassium sulfate	K_2SO_4	Cement-brick reactions
Calcium carbonate	$CaCO_3$	Mortar or concrete backing
Sodium carbonate	Na_2CO_3	Mortar
Potassium carbonate	K_2CO_3	Mortar
Potassium chloride	KCl	Acid cleaning
Sodium chloride	$NaCl$	Sea water
Vanadyl sulfate	$VOSO_4$	Brick
Vanadyl chloride	$VOCl_2$	Acid cleaning
Manganese oxide	Mn_3O_4	Brick
Iron oxide	Fe_2O_3 or $Fe(OH)_3$	Iron in contact or brick with black core
Calcium hydroxide	$Ca(OH)_2$	Cement

Fig. 10-6 Common sources of efflorescence

rescence is a deposit of water-soluble salts upon the surface of a masonry wall, usually white in color. Efflorescence tends to spoil the appearance of the finished wall, Figure 10-5. It occurs many times just after the structure is completed, when architect and owner are most concerned with the finished appearance of the building.

Efflorescence appears as a white stain on brick walls. It is believed that the calcium chloride used to accelerate the setting time of mortar is responsible for this condition. If the weight of the cement in the mixture does not contain more than 2% calcium chloride, efflorescence should not occur. Excessive amounts of calcium chloride may be responsible for efflorescence.

Efflorescence occurs when certain conditions are present. The wall must contain soluble salts and moisture must be present in the wall for sufficient time to take the salts into solution and carry them to the surface of the wall. If either of these conditions can be eliminated, efflorescence will not occur.

In most cases, the salts originate in the wall interior in either the masonry unit or the mortar itself, Figure 10-6. The entrance of water or moisture in the wall causes a reaction with the wall resulting in efflorescence. Sometimes chemicals in the atmosphere and the masonry combine and cause this action. Another source of salt is due to the masonry coming in contact with ground moisture, such as in basement walls or retaining walls. If the walls are not protected by some type of moisture barrier, the salts may be carried up through the first several courses of masonry above the ground level.

Preventing Efflorescence

The following are several ways to prevent efflorescence:

- Reduce salt content by using washed sand.
- Use clean mixing water in the mortar.
- Keep materials on the job covered and stored off the ground at all times.
- After the wall is built, cover it with a canvas cloth or some other suitable waterproof covering.
- Overall good-quality workmanship and attaining a good adhesive bond between the unit and the mortar is the best way to eliminate the problem of efflorescence.

The appearance of efflorescence is dependent upon excessive moisture in the masonry work.

Fig. 10-7 The mortar shown here has been mixed with the proper water content. Note how it holds its body when pulled with the trowel.

Removing Efflorescence

Efflorescence can be removed by washing the wall with water. If that fails, use a diluted solution of part muriatic acid to 9 parts water and rinse thoroughly with clear water.

> **Caution**: When mixing the solution of muriatic acid into the water, make sure safety glasses or goggles are worn. Otherwise, eye injury could result

This is only a temporary cure, however, and the stain may reappear after a period of time. Moisture must be kept from entering the wall.

WATER CONTENT OF MORTAR

Regardless of what type of mortar is used on a structure, one of the most important considerations is the water content.

Water content is possibly the most misunderstood aspect of masonry mortar. This is probably due to the confusion between mortar and concrete requirements. Many architects and builders assume that mortar requirements are the same as concrete. Therefore, they may make inaccurate specifications for mortar. The makeup of mortar and concrete is very different, especially in the water-cement ratio. Many specifications are incorrect. They require mortar to be mixed with the minimum amount of water to produce workable material and prohibit retempering of mortar. These specifications result in mortar which has maximum compressive strength but which has less than maximum tensile bond strength. A strong bond between the unit and the mortar is not achieved. Mortars should always be mixed with the maximum amount of water to provide maximum tensile bond strength of the mortar.

Mortar used to lay brick is somewhat thinner than mortar used to lay a concrete block, Figure 10-7. This is because the concrete block is heavier and has a tendency to sink into the mortar. There is also a difference in the water content of mortar used with a soft, sand brick and that for hard, dense brick. This is because the absorption of the two units will differ and the setting time will, therefore, differ.

Retempering mortar (adding water and remixing) is acceptable, but must be done correctly. Only that water which has been lost from evaporation should be added. In the past, masons commonly discarded mortar if it had become stiff from dryness. Modern-day specifications allow the mortar to be retempered as often as necessary, as long as it is used within 2 hours of the inital mixing. Pace the amount of mortar being mixed on the job with the amount of mortar being used. When tempering mortar, always wet the edges of the mortar pan or board with water. This helps prevent the mortar from drying out too quickly. Wetting the mortar board or pan before filling it with mortar also helps to reduce evaporation of water.

Mortar must be workable to permit the mason to be productive. It must also have good strength and durability. Mixing the correct amount of water initially is the best way to achieve mortar with the proper consistency. Since masons spend approximately 50% of their time working with mortar, it is important that they understand the role of admixtures and water in mortar.

ACHIEVEMENT REVIEW

The column on the left contains a statement associated with the composition and mixing of mortar. The column on the right lists terms. Select the correct term from the right-hand list and match it with the proper statement on the left.

1. Mortar used for most masonry work such as exterior walls, chimneys, parapet walls, and exposed walls above the finished grade; tests 750 psi
2. Mortar with a high compressive strength and greater than average durability; used for masonry below grade, sewers, manholes, and retaining walls; tests 2500 psi
3. Low-strength mortar used for interior walls; tests about 350 psi
4. High compressive strength mortar used to reinforce masonry walls and where high bond strength in the wall is required; tests 1800 psi
5. Materials added to mortar mixture to achieve particular results such as color or better workability
6. Chemical used to speed the setting time of mortar
7. Deposits of water-soluble salts on the surface of masonry work; whitish in color
8. Principal cause of efflorescence
9. Chemical used to remove efflorescence from masonry work
10. The process of adding water to mortar to make it more workable after it has become stiff from dryness
11. Mortar that has been mixed with a minimum amount of water acquires a high compressive strength but is weak in this strength measurement

a. Type S mortar
b. Calcium chloride
c. Tensile strength
d. Type N mortar
e. Moisture
f. Type O mortar
g. Tempering
h. Muriatic acid
i. Type M mortar
j. Admixtures
k. Efflorescence
l. Bond

Unit 11
Mixing Mortar

OBJECTIVES

After studying this unit, the student will be able to

- proportion mortar ingredients for specific mixes.
- mix mortar manually with hand tools and equipment.
- mix mortar with a power mixer.
- follow correct safety practices when mixing mortar.

Student masons or apprentices must be able to mix mortar properly for different job requirements. As previously mentioned, masons usually do not mix their own mortar or *mud,* as it is known in the trade. However, on small jobs or repair jobs, it is often economically practical for masons to mix their own mortar. Small jobs or repairs include repairing a window sill, pointing up cracks in a wall, or replacing the last course of brick on a retaining wall.

When only a small amount of mortar is needed, it can be mixed in a contractor's wheelbarrow or a small mortar box. For larger jobs, a mechanical mixer is used. Setting up the job with the necessary materials beforehand offers a savings when the work is started. This unit presents procedures for mixing mortar both manually with a shovel and hoe and mechanically using a power mixer.

> **Caution** Persons mixing mortar should wear safety glasses to protect against splashing mortar. When mixing in a poorly ventilated area, wear a dust mask.

ASSEMBLING AND STORING MIXING MATERIALS

The materials needed for mixing mortar should be placed close to the area where the work will be done. However, the materials should not be so close as to interfere with the laying of the masonry walls. Keeping materials close to the working area speeds up the work process.

All of the mixing ingredients should be covered with a plastic or canvas tarpaulin to protect them from the elements, Figure 11-1. It is not necessary to protect the sand from moisture in warm weather. During winter, however, masons must keep the sand pile covered to prevent freezing. If sand freezes, it must be completely thawed before it is usable. Cement and lime must be kept dry. They may be stored in a shed or placed on wooden pallets or boards off the ground to ensure complete protection from moisture.

Place mixing materials close together and have a source of water available for mixing materials and cleaning mixing equipment. A 50 gallon (gal) steel drum filled with clean water is placed near the mixing area on many jobsites. A length of rubber hose is required to fill the water barrel. The hose is also useful in cleaning mortar from the mixing equipment.

STANDARD PROPORTIONS FOR MORTAR MIXTURES

Proportions of ingredients in mortar are based on volume measurements. Mortar materials are manufactured, bagged, and delivered to the jobsite accordingly. Portland cement is available in 94 lb bags containing 1 cu ft. Mason's hydrated lime comes in bags

Fig. 11-1 A roofed frame protects the materials and worker. Such protection is especially valuable in cold weather.

which contain 50 lb. Sand is sold by weight but is measured by the cubic foot (cu ft) or yard (yd).

In actual practice, portland cement-lime mortar is mixed on the job by combining a certain number of parts of sand, lime, and portland cement. The ingredients may be measured by the shovel and then combined in the mixing box or mechanical mixer. The cement and lime may be measured by the bag.

Specifications of some jobs require that the mortar ingredients be measured very carefully by using a material batcher. This is done by counting the number of shovels of sand needed to fill a box 1′ long x 1′ wide x 1′ deep (inside measurements), Figure 11-2. After the number of shovels of sand for 1 cu ft has been determined by filling the batcher, a matching amount is placed directly in the mixing box or mixer for each cubic foot required. The number of shovels required to fill the batcher should be rechecked each day to allow for any moisture or other factors which may cause the measurement to vary.

Masonry cement mortars require that only sand and water be added to the mix. Usually 1 shovel of masonry cement is added to 3 shovels of sand, or 18 shovels of sand to one 70 lb bag of masonry cement. This combination should be equal to a Type N mortar by ASTM standards and is a standard proportion for masonry cement mortar throughout the United States.

Different proportions and combinations of materials for mortar can be reviewed by referring to Unit 10, Figure 10-4.

Fig. 11-2 Cubic foot box used to measure dry mortar ingredients

MIXING MORTAR MANUALLY

On small jobs, mortar may be mixed manually rather than with a mechanical mixer. To make mixing mortar with hand tools and equipment as easy as possible, specific steps and procedures should be followed.

The necessary materials must first be assembled. The ingredients for the mix in this example are portland cement (Type 1), lime, sand, and water. The proportions of the mix are expressed as *1:1:6 (Type N)*, or 1 part portland cement, 1 part lime, and 6 parts sand. The mixing process may be followed in Steps 1 through 9, Figure 11-3.

1. Before mixing materials, the mortar box should be blocked up and leveled. There is no definite height at which the mortar box should be raised; It is determined by the existing ground level. Leveling prevents the water used in the mix from running to one end of the mortar box. Leveling also positions the mortar box more conveniently for the mason to use the hoe.
2. Put one-half the desired amount of sand needed for the batch in the mortar box. Measure all proportions for this mix with a standard dirt shovel. (This is standard practice on the job unless otherwise specified.) Spread the sand evenly over the bottom of the mortar box.
3. Use a shovel to spread the desired amount of portland cement and lime over the sand. If a large batch of mortar is being mixed, the portland cement and lime may be added by the bag rather than by the shovel. Add the remaining

half of the sand to the mortar box over the portland cement and lime. Adding the sand in halves helps to distribute materials evenly throughout the mix.

4. Blend the dry materials together with the shovel or hoe and push them to one end of the mortar box.

STEP 1

STEP 2

STEP 3

STEP 4

STEP 5

STEP 6

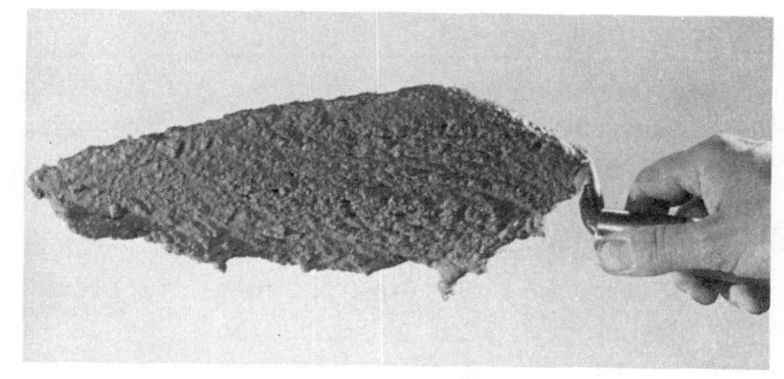

STEP 7

STEP 8

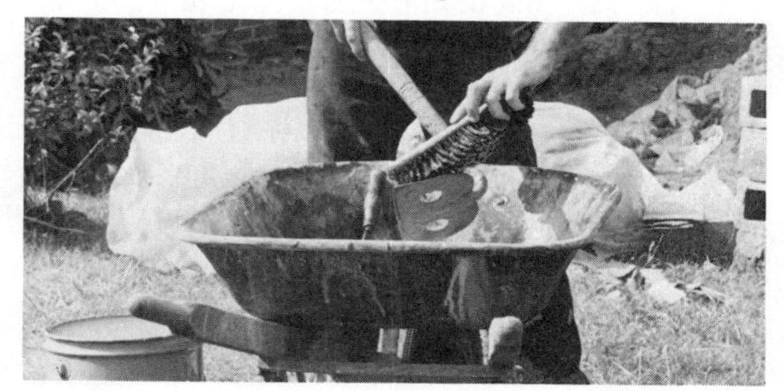

STEP 9

Fig. 11-3 Steps in mixing mortar

5. Add about 2 1/2 gal of clean water to the other end of the mortar box from a 5 gal bucket. Do not use a hose to add water since the water cannot be measured accurately.

   The total amount of water for a mix is determined by the moisture content of the sand.

6. Combine the dry mix and the water with a chopping action. Pull and push the mortar back and forth through the mix with the hoe at a 45° angle until all of the mortar is well mixed. Add water as needed to complete the mixing process. Tilting the hoe helps eliminate strain on the arms and wrists.

> **Caution**: Wear safety glasses or some other eye protection when mixing mortar. If mortar splashes in your eyes, do not rub them, as the sand may scratch your eyes. Wash your eyes immediately with clean water. If you cannot remove all of the mortar by rinsing with water, see a doctor as soon as possible for further treatment.

7. After the mortar has been mixed, pull all of the mortar to one end of the box with the hoe. This will help prevent the mortar from drying out before it is removed from the box and placed in the wheelbarrow. The mortar can be checked at this time by a simple test to see if it has the proper workability and adhesiveness. Pick up a small amount of mortar with the trowel and set it firmly on the trowel by flexing the wrist downward with a sharp, jarring movement. Turn the trowel upside down. If the mortar is of a proper consistency, it will adhere to the blade of the trowel.

8. Wet the wheelbarrow with a small amount of water before shoveling the mortar into it. Do this so that the mortar does not stick to the sides of the wheelbarrow. Move the mortar to the wheelbarrow with a flat shovel. When only a small amount of mortar is needed for a job, mix it in the wheelbarrow instead of the mortar box.

9. The final step in mixing mortar is to clean the mixing equipment with water. Do this as soon as possible after use. Mortar setting and drying on mixing tools could cause a buildup of mortar and permanent damage to the mixing equipment. A hose with a spray nozzle does the best job of cleaning because the water is under pressure. To scrub stubborn spot or particles of mortar from the tools, use a stiff brush.

MIXING MORTAR WITH THE POWER MIXER

When using a mechanical mixer (gasoline-powered or electrically driven), place all materials near the mixing area, as in the manual operation. The mechanical mixer should be in the center of the assembled materials. On many jobs, to avoid excessive wear on the tires, prop up the mixer with blocks. Secure the mixer carefully, since on a large job, it may stay in one location for over a year.

When mixing mortar with a power-driven mixer, the following steps should be followed:

1. Add a small amount of water to prevent the materials from building up and sticking to the sides of the mixing drum, Figure 11-4. This also helps to prevent the mortar from caking on the machine paddles as they whip through the mix.

> **Caution**: Wear safety glasses or goggles to protect the eyes.

2. Add one-half to one-third of the sand needed for the total mix. Add sand to the mixer while the paddles are turning to prevent excessive strain on the motor, Figure 11-5.

Fig. 11-4 Adding water to the mix to prevent mortar from sticking to the drum

Caution: Be extremely careful not to place the shovel inside the mouth of the mixer where it may become caught in the turning paddles. You could be severely injured by the uncontrolled shovel handle. Place a safety grate in the mixer when adding materials to help prevent accidents.

3. Add the necessary amount of portland cement, lime, or masonry cement to the mixing drum, Figure 11-6. Usually this is done by the bag. Lay the bag on the safety grate and cut it with a penknife, an old trowel, or by pulling the string on the corner of the bag to release the material. Some mixers have a sharp piece of metal on the top of the grate to puncture the bag.

Caution: Hands or fingers should never be put inside the mixer while it is running. If part of the torn bag falls into the mixer, do not try to remove it while the mixer is running.

4. Add the rest of the sand to the mixer. Add additional water at this time as needed to bring the mortar to the desired consistency. Allow the mixer to blend all materials thoroughly for 3 to 5 minutes. Prolonged mixing of mortar should be avoided. Excessive amounts of air will become trapped in the mix and the mortar will be spongy. Mortar of this consistency will not have the proper body and may cause problems in supporting the masonry unit.

5. At the completion of the mixing process, grasp the mixing drum by the dumping handle while the mixer is still running and dump the mortar into the wheelbarrow or mortar box, Figure 11-7. The turning of the blades clears most of the mortar from the mixer.

Caution: Do not scrape remaining mortar from the mixer drum until you take the blades out of gear and shut off the mixer.

6. Start the mixer again. Put enough water in the mixing drum to clean the inside while the paddles are turning. When no more mixing is to be done, scrub the mixer with a stiff-bristle brush and flush with enough water to remove all of the mortar, Figure 11-8.

Caution: Never attempt to scrub the mixer with the brush when the mixer is running.

Fig. 11-5 Adding the sand to the mixer. Sand is thrown on the grate, never into the mouth of the mixer.

Fig. 11-6 Adding the masonry cement to the mixer. One bag of cement makes one batch.

Fig. 11-7 Dumping the mortar from the mixer into the wheelbarrow

Fig. 11-8 Rinsing the mixer with water

Power mixers are necessary on larger jobs because they complete the job with relative speed and save labor since they require only one operator. However, the use of good safety practices cannot be overemphasized when operating a power mixer. Power mixers also assure a uniform mixture of materials. Usually, if the same person mixes a batch of mortar, the proportions and consistency should remain relatively the same.

Power mixers can be obtained in single-batch or double-batch sizes. A *batch* is defined as 1 complete mix of mortar, usually consisting of 1 bag of portland cement or masonry cement and other necessary materials.

PROBLEMS ENCOUNTERED IN MIXING MORTAR

Three problems masons face when mixing mortar are proportioning materials improperly, using poor-quality materials, and working in cold weather.

Proportioning Materials Improperly

If an excessive amount of water is added to a mortar mixture (known as *drowning* the mortar), the proper water-cement ratio is destroyed. To correct this problem, cementing materials, lime, and sand should be added in the correct proportions until the desired consistency is reached.

Producing mortar of the proper consistency for specific job conditions comes only with experience. The job will be easier for the trainee if water is added very slowly to the mix. It is considerably more trouble to add all the necessary ingredients again in their proper proportions. The same problem results when too much water is used to temper mortar.

Adding too much sand is one of the more common problems encountered when mixing mortar. This is caused primarily by poor measurement of materials. If a shovel is used for proportioning, the same type of shovel should be used all of the time with the same amount of material placed on the shovel each time. Usually, a standard dirt shovel is used. The amount of sand that a shovel holds will vary slightly depending upon the moisture content of the sand. The batcher measuring device is the most accurate.

Masons should never add sand to stiffen a mortar mixture without adding the proper proportions of other ingredients at the same time. This is a common problem in the trade, but should be avoided. If only sand is added, the compressive and tensile strength of the mixture changes drastically. The result is inferior mortar. It also causes variations of color which are noticeable in the finished mortar joint. Overly-sanded mortar is harsh, difficult to use, and forms a very weak bond. All of these factors result in a poor job.

Mortar which contains a high percentage of cementitious materials, known as *fat mortar,* is sticky and hard to dislodge from the trowel. Mortar which is lacking in cementitious materials is called *lean mortar.* Avoid both of these conditions.

Using Poor-Quality Materials

Cement which is old or which contains hard knots or lumps should not be used. The knots or lumps are caused by moisture penetrating the material and hydration taking place. The mason must remove all the knots and hard lumps from the mortar before the masonry unit can be laid in the wall. This is a tedious practice which wastes labor. It is more economical to discard the hardened mortar. Proper storage prevents this condition.

Working in Cold Weather

Cold weather presents special problems when mixing mortar. The use of admixtures or antifreezes is discouraged since they sometimes decrease the strength and affect the color of the mortar. If used, calcium chloride sometimes causes corrosion of metal ties.

When the temperature falls below 40°F, the mortar materials may require heating. Sand must be heated slowly and evenly to prevent scorching, which which turns sand slightly red. Heating is usually done by piling the sand over a metal tube or pipe and lighting a fire inside the tube.

Mixing water should also be heated in cold weather. The water should never be heated above 160°F, however, because of the danger of a flash set when portland cement is added to the mix. A *flash set* occurs when the portland cement sets prematurely due to excessive heat. After combining all of the ingredients to the mix, the temperature should be between 70°F and 105°F. If mortar temperatures exceed this limit, premature hardening may occur which results in greatly reduced compressive and bond strength. If it is less than 70°F, mortar sets too slowly.

When a problem concerning mortar occurs on the job, analyze and correct it as soon as possible. Good-quality mortar requires strict adherence to specific proportioning of materials. This care results in strong, watertight masonry walls.

MORTAR FOR TRAINING PURPOSES

If mortar is being used for training in a shop, only hydrated lime (no cement) is added to the mix. Lime is available in 50 lb bags. For good-quality mortar for training purposes, mix 1 part lime to 2 1/2 to 3 parts sand, depending on the fineness of the sand.

Lime is used without cement in a training mortar because the mortar does not harden and may be reused. Lime gives the mortar the necessary plasticity and workability. It may be necessary to add a small amount of lime to the rescreened mortar from time to time to increase the plasticity and workability. The amount to be added must be determined by the trainee according to the needs of the job.

When mixing mortar, the student mason should observe certain safety precautions.

- Always protect eyes and hands from injury.
- Wear safety glasses or goggles.
- Do not place hands in a power mixer when the machine is operating.
- Report all accidents at once to the class instructor or job supervisor.

ACHIEVEMENT REVIEW

A. There are specific procedures concerning mortar mixing which the trainee should have mastered at the conclusion of this unit. Mixing good-quality mortar requires a knowledge of the material and proper technique to form the finished mix. Review the following procedures until each one is learned.

Mixing by Hand

- Level the mortar box.
- Measure ingredients for the desired mix.
- Dry mix the ingredients.
- Add water.
- Mix the mortar with the hoe.
- Clean mixing tools.

Mixing Mortar by Machine

- Add water before combining ingredients.
- Proportion ingredients in the mixer.
- Add additional water.
- Mix 3 to 5 minutes.
- Unload mixer.
- Clean mixer.

B. Select the best answer from the choices offered to complete each of the following statements. List your choice by letter identification.

1. The trade name for mortar is
   a. cement.
   b. concrete.
   c. mud.
   d. gypsum.
2. Proportions of masonry mortar are based on
   a. volume.
   b. square feet.
   c. weight.
   d. liquid measurement.
3. Usually, the type of portland cement used to mix portland cement mortar is
   a. Type 1.
   b. Type 2.
   c. Type 3.
   d. Type 4.
4. The total amount of water needed for a mixture of mortar is determined by
   a. the amount of dry material being mixed.
   b. the thickness of the mortar joint.
   c. the moisture content of the sand.
   d. the type of mortar being mixed.
5. After the mortar has been mixed, it is tested by filling a trowel with mortar and turning it upside down to see if the mortar remains on the trowel. The property of the mortar being tested is
   a. water retention.
   b. plasticity.
   c. strength.
   d. adhesive quality.
6. When mixing mortar with the mechanical mixer, water is always added to the mixer first. The main reason for this is
   a. to help the mortar mix more thoroughly.
   b. to speed the mixing process.
   c. to prevent the mortar from sticking to and building up on the mixer.
   d. to increase the workability of the mix.
7. If mortar is mixed for an excessive amount of time in a mechanical mixer, it becomes very spongy. The correct mixing time is
   a. 8 minutes.
   b. 3 to 5 minutes.
   c. 2 minutes.
   d. 10 minutes.
8. If the mortar has been mixed with too much water, the mason
   a. adds more sand.
   b. adds more cement.
   c. adds more lime.
   d. adds in proportion all of the materials that were originally in the mix.
9. Mortar which contains a high percentage of cement is sticky and difficult to dislodge from the trowel. The trade term for this type of mortar is
   a. lean mortar.
   b. fat mortar.
   c. sharp mortar.
   d. mud.

10. In cold weather, masonry materials must be heated so that proper curing of the mortar may take place. Heating the materials is recommended after the temperature drops below
    a. 32°F.
    b. 40°F.
    c. 45°F.
    d. 50°F.

SUMMARY, SECTION 3

- Portland cement added strength to the traditional lime-based mortar and changed the entire building industry.
- Lime-based mortar has many advantages over other types, including high water retention, workability, plasticity, greater bond strength, elasticity, flexibility, economy, high resistance to moisture from weathering, and the ability to reseal cracks in joints by autogenous healing.
- Sand for masonry mortar should be clean and free from organic matter.
- Water used to mix mortar should be clean and free from alkalis, salts, acids, and organic matter.
- When selecting a mortar, consider the needs of the project.
- The different types of mortars include Types M, N, S, and O. The different types are identified by the proportions of sand, lime, and portland cement in the final mix.
- Standard masonry cement is obtained prepackaged and requires only the addition of sand and water. It is very popular with masons, as there is less material to combine when preparing mortar on the job.
- Admixtures are available for use in mortar. Their benefits vary, but there is the possibility of the admixture affecting the strength of the mortar. Admixtures should be used only when there is a definite need. If they are used, the manufacturer's instructions must be followed to the letter.
- Efflorescence is one of the major problems encountered in the masonry trade. It is caused mainly by moisture coming in contact with soluble salts either in the mortar or the masonry unit. It can be prevented somewhat by keeping the masonry units dry.
- Always mix mortar with the maximum amount of water to maintain proper workability.
- Place all necessary mortar mixing materials as close together as possible without hindering the work.
- When mixing mortar manually or by machine, proportion the ingredients as accurately as possible.
- Always observe good safety practices.
- Do not use old or defective cement.
- Mortar materials can be heated in cold weather but strick adherence to designated temperatures must be observed or the strength of the mortar will be affected.

SUMMARY ACHIEVEMENT REVIEW, SECTION 3

Complete each of the following statements which refer to material found in Section 3.

1. The ingredients of mortar include ________________.
2. Quicklime is changed to hydrated lime by a chemical reaction caused by the addition of ________________.
3. The ability of mortar to hold moisture when applied to the masonry unit is called ________________.
4. The use of portland cement in mortar gives the mortar ________________.
5. Sand is defined as the deterioration of ________________.
6. The siltation test is designed to detect the presence of ________________ in sand.
7. The total number of portland cement-lime mortar types discussed is ____________.
8. So that mortar may better resist freezing and thawing, ________________ are added.
9. Efflorescence can be removed temporarily from masonry work by using either plain water or muriatic acid and water. The cause of efflorescence is ________________.
10. Retempering of mortar should only be done to replace water lost through ________________.
11. When using a material batcher for measuring mortar ingredients, the unit of measurement is ________________.
12. In cases where mortar is to be mixed by hand, it is mixed with a (an) ___________.
13. Water is added to the mix slowly to prevent ________________.
14. Before scraping excess mortar from a mechanical mixer, the mason should be sure that the mixer is ________________.
15. Portland cement setting prematurely due to excessive heat is called a (an) ________________.
16. If only one type of portland cement-lime mortar could be chosen for a variety of conditions, the best choice would be ________________.

Section 4
Essentials of Bonding

Unit 12
Introduction to Bonding

OBJECTIVES

After studying this unit, the student will be able to

- discuss the different meanings of *bonding.*
- describe various types of bonds.
- describe various brick positions and their application.

BOND

The term *bond* has several different meanings to the masonry student or apprentice. The following three types of bonds are important for masons to know:

- The adhesion of the masonry unit with the mortar joint is known as a *mortar bond.*
- The interlocking of masonry units to each other to distribute the weight of the wall is called a *structural bond.*
- The arrangement of masonry units to form a pattern, design, or texture is called a *pattern bond.*

All bonds should be structurally strong, provide a good adhesive bond between the mortar and the masonry unit, and maintain the pattern or design specified. These qualities are present in varying degrees, depending on the type of bond used.

Once the bond type has been established and construction of the wall has begun, the bond must be carried true for the entire length and height of the wall. *True,* in this sense, means that in every other course the unit should be in line vertically with the unit underneath it or so that the pattern or design is not changed. All masons share the responsibility of building a true, strong bond if the finished job is to be an acceptable one.

MORTAR BOND

Mortar is the first material the mason uses when laying any masonry unit. Therefore, establishing a good bond between the unit and the mortar should be a prime consideration. The adhesion of mortar to the masonry unit, wall ties, or other means of reinforcement is known as the *mortar bond.*

The stronger the mortar bond is, the less chance there is of the wall leaking due to space forming between the mortar joints and masonry units and shrinkage of the joints, Figure 12-1. When reinforcing is used to strengthen the wall, a good bond must be formed between the steel and the mortar. Lime-based portland cement mortar provides a very high bond strength. Mortar must not be allowed to dry too quickly or a very poor bonding between the masonry units results and the strength of the wall is affected. Spreading mortar too far ahead of laying the masonry units results in a poor mortar bond. Spreading mortar so that it does not dry too quickly is a skill that every mason must master.

Fig. 12-1 Bricks from an old building demonstrate that lime-based mortar adheres much better to brick (mason's left hand) than mortar with a high cement content (mason's right hand). Notice that the brick in the right hand has separated completely from the mortar.

Once the masonry units have been laid and the initial set has taken place, the units should not be moved or shifted in the wall. If the unit is moved after the mortar has partially set, the bond will be seriously weakened or broken. In these cases, the broken bond between the mortar and masonry unit never reunites completely. The high water retention of lime-based mortar allows more time to lay the unit before the mortar drys on the wall.

THE STRUCTURAL BOND

Structural bonding is the method by which individual units are interlocked or tied together when laid so that the entire mass acts as a single structural unit. Structural bonding of brick and concrete block may be accomplished in three different ways.

Overlapping of Units

Masonry units may be lapped over one another to provide structural strength. Lapping one unit halfway over the next unit provides the best distribution of weight, Figure 12-2.

Fig. 12-2 A 4-inch brick wall laid in one-half lap stretcher bond. This bond provides excellent distribution of weight.

Bonding Ties

Metal ties embedded in the mortar or steel wire reinforcements embedded in the horizontal mortar bed joint are used to strengthen and bond walls together. Modern building codes permit the use of rigid steel bonding ties, Figure 12-3. At least 1 metal tie should be used for every 4 1/2 square feet (sq ft) of wall surface. Ties in alternate courses should be staggered so that no 2 ties form a continuous vertical line. The maximum distance between ties in a horizontal position is 36". In a vertical position, ties should be installed every 16" for best results. Since there is a high degree of moisture present in all masonry walls due to condensation, ties should be coated to prevent deterioration. The most popular method is to galvanize the ties.

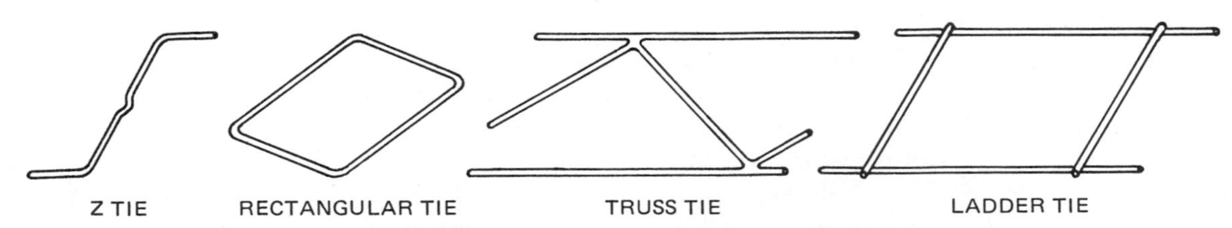

Fig. 12-3 Various ties used to bond masonry walls together

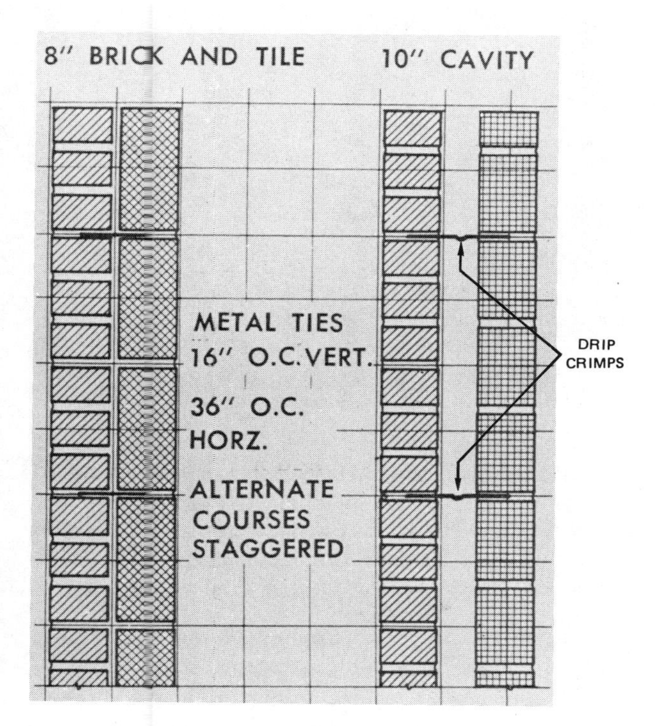

Fig. 12-4 Masonry wall with metal ties. Notice the drip crimps which allow moisture to drip into the cavities.

Structural bonding with wall ties is also used when building *cavity walls* (walls which have a cavity or air space of a minimum 2″ between them). This type of tie may have a *drip crimp* in the center. This drip crimp allows the moisture to drip from the center of the tie into the cavity rather than travel across the cavity to the other wall, Figure 12-4.

Structural bonding can also be accomplished by use of a header brick, considered the strongest tie in masonry. A *header brick* is a brick laid in such a position that it crosses over and rests on both sides of an 8″ wall. Its name is derived from the fact that the head or end of the brick is exposed on the face of the wall. Header ties are used in the construction of a solid masonry wall or cavity wall, Figure 12-5.

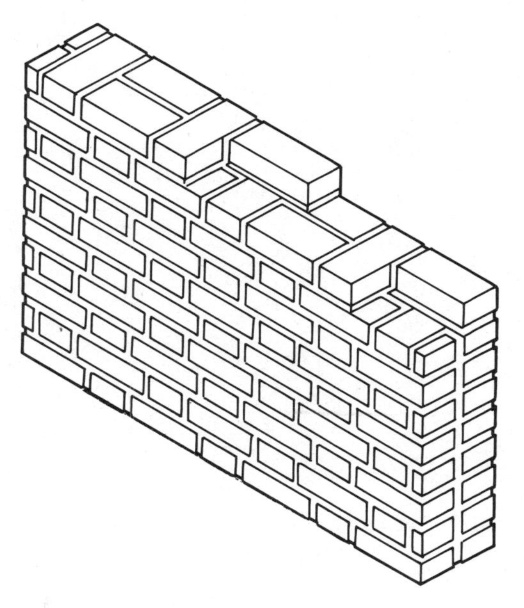

Fig. 12-5 Solid brick wall bonded with header bricks

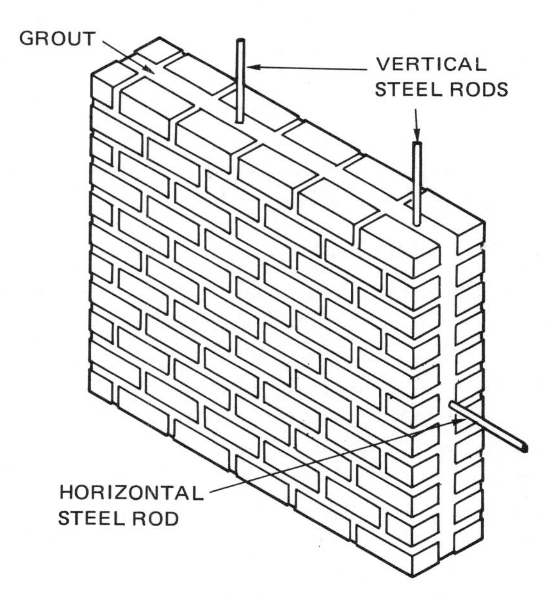

Fig. 12-6 Reinforced brick wall

Reinforcement Rods

The third method of structural bonding is accomplished by building a double wall and leaving a space in the center, usually 2″ to 4″ wide. Steel reinforcement rods are installed in the air space and mortar or concrete grout is mixed and poured in the space to fill all voids or holes, Figure 12-6. This type of mortar or concrete is known as *grout*. Although the grout should be thin enough to be poured, it must also be strong. There are specific proportions for the mixture of grout. This type of masonry work is known as *reinforced masonry*. This is a good method to use in structures which will have a great lateral load on the wall.

The method of structural bonding which is used depends on the requirements of the building, the wall type, and other factors. The metal tie method is

Fig. 12-7 Recessing and projecting of brickwork can be used to add variety and interest to structures.

Fig. 12-8 This section of brick from a building on the campus of the University of Virginia shows the colonial grapevine joint. The building, dating from the late eighteenth century, was constructed with a lime-based mortar, since portland cement had not yet been developed. The structure is still in good condition.

highly recommended for exterior walls. Advantages include greater resistance to rain penetration and ease of construction. Metal ties also allow slight differential movement of the facing and backing walls with a minimum of cracking. On walls with headers, moisture may collect on the inside of the wall where the header is installed. Any shifting of the wall also may cause cracks, as the brick header does not have the flexibility of the metal tie.

THE PATTERN BOND

Pattern bonding is defined as the arrangement of masonry units in a wall to form a pattern or recurring design. *Pattern* can also refer to a change in color and texture of the units. It may be possible to secure many different patterns with the same structural bond by mixing various colors and textures. For example, placing dark brick headers in specified positions throughout a wall has an interesting effect.

Recessing bricks or concrete block from the face of the wall and then returning the next course to the original wall line causes shadows and adds depth to the wall by suggesting a three-dimensional appearance. This method can be used to add interest and character to a structure, Figure 12-7.

Joints can be finished in a variety of ways with the use of striking tools. The raked joint adds depth to any wall. The V-joint highlights a rough-textured brick. The grapevine joint creates a pleasing effect with an irregular line running through the bed and head joints, Figure 12-8. The effect is especially appealing when used with pink colonial sand bricks.

There are five basic structural bonds which form pattern arrangements in masonry work. Included are the *running bond,* more commonly called the *all stretcher bond, common* or *American bond, Flemish bond, English bond,* and *stack bond.* These bonds will be explained further in Unit 13. There are, of course, more than five bonds that can be used, but the majority of all masonry work specifies one of these five. Through the use of bonds, mixing of color and texture, recession and projection of the masonry units, and special treatment of mortar joints, an almost unlimited number of patterns can be developed.

Brick Positions

To create the various bonds and patterns found in masonry work, the masonry unit must be laid in a particular position. These positions of bricks have been given trade names over the years by the masonry industry. Architects specify the arrangements of patterns and bonds by indicating the position of the bricks or blocks when laid in the wall.

There are six different positions in which a brick or masonry unit may be laid, Figure 12-9.

The *stretcher* is laid in a horizontal position with the longest, narrowest side exposed at the front of the wall. This is the most frequent position in which bricks are laid.

The *header* is laid with the 4" face of the brick exposed at the front of the wall. The widest part of the header is laid down in the mortar bed joint. Headers are used primarily for tying two separate masonry

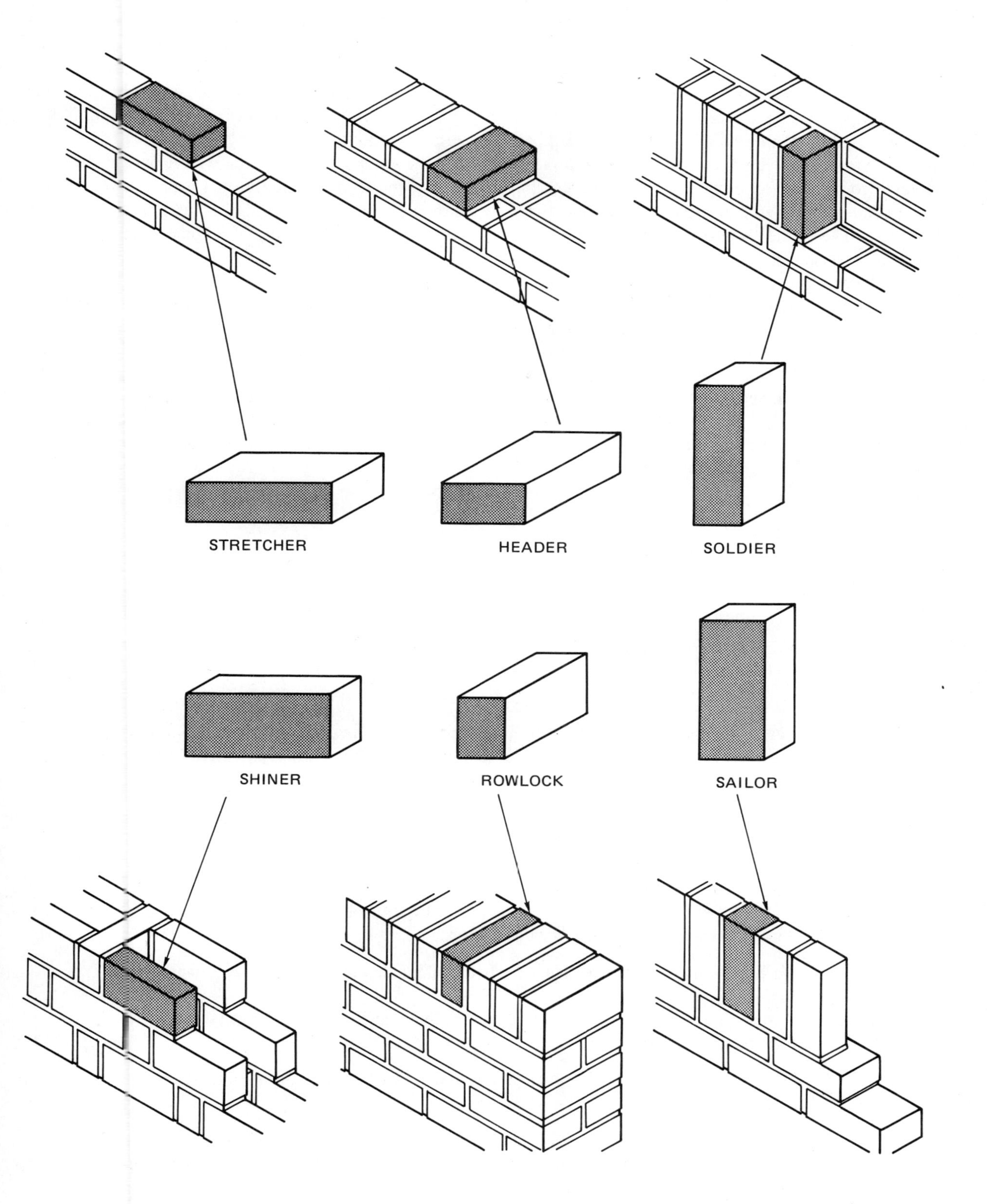

Fig. 12-9 Brick positions as they appear in a wall

walls together. They can also be used for capping walls, contributing to different pattern arrangements, and on flat windowsills.

The *soldier* is laid in a vertical position on its narrowest and shortest side, with the longest side of the brick exposed on the front side of the wall. Soldiers are most frequently used over doors, windows, or openings to simulate an arch. Do not, however, confuse a soldier course with an arch, as they are entirely different. The soldier is also used in different types of pattern arrangements.

The *shiner* is laid with the widest side of the brick in a horizontal position exposed on the front of the finished wall. It is also used for leveling when a material 4" high is needed, or for decorative purposes when constructing certain pattern bonds. The most common use of a shiner is in flat paving work, such as brick walks.

The *rowlock* is laid in a vertical position with the end of the brick facing the front of the finished wall. This is different from the header in that it is laid on its narrowest edge. The most common use of the rowlock is brick windowsills. Rowlocks are also used to cap walls, in pattern arrangements when building complicated bonds, and in the ornamental cornices of brick buildings.

The *sailor* is laid with the widest part of the brick facing the front of the finished wall in a vertical position. Sailors have very limited use, but are found in different pattern bonds. The most common use of the sailor is in brick walks and to form bond patterns in a wall.

It has been said that much of the old charm and appeal of masonry work has been lost since prefabricated materials have become popular. However, through the use of various bonds and patterns, the mason can exercise creativity and add interest to structures. However, masons should remember that as the design becomes more complicated, the cost involved in creating the structure rises.

ACHIEVEMENT REVIEW

A. Select the best answer from the choices offered to complete the statement or answer the question. List your choice by letter identification.

1. Mortar bond is related to which of the following terms?
   a. Pattern
   b. Adhesiveness
   c. Interlocking units
   d. Tying

2. The term *bond* has several different meanings to the mason. Which of the following is the most important type of bonding relating to distribution of weight?
   a. Mortar bond
   b. Structural bond
   c. Pattern bond
   d. Finish and texture bonding

3. Which of the following describes the term *breaking the set*?
   a. Cutting the masonry unit
   b. Laying out the bond without mortar
   c. Moving the masonry unit after the mortar begins to harden
   d. Arrangement of the masonry unit to form a pattern

4. Masonry units may be overlapped when building a wall. The strongest overlap is the
   a. one-quarter lap.
   b. three-quarter lap.
   c. half lap.

5. The maximum distance between horizontal wall ties in a masonry wall should be
   a. 36".
   b. 16".
   c. 24".
   d. 48".

6. The maximum distance between vertical wall ties in a masonry wall should be
   a. 24".
   b. 36".
   c. 48".
   d. 16".

7. Structural bonding can be accomplished by tying two walls together with a brick. When a brick is laid in this position it is called a
   a. stretcher.
   b. snap header.
   c. soldier.
   d. header.

8. Mortar proportioned so that it is thin enough to pour into a cavity wall and fill all voids is called
   a. rich mortar.
   b. grout.
   c. portland cement.
   d. lean mortar.

9. A wall can be given the appearance of pronounced shadows and a sense of depth by
   a. staggering the brick alignment in the bond.
   b. rough-textured brick.
   c. recession and projection of brick.
   d. use of a darker mortar in the joints.

10. It is possible to design and construct many different patterns while using the same structural bond. This is done by
    a. mixing colors and textures of the bricks.
    b. cutting the bricks.
    c. tooling the joints differently.
    d. changing the color of the mortar joint.

B. Study the different brick positions shown in Figure 12-9. Make a drawing showing each of the brick positions as they may be used in a brick wall.

Example:

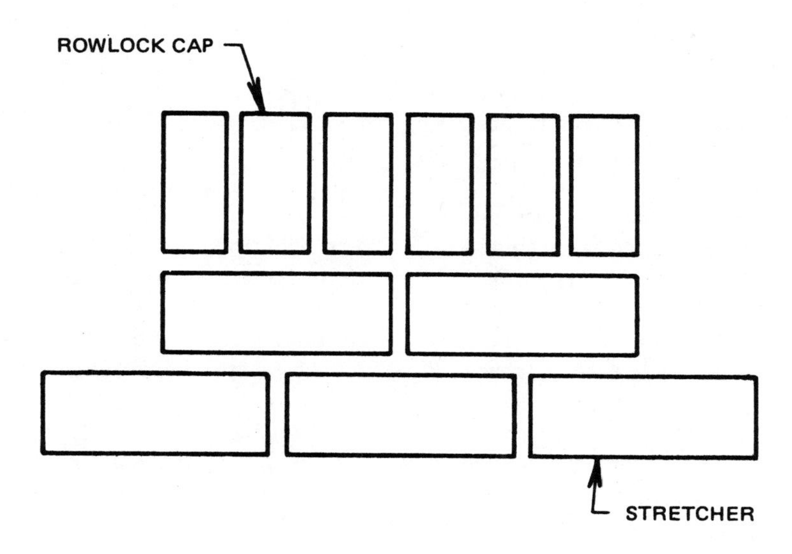

Unit 13
Traditional Structural and Pattern Bonds

OBJECTIVES

After studying this unit, the student will be able to

- describe the five basic bonds and how to start each.
- lay out and construct a brick panel which includes the five basic bonds.
- dry bond masonry projects in preparation for construction.

As mentioned in Unit 12, there are five basic bonds with which the mason should be familiar. They are the *running* (or *all stretcher*) *bond, common* (or *American*) *bond, Flemish bond, English bond,* and the *stack bond.* Although the mason may not build all of these bonds frequently, a thorough knowledge of the construction of each is necessary, since every mason will lay out and construct each bond at some time. Remember that the arrangement of texture and color can accentuate all bonds and pattern designs.

The most important step in building any bond is to start the course correctly. Starting a course involves using different pieces or cuts of brick on the corner jamb of the structure. The mason should be aware of the requirements of brick pieces for each bond.

The bond must be built true to the pattern or layout course. Breaking up an established pattern in a decorative bond destroys the beauty and meaning of the bond. Architects and builders are also very conscious of good bond pattern development, since even people unfamiliar with masonry construction can detect broken bond patterns when the job is completed.

THE RUNNING OR ALL STRETCHER BOND

The simplest of all bonds to build is the *running* or *all stretcher bond.* This bond consists entirely of stretcher bricks (full bricks) with the exception of pieces which would occur at windows, doors, or other openings. Since there are no headers, metal ties are used when the wall must be tied to another masonry backing or framework such as the sheathing of a frame house. The running or all stretcher bond is simple, fast, and economical to construct. It is, therefore, a favorite of builders for masonry construction.

The majority of brick masonry houses constructed today are brick veneers, Figure 13-1. Brick veneer walls are usually 4″ thick and built against a frame or a comparable substitute. The running bond is a natural choice for veneer work, since the standard brick is 4″ thick.

The running bond is constructed with either the *half lap* (half bond) or *one-third lap* (one-third bond). The half lap is used when laying standard 8″ bricks, since the return brick on the corner measures 4″ and the standard brick will, therefore, bond evenly over the next brick, Figure 13-2.

The one-third lap is generally used when laying a 12″ long Norman brick, Figure 13-3. It forms a weaker bond than the half lap because the weight is not distributed as evenly and the half lap covers more area. The one-third lap is, however, a very sound lap which forms a good bond.

THE COMMON BOND

The *common* or *American bond* is a variation of the running bond, with a course of headers at regular intervals tying the units together. The headers may be installed on the fifth, sixth, or seventh course, depending on the size of the masonry or back-up units, Figure 13-4. As a rule, the common bond is only

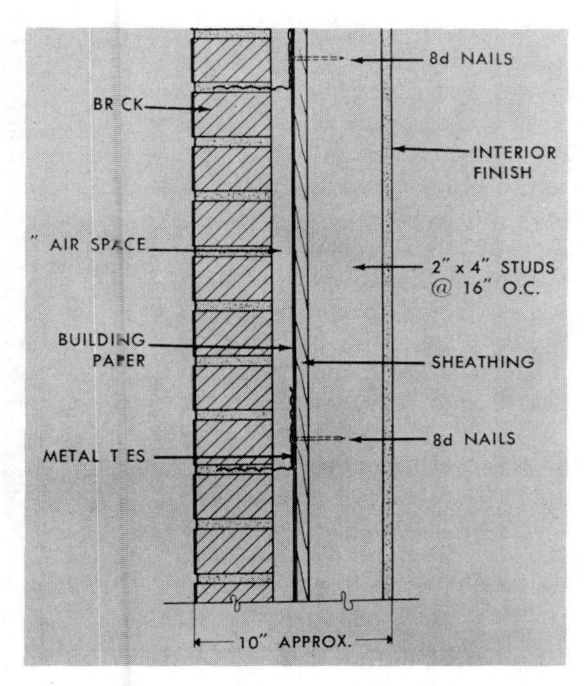

Fig. 13-1 Brick veneer wall section. Note the wall ties in the framework.

used when a solid wall is being built. The common bond was the most often used bond until the demand for veneer masonry became so great.

To start the corner for a header course, lay a 6" piece of brick (often called a *three-quarter*) on the corner. Two three-quarters are required as one must be laid on each side of the corner. The header brick is laid against the three-quarter and carried throughout the course. This is known as a *full header course,* Figure 13-4. The three-quarter on the corner allows the header to center over the head joint in the stretcher course below. Every other header laid will then be centered over the head joint of the stretcher beneath it.

If a variation of the header is desired for a pattern or design, a *Flemish header* may be used. The Flemish header is also started from the corner with a three-quarter piece. Instead of using headers throughout the course, however, a stretcher is alternated between every header. The Flemish header does not form as secure a bond as the full header since only every other brick is tying the wall together, Figures 13-5 and 13-6.

When the mason reaches the height of the wall at which the header is to be installed, the wall is said to be *header high.* When using the common or American

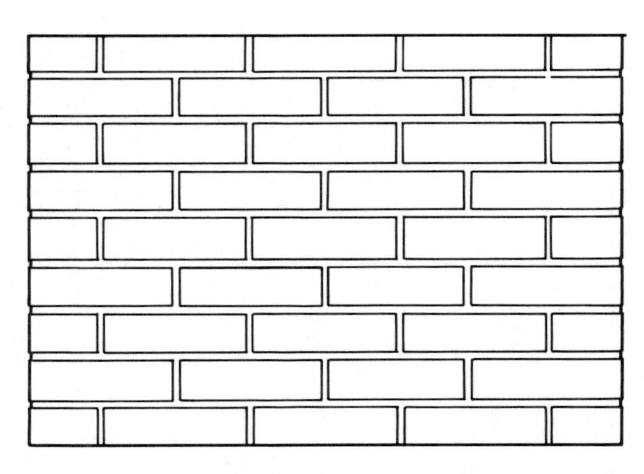

Fig. 13-2 Running bond (all stretcher) wall laid with a half lap over brick underneath.

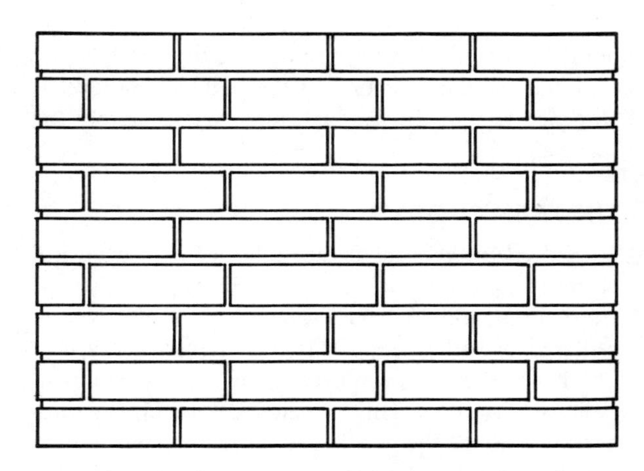

Fig. 13-3 Brick wall laid in running bond with one-third lap

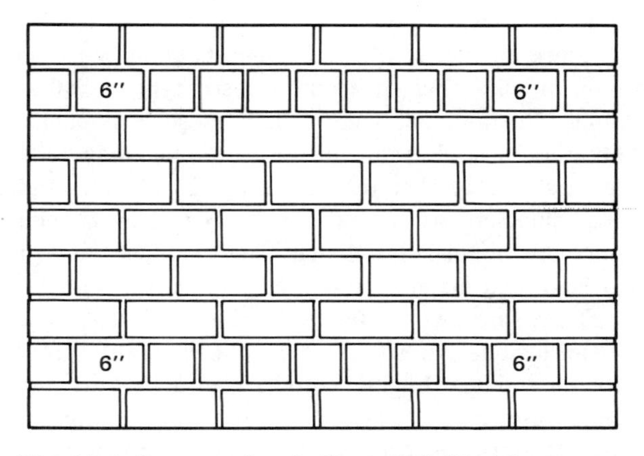

Fig. 13-4 Common bond. Note the full header courses every sixth course.

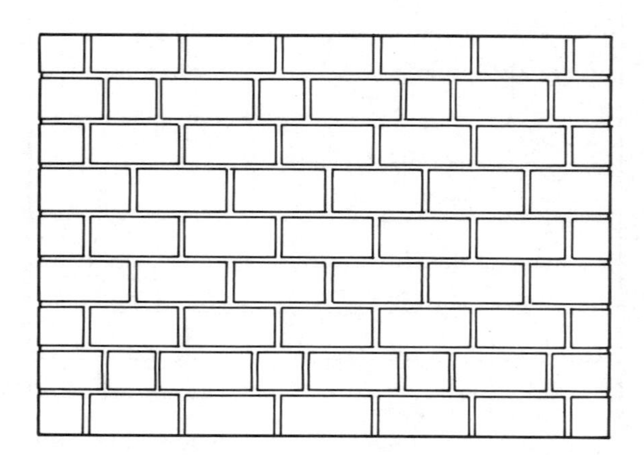

Fig. 13-5 Common bond. Flemish headers are on the sixth course.

Fig. 13-6 Brick wall laid in common bond with a Flemish bond header every seventh course.

bond on a wall, it is important that the front of the wall and the back portion of the wall be kept level at all times so that the header course is level across the two walls.

THE FLEMISH BOND

The *Flemish* bond is one of the truly beautiful pattern bonds. Each course of bricks consists of alternate stretchers and headers, with each header centered over the stretcher that is directly beneath it. The headers, located on every other course, are in an even vertical line. This presents a very pleasing appearance. Since the wall is tied on each course, due to the fact that each header is tied into the backing course, a wall with a Flemish bond will always be built of solid brick. The only exception to this rule occurs when a 4″ wall is built with a Flemish bond and *dummy headers* (or *snap headers*) are cut and used on the wall.

Since the wall with the Flemish bond must be built of solid brick, it is a more costly type of wall to build. Operators of many buildings such as banks, libraries, schools, and museums prefer the architectural beauty that the Flemish bond provides and are willing to pay the extra cost. The Flemish bond was very popular in colonial America when the cost of labor and materials was not a factor as it is today.

There are two different methods of starting a Flemish bond from the corner, known as the Dutch corner and the English corner, Figure 13-7. In the Dutch corner, a 6″ piece of brick (three-quarter) is used to start the corner or is laid against the head of the return brick on the corner, depending on the structure. The header is then laid against the three-quarter in the same way as the full-header bond is laid. The English corner uses a 2″ piece (also known as a *plug*) against the corner return brick, Figure 13-8. Very small pieces of brick are normally not used in masonry. It is very seldom that a piece smaller than a half brick is specified by an architect unless a true colonial job is to be built and then a 2″ piece is used.

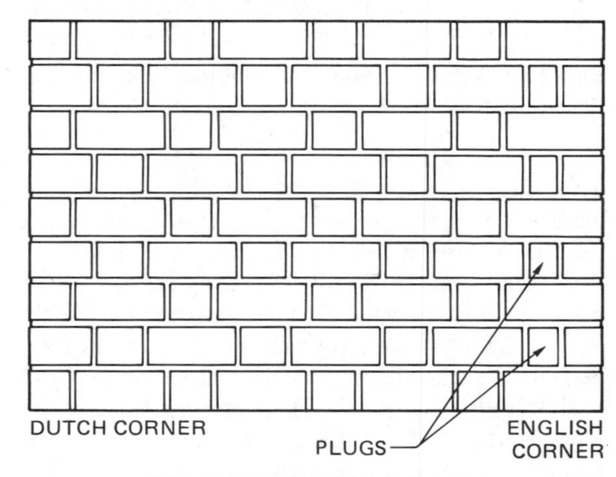

Fig. 13-7 Flemish bond with Dutch and English corners.

The Dutch corner is considered a more modern design than the English corner. The Dutch method is more commonly used since it is easier to cut and is more firm in the mortar joint, thereby being more economical to build. Both methods serve the same purpose of breaking the bond of the brick which is located underneath it.

Fig. 13-8 Corner of a brick building built with Flemish bond. Note the 2-inch starter pieces near the corner.

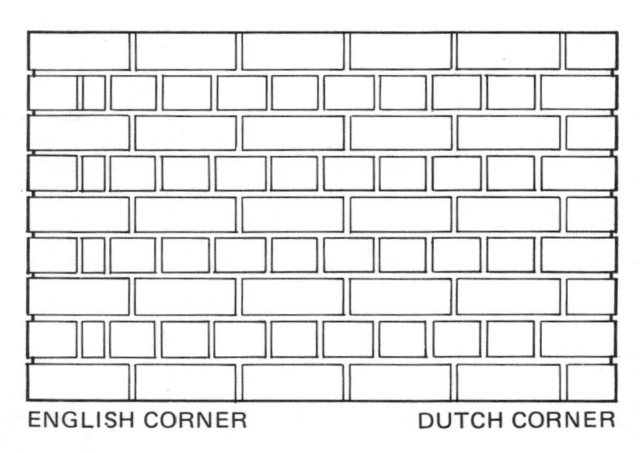

Fig. 13-9 English bond with English corner (1-inch piece) and Dutch corner (6-inch piece)

ENGLISH BOND

The *English* bond is composed of alternate courses of headers and stretchers, Figure 13-9. Do not confuse this bond with the Flemish bond, which has a header and stretcher alternating on the same course. In the English bond, the headers are centered on the stretchers and the joints between the stretchers in all courses are in line vertically. This bond is usually used when a structure with an 8" solid brick wall is being built. If a 4" wall is specified, snap headers must be cut and used in place of the header bricks.

The English bond is seldom used in masonry today due to the expense of constructing a solid brick wall and the high cost of laying an intricate pattern. It was a favorite in colonial period masonry and is still preferred when colonial reproduction is required, Figure 13-10.

The mason must be especially precise in keeping the head joints in true plumb alignment. If the bond staggers even slightly, the effect of the pattern is seriously affected. The cross joints between headers must be kept full of mortar to prevent any water from leaking through the wall. Any bond that has many headers is more subject to moisture leakage than a running bond, since moisture which collects in the center of the wall leaks out at the header level.

The English bond can be started from the corner or jamb with a 2" piece of brick (English corner) or a three-quarter (Dutch corner), just as a Flemish bond is started.

Fig. 13-10 Example of an English bond wall. Note the diamond pattern effect of this bond.

THE STACK BOND

The *stack* bond, Figure 13-11, is used basically for decorative purposes. In this bond, there is no overlapping of masonry units when they are laid in the wall. This pattern must be bonded or tied with ladder or truss-type steel reinforcing wire, whether the wall is *load bearing* (a wall which supports part of the weight of a building, such as steel beams, concrete floors, or joists) or *non-load bearing* (a wall which supports only its own weight).

Masonry units in stack bond must be of the same length and height since the vertical alignment of the head joint must be perfectly straight. A truer size of

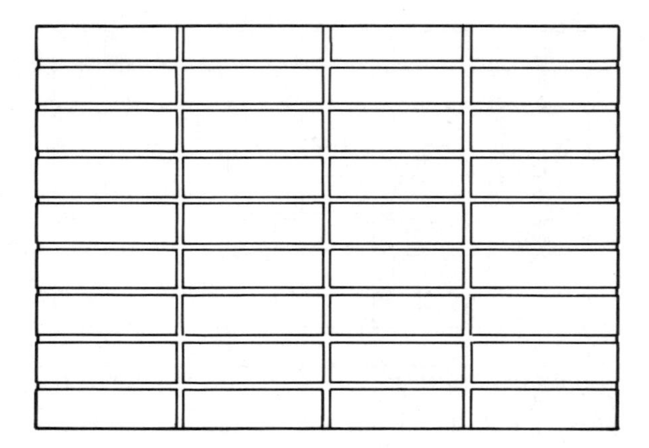

Fig. 13-11 Stack bond. Notice the plumb vertical head joints.

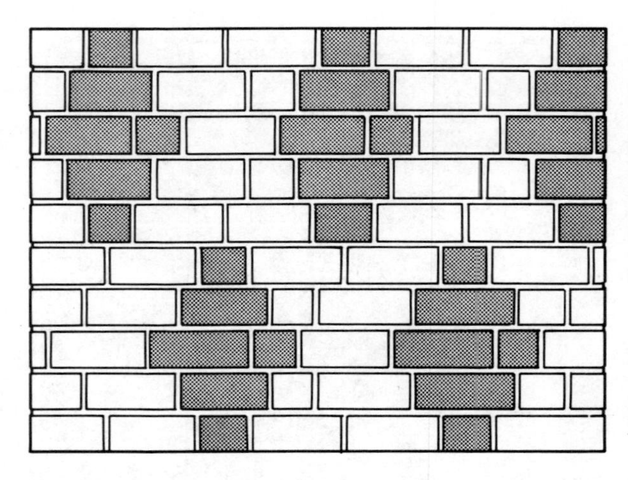

Fig. 13-12 Double stretcher garden wall bond with units in diagonal lines

face brick is usually selected when building a project in the stack bond to help assure that the joints will be plumb. The mortar head joint will compensate for only a minor difference in brick size before it becomes noticeable in the wall. The stack bond takes more time for the mason to construct than a conventional running bond because of these factors.

If a stack bond is specified, it is for design purposes. Structurally, the stack bond is the poorest of all of the bonds discussed since there is no overlapping of the masonry units. A slight movement of the units caused by expansion or contraction of the wall could cause cracks to appear in the vertical head joints.

However, a wall in the stack bond is a safe and acceptable method to use when constructing a masonry wall. If specifications are followed and the units are tied together with ladder or truss-type wall ties, the wall will be structurally sound. Many architects design and recommend the use of stack bonded walls in modern buildings.

GARDEN WALL BOND

An adaptation of the Flemish bond, the *garden wall bond,* is often used in walls to enclose or screen a garden, court-yard, or estate. There are two versions of the garden wall bond. A bond with 2 stretchers and a header alternating on the same course is known as a *double stretcher garden wall bond,* Figure 13-12.

A garden wall bond with 3 stretchers and a header alternating on each course is known as a *three stretcher garden wall bond,* Figure 13-13. The diamond effect created appears larger in the three stretcher garden wall bond.

Fig. 13-13 Three stretcher garden wall bond with brick in dovetail fashion

Combinations of dark headers and stretchers emphasize the pleasing diamond pattern, making the garden wall bond popular for decorative work. Projecting or recessed bricks also achieve special effects in the pattern. However, each brick must be properly placed or the pattern will be ruined. Masons should follow an elevation view of the project to assure that the pattern is correct.

DRY BONDING THE MASONRY UNIT

Bricks

Dry bonding is another important type of bonding that the mason must be able to perform. This involves laying out the masonry unit dry (without using mortar) to determine if a unit will fit in a given area without cutting small bricks. Dry bonding by the proper method is the sign of a true craftsperson.

It is always good practice to lay out the bond dry before spreading mortar and actually laying bricks. Since bricks are products which have been burned under intense heat, the sizes are going to vary somewhat even though they are initially constructed in standard size. Architectural plans for wall length cannot be changed to accommodate brick length.

How to Dry Bond

To dry bond, select 1 brick of each size that will be used in the finished wall. Three different sizes of bricks may be necessary to lay out the course. Starting at the beginning point, usually the corner, lay a brick even with the edge. As this is being done, insert your index finger sideways between the head of the bricks as a gauge for the head joint, Figure 13-14. Lay the first course of bricks across the project to the opposite corner. Using the index finger as a gauge, the head joint will usually measure 3/8", which is the standard preferred head joint. It may be necessary to open the head joints further to work with the bond or tighten the joints to prevent cutting the bricks. If the head joints are too big to open up, reverse the corner and tighten the joints to make up for the difference.

After the bricks have been laid out dry across the entire course and adjusted to accommodate all full-sized bricks the mortar is spread and the bricks are laid permanently into position. Do not remove all the bricks from the dry bonding position before spreading the mortar. If all the bricks are removed at once, the mason cannot possibly remember all the proper positions and the dry-bonding technique would be useless.

Spread mortar for 2 or 3 bricks at one time and lay them in their proper position without disturbing the others. Continue to pick up and lay bricks across the course until all have been laid. If all the bricks must be removed across the entire course before spreading the mortar, the end of every third brick could be marked with a pencil or crayon on the base where it is to be laid.

At no time should a piece of brick smaller than 6" be laid in the wall unless, of course, the bond specifies its use. If it is not possible for the wall to be entirely of full-sized bricks, cut as many three-quarters as are needed to make up the difference.

For example, suppose that a mason dry bonds a brick wall and finds that full bricks will construct it. The corners are laid up and the course filled in with bricks. However, upon reaching the center of the wall, the mason discovers that either due to a mistake in dry bonding or variations in brick sizes, a *bat* (half brick) is in the middle of the wall, Figure 13-15A. Since tearing down both corners would be expensive and impractical, the mason should cut 3 bricks of equal size to replace the half brick, Figure 13-15B. If the bricks are cut neatly, it is difficult to distinguish them from full bricks. A wall of full bricks is preferable, but faced with this field problem, a mason must deal with it in the most practical way.

Fig. 13-14 When dry bonding brick, use the index finger as a gauge for the head joint. The joint should be about 3/8 inch.

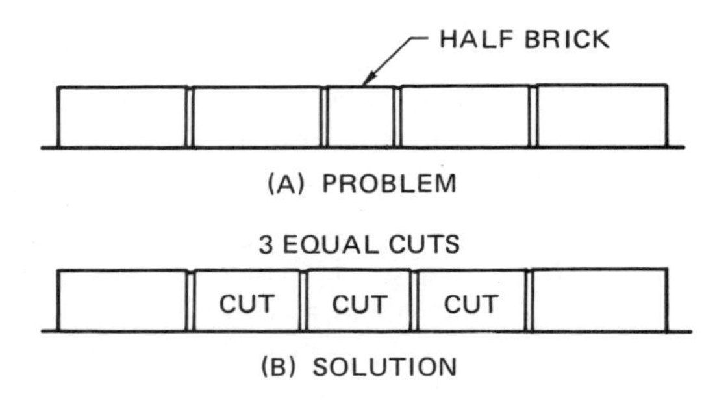

Fig. 13-15 Proper method of replacing a bat (half brick) in the wall

ACHIEVEMENT REVIEW

The column on the left contains a statement associated with bonding. The column on the right lists terms. Select the correct term from the right-hand list and match it with the proper statement on the left.

1. Number of basic bonds the mason should know and be able to build
2. Bond built completely of whole bricks or stretchers
3. Double wall with air space in the center
4. Reason that the half lap of one brick over another is considered the strongest type of bond
5. Variation of the running bond with a course of headers being laid at the fifth, sixth, or seventh course
6. 6" piece of brick
7. Header course with alternating stretchers on the same course
8. Header course that is composed completely of headers
9. Height of the wall where the header is to be installed
10. Bond consisting of alternating headers and stretchers on every course
11. Dummy headers that have been cut for use in a 4" wall
12. Method of starting a corner by using a 2" piece of brick against the return brick
13. Method of starting a corner by using a 6" piece of brick
14. Bond constructed of 3 stretchers alternating with a header on the same course
15. Bond constructed of alternate courses of headers and stretchers
16. Bond in which there are no overlapping bricks and all of the head joints are in true plumb alighment
17. Procedure of laying out bricks to check the bond without using mortar
18. Common term for a half brick
19. Mortar joint between headers
20. Common term for a 2" piece of brick

a. Snap headers
b. Full header
c. Three-quarter
d. Dry bonding
e. Bat
f. Cross joint
g. English corner
h. Header high
i. Flemish header
j. Cavity wall
k. Common or American bond
l. Garden wall bond
m. Stack bond
n. Dutch corner
o. Best distribution of weight
p. Five
q. Flemish bond
r. Plug
s. English bond
t. Running bond
u. Nine
v. Economy

PROJECT 4: DRY BONDING AND CONSTRUCTING A WALL

OBJECTIVE

- The masonry student will dry bond and build a 4" brick wall 10 courses high and 4' long in the running bond.

EQUIPMENT, TOOLS, AND SUPPLIES

Mortar pan or board
Mortar
Face bricks
Chalk box
Tempering water
Mason's trowel
Brick hammer
Convex striker
Plumb rule
Modular rule
Brush

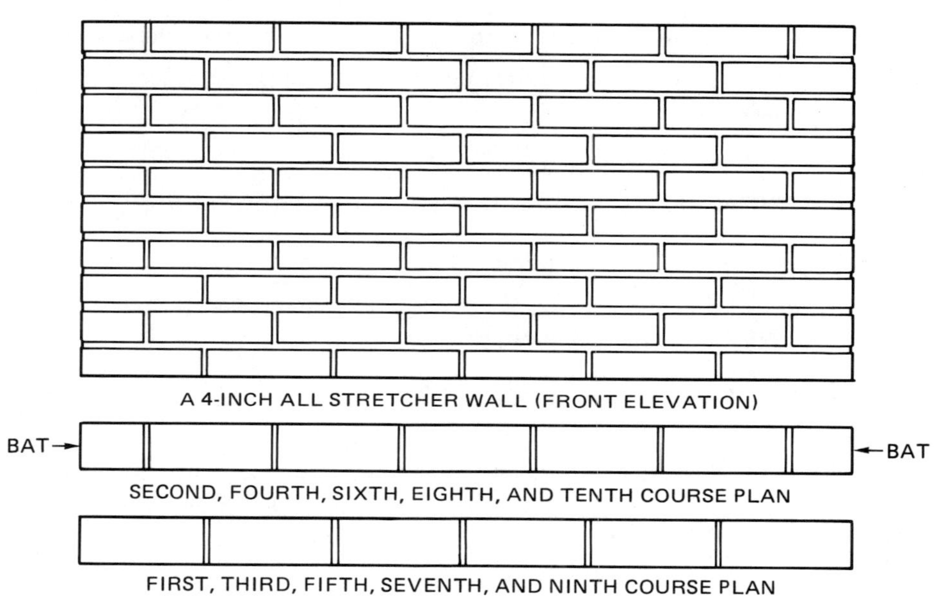

SUGGESTIONS

- Space materials being used approximately 2′ from wall.
- Keep all tools not being used away from the immediate work area to avoid accidents.
- Have a concrete block handy in which to set plumb rule when not in use.
- Be careful not to splash mortar over the wall.
- Use as little pressure as possible when leveling and plumbing. The higher the wall is built, the more delicate the wall is until the mortar sets.
- Return all mortar that falls on the base to the pan and temper the mortar as necessary. Keep the work area neat and clean.
- Observe proper safety practices at all times.
- If any mortar splashes in the eyes, wash out immediately with water. Avoid rubbing the eyes.

PROCEDURE

1. Strike a chalk line a little longer than 4′ on the base for a reference point.
2. Lay out 6 bricks dry using the index finger as a guide for the head joints.
3. Bed up a brick on each end of the wall, level, and plumb.
4. Lay bricks between the two end bricks. Level and plumb. Be sure to straighten the top edge of the course with the level.

5. Do not be concerned if the wall is slightly over 4′ in length. Be certain, however, that all of the head joints are 3/8″ in width and well filled with mortar.
6. Lay a bat on each end of the wall on the second course, as shown on the plan. Run the second course. Follow the same procedure of leveling and plumbing.
7. Check the height of the wall with the modular rule. Use the number 6 on the rule as your checking point.
8. Build the wall following the described procedures until it is 10 courses high.
9. Strike the mortar joints as needed with a convex jointer and brush the wall.
10. Double-check the wall to be certain that it is built according to directions and is ready for inspection.

PROJECT 5: BUILDING AN 8″ WALL IN COMMON BOND

OBJECTIVE

- The masonry student will lay out and build an 8″ wall in the common bond and lay a header course to tie the wall together.

EQUIPMENT, TOOLS, AND SUPPLIES

Mortar pan or board
Mortar
Face bricks
Mortar hoe to temper mortar
Tempering water
Brick tongs
Chalk box
Mason's hand tools (trowel, convex or round striker, plumb rule, brick hammer, modular rule, square, and brush)

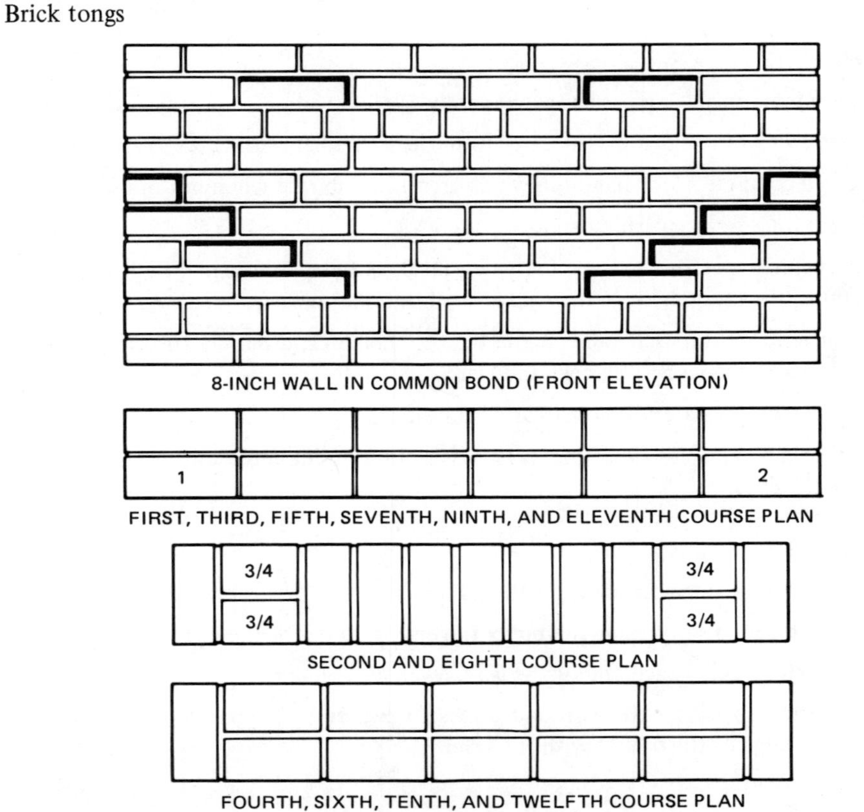

SUGGESTIONS

- Cut the header starter pieces 6″ long.
- Be sure that the header is laid level and straight across both walls.
- Use solid joints, particularly on header courses.
- Check mortar joints constantly and strike when the thumbprint is firm.
- Prevent excess mortar from dropping between the walls.
- Build to the brick header a height on one side of the wall, then build this height on the other side to keep the wall in alignment.
- Keep the work area clean and free from hazards.

PROCEDURE

1. Assemble materials in the work area.
2. Set the mortar pan back from the wall approximately 2′ to provide sufficient working space.
3. Strike a chalk line a little longer than 4′ on the base as a reference point.
4. Dry bond the project.
5. Spot (lay) a brick in mortar on each end of the project level and plumb. Square jambs (sides).
6. Lay the course level and plumb with the spotted bricks.
7. Back the wall with another course of bricks, being certain that the wall is the same width as the header bricks. This measurement will be approximately 8″ in width but must be the exact size of the header bricks being used. Use a header brick laid dry to gauge the width.
8. Lay the second course of bricks as shown on the plan. Be sure to follow the course layout for the header course. Follow the same procedure of leveling and plumbing.
9. Continue building the wall until completed as the plan indicates.
10. Check the height of the wall with the number 6 on a modular rule as needed while constructing the wall.
11. Strike the work as needed with a convex or round jointer and brush the wall upon completion.
12. Recheck the work for accuracy according to plan. The work is now ready for inspection.

Unit 14
Bonding Concrete Block and Rules for Bonding

OBJECTIVES

After studying this unit, the student will be able to

- lay out a block wall in the running bond.
- lay out a block wall in the stack bond.
- apply rules for bonding brick and block.

Concrete masonry units, like bricks, can also be laid in different bonds. As a rule, they are laid in either the running bond or the stack bond. Various effects can be achieved depending on the bond, the finish of the mortar joint, and the choice of the many types of decorative blocks now available. You may wish to review the types of blocks described in Unit 4.

RUNNING BOND

The procedure for establishing a running bond in block is the same as that described for brickwork. That is, lapping half a unit over the unit underneath. As a rule, this allows the best distribution of weight and forms the strongest wall.

Laying Out the Corner in Running Bond

The size of the block used to start the corner depends on the size of the block being laid. *Size of the block* refers to its width, as all standard blocks are the same length: 15 5/8″, with 3/8″ for the mortar joint for a total of 16″.

Eight-inch concrete blocks are the only blocks that lap perfectly over each other without the necessity of cutting the corner block. This is because the corner end of the 8″ block is exactly half the length of the 16″ stretcher block. Each course laid will reverse and alternate as the wall is built, Figure 14-1.

When the concrete block is wider than 8″, special L-shaped corner blocks are available. The L-shaped corner block enables the adjacent block to make the correct lap of 8″ over the block beneath. Figure 14-2 shows how typical L-shaped corner blocks appear when then are laid in position.

L-shaped corner blocks are usually available only in 10″ and 12″ sizes. All other size block walls must be started with a cut block on the corner to produce the 8″ lap bond.

Figure 14-3 illustrates three different block sizes and the proper corner block lengths for each size. The cut piece is different for each size block shown. To determine what length to cut a block corner, add the width of the block being laid to the standard figure of 8″. For example, a mason working with 4″ block cuts a block corner 12″ long–the standard lap of 8″ plus the width of the 4″ block makes a total of 12″, Figure 14-4.

STACK BOND

A concrete block wall built in the stack bond is constructed the same as a brick wall. Full stretchers can be laid or, for decorative purposes, all half bricks may be used. Concrete blocks are very accurately sized and it is much easier to maintain them in true vertical alignment than brick units. On many jobs, the blocks are cut with the masonry saw. After the first course is laid out, cuts for the entire wall can be made since the size of the cuts from the first course to the top course never changes. This speeds the job by allowing the cuts to be made beforehand and the blocks to be set to dry in the work area.

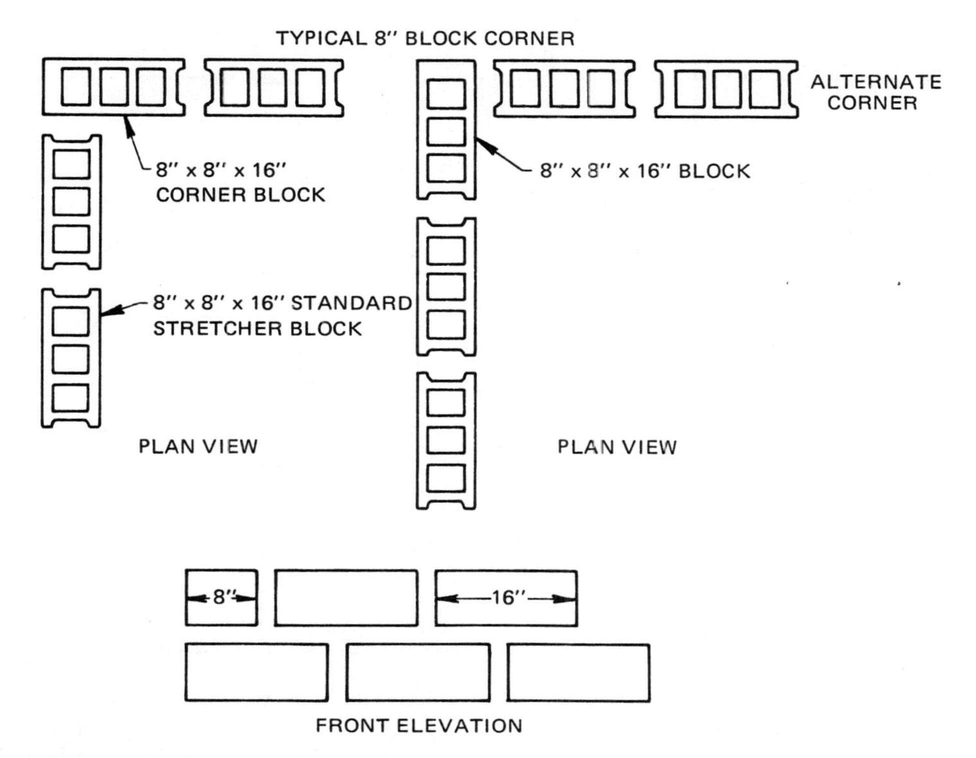

Fig. 14-1 Bonding an 8-inch corner block. The plan views show alternating courses; front elevation shows view facing the wall.

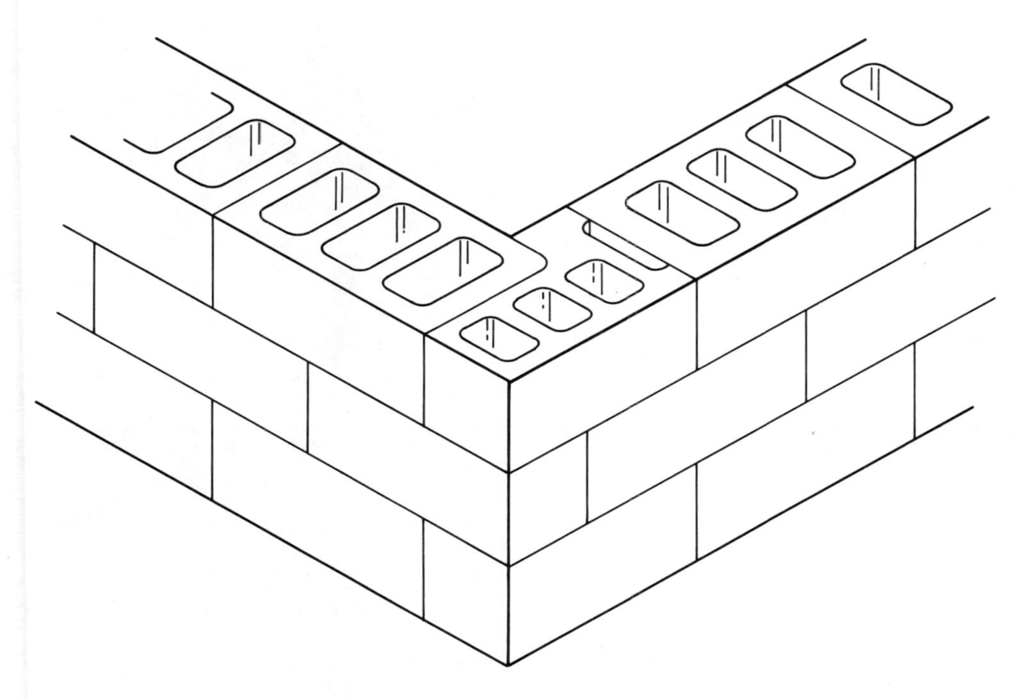

Fig. 14-2 12-inch, L-shaped corner block laid in position

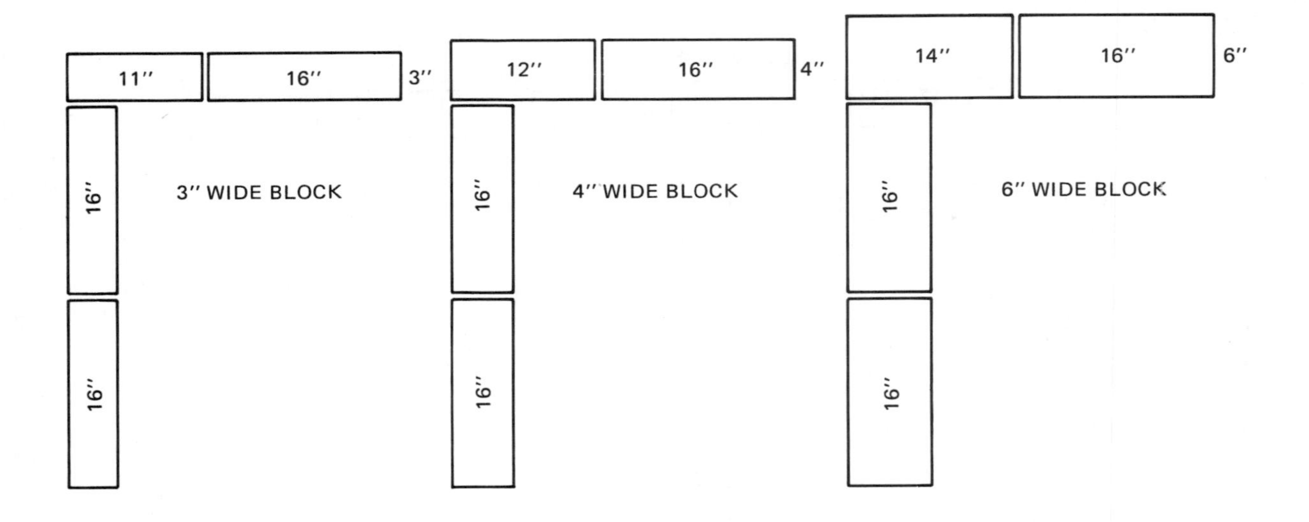

Fig. 14-3 Corner layout for 3-inch, 4-inch, and 6-inch block

PLANNING THE BOND

Block walls and bonds must be carefully planned for economy and efficiency. Because of the time factor and the cost of labor involved, the cutting of block should be kept at a minimum and all openings and heights should be planned specifically for the size of the unit being laid. Concrete block buildings should be laid out by the architect on an 8″ grid since either a half or whole block can be used around windows, doors, and other openings. The fewer pieces there are in the wall, the more efficient the project is.

A mason (or on larger jobs, a group of masons) is selected to work ahead of the rest of the crew and lay out major features of the structure. The layout mason determines if any dimensions can be changed so that the masonry can be built with as few cut units as possible. The bond is also laid out for the crew in advance, Figure 14-5. Since they do not have to figure out the bond pattern, the crew's efficiency is greatly improved.

Masons can use several tools to mark off concrete block bond on the base before laying any of the blocks in mortar:

- Mason's folding rule marks should be placed every 16″ on the layout line. These are known as *checkpoints* in the trade.
- The steel tape is used to mark off the bond over long distances.

Fig. 14-4 Student mason laying out 4-inch block corner with a 12-inch cut block

- Use the 48″ plumb rule as a gauge by laying it down on the layout line and marking the end of it. Then shift the plumb rule one length and mark again. Three concrete blocks equal 48″, the length of the plumb rule.

To begin the project, lay the first course and establish the bond for the wall. Study the floor plans and determine locations of windows, doors, and other openings With a steel tape or rule, mark these openings on the first course with a crayon or pencil. Windows and doors may often be shifted so that fewer block require cutting. If this may be done without changing any critical dimensions, a considerable lowering of costs is possible. However, the supervisor should always be consulted before any changes are made.

The height of concrete block walls should be laid out in the same way, Figure 14-6. Since most windows, doors, and openings are estimated by using the modular grid, the space between the top of openings and the floor level should be made up of full blocks. To accomplish this, it may be necessary to start with less than a full-sized block on the first course. (Blocks less than 8″ in height are referred to as *rip blocks.* The 4″ block is the most commonly used.) It is impossible, in some cases, to change wall dimensions to accommodate only whole block. If this is the case, a cut block must be used. However, masonry work should never be started without careful consideration and planning. Without careful planning, economy and appearance are likely to be sacrificed.

BONDING BRICKS AND CONCRETE BLOCK TO FORM A WALL

Another form of bonding involves the combination of two different masonry materials to form a single structure. This is known as *composite masonry.*

Many different masonry materials may be bonded together, the most common being bricks and concrete

Fig. 14-5 First course of block laid out and wall stocked with materials ready for masons

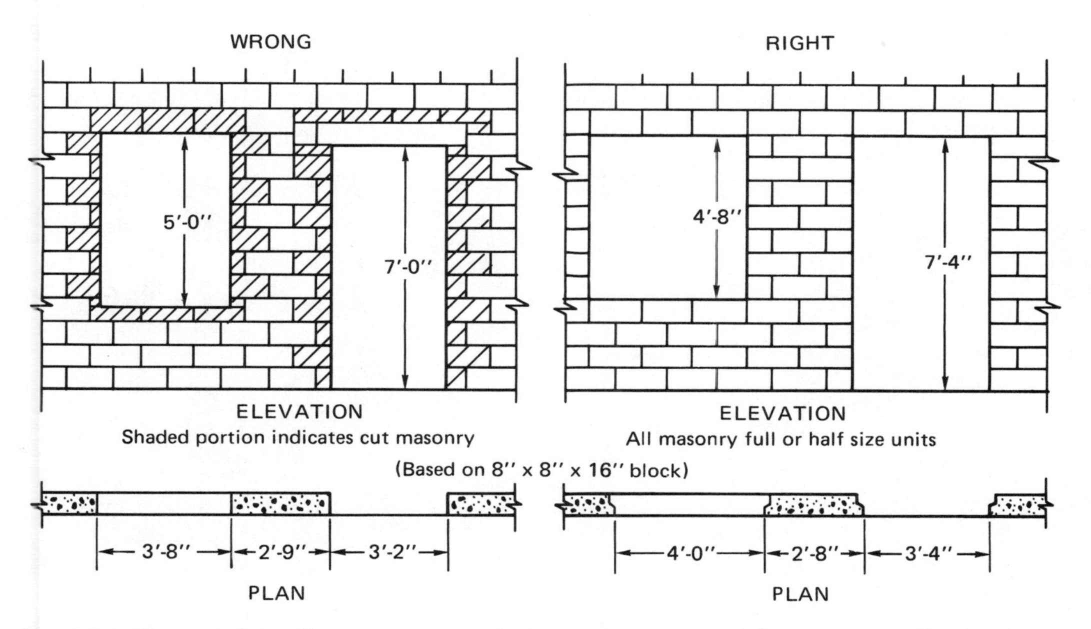

Fig. 14-6 Planning the wall using concrete block. Notice that the illustration showing the correct procedure calls for only full-sized block in the space between the top of the openings and the ceiling.

block. There are two basic methods of bonding bricks and concrete block together. One method has a specified number of courses of bricks laid to a given height. The bricks are then backed with concrete block which are the same height as the brickwork. Each single vertical section of masonry 1 unit in thickness is called a *wythe* or *withe.* A brick header may be utilized to tie 2 wythes together into 1 single wall, Figure 14-7.

The second method involves construction of the brick wall backed by concrete blocks and tied together with metal ties or reinforcement wire, Figure 14-8.

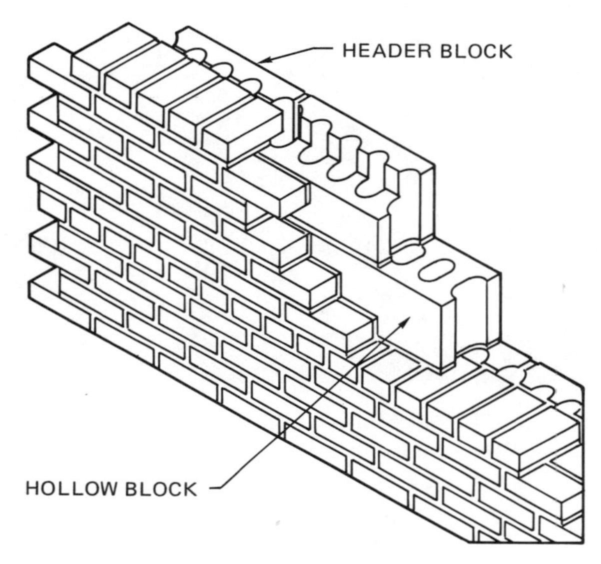

Fig. 14-7 A brick and block composite wall bonded with a brick header

Fig. 14-8 Brick and block composite wall tied together with metal wire joint reinforcement.

RULES FOR BONDING

When bonding masonry materials, masons should follow certain rules. Following these rules will ensure that the structures built are strong, well designed, and economically constructed with no more pieces than necessary. The following three particular rules will help the mason achieve high-quality bonds. These rules are:

- Keep the bond plumb.
- Keep walls level with backing materials.
- Maintain the set of the mortar bond.

Keeping the Bond Plumb

In high-quality masonry work, every head joint must be kept exactly plumb with those on alternating courses below. This is known as creating a *plumb bond.* The distribution of weight is more uniform in a plumb bond, resulting in a stronger wall. In decorative bond patterns, it is particularly important that the pattern be laid accurately on each course or the effect of the pattern will be lost.

When the architect has specified a very exact plumb bond, the mason sometimes plumbs the head joints with the plumb rule at given points on specific courses. This will assure that the joints are kept in true vertical alignment. It is the responsibility of the mason who is laying out a bond to maintain the bond until the wall is completed.

When masonry units are laid or stacked directly over one another with the head joints in plumb vertical position, they are laid, in the trade term, *jack-over-jack.* Only in the stack bond is this practice acceptable.

Fig. 14-9 Notice how mortar does not stick properly to brick after initial set has taken place. This brick must be relaid in fresh mortar to obtain a good mortar bond.

Keeping Walls Level

Backing courses should always be level with the face of the wall. This allows the proper placing of headers or wall ties to bond 2 wythes together to form a solid structural unit. The backing wall should be kept level as each course is laid, rather than being constructed all at once with 1 mortar joint. This is a common practice in the trade, but can cause problems in maintaining a level wall.

Maintaining the Set of the Mortar Bond

A masonry unit should not be moved or shifted once it is laid in mortar and takes the initial set. (*Set* means the mortar adheres to the unit.) If the unit is disturbed, the bond between mortar and unit is said to be *broken* or *lost*. The lack of a seal may result in a leaky mortar joint. If a unit must be moved after taking the set, it must be removed from the wall, fresh mortar spread, and the unit relaid. Mortar spread too far ahead will dry too quickly and lose its strength, resulting in a poor bond, Figure 14-9.

These rules of bonding apply to all masonry units. If the rules are followed strictly, the final result will be strong, waterproof masonry construction.

ACHIEVEMENT REVIEW

Select the best answer from the choices offered to complete each statement. List your choice by letter identification.

1. The only concrete block that can be laid one over the other without cutting the corner block is the
   a. 12" block.
   b. 8" block.
   c. 6" block.
   d. 4" block.
2. The correct starting piece for a 6" concrete block wall is the
   a. L-shaped corner block.
   b. 14" piece.
   c. 12" piece.
   d. 11" piece.
3. Concrete block should be laid out by the architect on a grid that would be suitable for structures with windows and doors. The correct grid for concrete block is the
   a. 6" grid.
   b. 4" grid.
   c. 8" grid.
   d. 12" grid.
4. Before changing the locations of openings of a building, such as windows and doors, the mason should always consult the
   a. architect.
   b. supervisor.
   c. fellow mason.
   d. contractor.
5. Many times it is necessary to use a rip block when starting a concrete structure to attain the correct height. The standard height for a rip block is
   a. 4".
   b. 6".
   c. 8".
   d. 2".
6. The term *plumb bond* indicates that
   a. all bricks are laid perfectly horizontally.
   b. wall ties are used in the wall.
   c. vertical joints are aligned in every other course.
7. A term used frequently to refer to a wall of a single thickness is
   a. stack.
   b. column.
   c. wythe.
   d. unit.
8. If the set of the mortar has been broken, the best procedure is to
   a. relay the units with fresh mortar.
   b. wet the bricks.
   c. tap the units until contact is re-established.

SUMMARY, SECTION 4

- The term *bond* has three meanings to the mason. The mortar bond is the adhesion of mortar to the masonry unit. The structural bond is the interlocking of masonry units to each other to distribute the weight of the wall. (The half lap is an extremely strong structural bond.) The pattern bond is the arrangement of masonry units to form a pattern or design.
- Patterns can also be achieved by projecting and recessing units on the face of the wall. Tooling the mortar joint in various ways also contributes to the design and beauty of the structure.
- The different meanings of bond apply to all types of bond in varying degrees.
- There are several different brick positions which architects specify by trade term.
- The most commonly used bond for brick or block work is the running bond. This is due to the simplicity of the running bond which requires no cutting or designing of complicated patterns. Of all the bonds discussed, the running bond is the most economical for the contractor to construct.
- Masonry work should be planned and dry bonded before bricks or block are laid.
- Eight-inch concrete block is the only size block that does not require cutting of the corner block.
- Concrete block is commonly used for walls built in the stack bond because the units do not differ in size.
- Jobs should be well planned so that the cutting of blocks is kept to a minimum.
- The mason's main responsibility concerning bonding is to maintain the bond as it was specified and laid out, and to be certain that it remains plumb throughout the job. Changes in the bond should not be made without approval of the supervisor.

SUMMARY ACHIEVEMENT REVIEW, SECTION 4

Complete each of the following statements referring to material found in Section 4.

1. The adhesion of mortar to the masonry unit is called ________________ .
2. A type of bonding in which the units are interlocked over one another and which is designed primarily for strength is called ________________.
3. The arrangement of masonry units to form a particular design is called ________________.
4. A standard brick laid in a horizontal position in a wall is known as a (an) ________________.
5. The most common brick position in windowsills is the ________________.
6. The most commonly used bond in brick veneer work is the ________________.
7. The bond which is a variation of the running bond with headers installed on the fifth, sixth, or seventh course to provide strength and design is the ____________.
8. Laying one masonry unit halfway over the next is called the ________________.
9. The best method of starting a header course from the corner of a structure is to cut and install a piece measuring ________________.
10. One of the most beautiful bonds in brickwork calls for alternating headers and stretchers on the same course. When this occurs in every course of a wall, the bond formed is the ________________.

11. The Flemish bond may be started from the corner one of two different ways. If a 2" piece is used it is called a (an) ______________ corner; and, if a 6" piece is used, it is called a (an) ______________ corner.

12. The Flemish bond can be adapted to present various appearances. If 3 stretchers are alternated with a header it is known as a (an) ______________.

13. One of the least used bonds in brickwork consists of alternate courses of headers and stretchers. This bond is called the ______________.

14. A bond that is specified by many architects for decorative purposes and in which there is no overlapping of masonry units is called a (an) ______________.

15. The method of laying masonry units dry (without mortar) to check the bond is called ______________.

16. Concrete blocks are usually laid in the running bond because ______________.

17. Concrete blocks measuring 10" and 12" in width are usually used for basement walls. To avoid cutting such large blocks to start the corner, special blocks are used. These blocks are known as ______________.

18. When building a brick wall which is backed by another masonry wythe, the most important consideration is to ______________.

19. A mason's plumb rule can be used to space concrete blocks on the layout course. The length of the plumb rule equals the length of ______________ blocks.

20. When laying out a concrete block corner built of 6" blocks, it is necessary to start the corner off with a piece measuring ______________.

Section 5 Laying Brick and Concrete Block

Unit 15 Laying Brick to the Line

OBJECTIVES

After studying this unit, the student will be able to

- attach a line block, nail, and line pin to a wall.
- set a trig.
- lay bricks to the line while spacing them the correct distance from the line.

The first tasks of the apprentice masons usually include spreading mortar and striking joints. These practices acquaint apprentices with the properties of mortar and proper tooling of joints. Beginning masons may also help stock materials near the work area and cut materials with the saw. In this way, they learn material requirements for various jobs and the ways in which the materials are handled. The various sizes of bricks and block are quickly learned.

After learning the particulars of these jobs, apprentices then learn to lay bricks to the line. Masonry contractors always require that apprentices learn the skill of laying to the line before learning to use the plumb rule. The line, which acts as a guide for the wall, is used as the best means by which apprentices can learn to lay bricks. After learning how to lay bricks, apprentices proceed to the more difficult task of building the corner. By progressing from the simpler jobs to the more difficult, apprentices learn as they work and master each task independently of others.

LAYING BRICKS

The Corner Pole

There can be difficulty for apprentices learning to lay bricks in the shop because there are not enough experienced masons to build corners so that trainees can finish the wall. To overcome this problem, a line can be attached to a corner pole so that the wall can be constructed without building corners. A *corner pole* (or *deadman*) is any type of post which is propped and braced into a plumb position so that a line can be fastened to it. A manufactured corner pole can be bought or a wooden pole can be constructed in the shop from a 2" x 4". Angle-iron frames or stacks of block can also be used.

On brick veneer homes, the manufactured corner pole can be braced against the frame of the house and the base masonry, Figure 15-1. This completely eliminates the need for a corner to be built before the bricks are laid. This is a very popular practice in the building trade, as labor costs are considerably reduced.

Use of the Line

When building walls longer than 4′, a line should be used as a guide for laying the units, since the plumb rule is only 4′ in length and, therefore, cannot accommodate a longer wall. The line serves to keep the wall level and plumb and greatly increases the speed by which the bricks may be laid. Most masonry work is laid to the line.

Note: In the practice situations which follow this unit, the plumb rule will be used only to spot check walls to be sure they are level and plumb and for work around window jambs and doorways.

Preparing the Work Area and Beginning the Job

Sweep or brush the area where the wall is to be built. It is important in the beginning to have a clean surface to achieve a good mortar bond with the base. Set the mortar pans approximately 2′ away from the wall line to allow for sufficient working space.

> **Caution:** When stacking the bricks on each side of the mortar pan, be sure to alternate the bricks on every other course to prevent the pile from falling over and injuring yourself or a fellow worker.

Strike a chalk line on the base if necessary to establish the wall line for laying out the first course. At times, a nail may be found in the concrete base to mark the wall line. It is usually placed there by the building engineer or carpenter foreman. Dry bond the wall to establish the proper spacing of the bricks.

Lay 1 brick in the mortar bed at one end of the wall or corner, lining it up with the mark established by the dry bond. Level and plumb the brick with the wall line. Repeat the steps at the other end of the wall. Be sure that the brick is laid to the proper height by checking with a rule or a gauge rod. This process is known as *spotting the brick.*

If a corner pole is to be used, it may be placed in position at this time.

Attaching the Line with the Line Block

Line blocks have a slot cut in the center to allow the line to pass through. Fasten 1 block against the left corner of the wall either by passing the line through the slot and tying a nail or large knot on the end to secure the line, or by making several wraps around the block with the line.

Fig. 15-1 Masons laying brick with a corner pole as a guide. Notice the line attached to the pole at the top of the bricks. The wall is being laid in the running bond.

The person on the left end of the line must hold the line block in line with the top of the course to be laid until the person on the right end of the line is ready to pull the line tight. Pull the line as tightly as possible without breaking it, being certain that it is passing through the slot in the block. When the line is pulled tight, wrap the line about 3 or 4 turns around the line block. Hook the line block on the right corner, Figure 15-2. Make sure that the line is perfectly level with the height of the course being laid.

> **Caution:** Use extreme care when working with line blocks. If a line block is used on a long wall, there is a danger of it being jolted loose and causing injury. For this reason, use line blocks only on short walls or with a corner pile.

Fig. 15-2 Attaching a line to a brick corner with a line block

When line blocks are attached to a corner pole and the line is tightened, be sure to check that the line is the correct height before laying any bricks. This is very important, since the height of the finished wall is directly affected by the height of the line.

Pulling and Attaching the Line with the Nail and Line Pin

Use caution when attaching the line to a corner with line pins or with a nail and line pin. Traditionally, a nail is used on the *peg end* of the line (end against which the line is pulled), and a steel line pin is used on the pulling and wrapping end. Many masons prefer to use line pins on both ends because the line pin, due to its shape, is less likely to pull out of the joint. If a nail is used on the peg end of the wall, make sure it is a large nail (at least size 10d Common). Not many things are more dangerous to masons than a flying nail.

As the mason faces the wall, the peg end of the line is always driven into the left-hand side of the wall. Most people are right-handed and it is easier for them to pull a line to the right. Left-handed persons must adjust to this traditional practice. Drive the nail or pin securely into the head joint, making sure that its top is level with the top of the course of brick to be laid. Always place the nail at a downward angle (45°) in the head joint and several bricks away from the corner. This prevents the nail from coming loose as the line is pulled.

The person setting up the nail end of the line is responsible for making certain that the nail will remain secure in the mortar joint when the other person tightens the line. Testing the nail is done by giving the nail a few sharp tugs after driving it in the joint. This practice can reduce accidents and time lost on the job.

After the line has been tested, the person at the nail calls *tight away* to the person located by the pin. This indicates that it is safe to pull the line tight. If it is a very long wall, the masons working at the center of the wall may relay the call to tighten the line or give 2 or 3 short tugs on the line as a signal.

The mason on the pulling end of the line drives or pushes the line pin securely into the head joint, also even with the top of the course. Great care should be taken to see that the pin is secure and in the correct position before pulling the line, Figure 15-3. The line is then pulled tight and wrapped around the line pin.

> **Caution**: When tightening the line, the left hand, which pulls the line, should be angled away from the wall in case the line breaks under the strain.

When the line pin is moved up for another course, immediately fill the holes where the nail and pin were located with fresh mortar. Waiting until the project is finished to fill the holes usually results in an added expense. There is also a noticeable difference in color where the holes were pointed.

Setting the Trig Brick

To prevent the line from sagging or being blown out of alignment by the wind, it may be necessary to set a *trig brick,* Figure 15-4. A trig brick is also set

Fig. 15-3 The mason pulls the line with the left hand and wraps the line around the pin with the right hand.

Fig. 15-4 Setting the trig brick. The mason is adjusting the brick to the line before laying brick on the wall.

Fig. 15-5 Trig brick in position with the line attached to the metal trig. Notice that the trig holds the line the correct distance (1/16 inch) away from the face of the brick.

Fig. 15-6 Position of the fingers when laying brick to the line. Notice that the mason's fingers grasp only the top edge of the brick.

to stabilize the line when there are many masons working on the wall. The first step in setting a *trig* (metal fastener or loop of line) is to be sure that the brick with the trig is set with the bond pattern of the wall. Be sure that the brick is level and plumb with the face of the wall. Check the trig brick for the proper height with a mason's rule or a course pole which is marked with the proper spaces for the masonry courses.

Sight down the wall every 3 or 4 courses to be sure that the trig brick is in line with the wall line. This sighting is usually done by the person located by the pin and another person who stands by the trig brick. One person holds the line in position with the trig brick, as the person sighting the line bends down and checks the top edge of the brick on one corner with the other corner to be sure they are in line. This method is usually accurate within 1/8". The trig brick may not always be exactly on number 6 of the modular rule, but should always be the same height as the corners. A straight wall can be maintained by these sighting techniques.

After the trig has been correctly set, place a dry brick on top of the trig to hold the line in position with the top outside edge of the trig brick, Figure 15-5.

The trig brick should always be set ahead of the line being raised for the next course. This is so that other masons working on the line do not have to wait for the line to be erected to proceed with their work. This can easily be done by setting the trig brick immediately after the line has been run. While one mason sets the trig brick, the other masons can be walling bricks for the next course and striking the mortar joints. This procedure calls for timing and teamwork and adds to the productivity of all masons working on the line.

Setting the trig brick (or *carrying the trig*) is a job which calls for a great amount of accuracy and responsibility. It must be laid perfectly plumb and level and kept in alignment with the wall. An incorrectly set trig brick will cause the wall to be incorrectly constructed even though the corners may be perfectly built.

Laying the Units

Spread mortar on the wall with the trowel. Try not to disturb the line. If the trowel is held on a slight angle as described in the unit on basic tools, it should not touch the line. Grasp the brick with the fingers and thumb. Position the fingers on the brick so that the line may pass between the fingertips and brick when the hand is released, Figure 15-6.

This process demands good coordination and timing which is developed only with practice. If the brick is held too long, the fingers or thumb may push the line out of alignment, preventing other masons from properly aligning bricks. Releasing the fingers or thumb from the brick too soon will result in the brick being laid unevenly on the mortar bed. If this happens, the brick will not be level and plumb and will require relaying in a fresh bed of mortar.

When bricks are protruding past the wall line, they are said to be *hard to the line.* When a brick is too far

Fig. 15-7 The correct spacing of brick is 1/16 inch from the line.

away from the line, it is said to be *slack to the line.* Either condition results in an unacceptable job. Bricks should be laid about 1/16" from the line, or as some persons in the trade say, one should be able to see a little daylight between the line and the brick. With practice, the mason will be able to leave the proper amount of space without using a rule. When laid correctly, the bottom edge of the brick should be in line with the top of the course beneath it, and the top edge even with the top of the line and 1/16" back from the line, Figure 15-7.

Laying bricks so that they touch the line (known as *crowding the line*) is one of the biggest problems the apprentice must overcome when learning to lay bricks. Developing this technique requires a great deal of practice.

Pressing the Brick into Place

It is sometimes permissible to tap the bricks into final position. However, to develop a good rate of speed in laying bricks, the trainee must perfect the technique of pressing the brick into place, Figure 15-8.

With the trowel, cut off the mortar which has squeezed out of the joints after pressing down and apply it to the head of the brick just laid. Time can be saved by applying the head joint in this manner, since the mason does not have to return to the mortar pan for each individual head joint. When applying the mortar for the head joint, hold the trowel blade at an angle so as not to move or cut the line. It is not possible to lay bricks to the line without disturbing the line to some degree, but with the correct movement, it can be held to a minimum.

Fig. 15-8 Adjusting brick to the line by pressing down with the hand.

Laying the Closure Brick

The last brick to be laid in a wall is the *closure brick,* usually located near the center of the wall. To eliminate any chance of leaking or moisture penetration, the mortar joints must be well filled. Apply mortar for the head joint (also known as *throwing* a joint) on each end of the two bricks adjoining the area where the closure brick will be placed. Then *butter* (apply mortar to) each end of the closure brick and lay it into position, Figure 15-9. When laying the brick, press it carefully so that the rest of the bricks are not disturbed. This practice is known as *double jointing.* Immediately fill any remaining holes to ensure a watertight joint.

POINTS TO REMEMBER

Most of the mason's time is spent laying bricks to the line. Since the procedure involves repetition,

Fig. 15-9 Laying the closure brick in the wall. Notice that the mason has double jointed both the closure brick and the two bricks surrounding it. This action protects the wall against leaks.

needless actions can result in a great deal of lost time. For efficiency on the job, remember the following points:

- A brick should always be picked up with the face out so that it is in the position in which it will be laid in the wall.
- A mason should be approximately 2′ from the stocked brick or within arm's reach. Unnecessary walking tires the mason and reduces efficiency.
- If the brick contains a depression or frog, it should be picked up with the frog facing down, since this is the way it will be laid in the wall. The mason should refrain from unnecessary turning of the bricks in the hand.
- Well-filled mortar bed and head joints should be formed. This greatly decreases the amount of time needed to strike joints later in the job. Solid joints also ensure stronger, waterproof walls.

ACHIEVEMENT REVIEW

Select the best answer from the choices offered to complete the statement. List your choice by letter identification.

1. A deadman is
   a. a method of bonding.
   b. a post to which the line is fastened.
   c. a poor safety practice.
   d. inefficiency on the job.
2. A line should always be used on a wall if it is longer than
   a. 6′.
   b. 2′.
   c. 4′.
   d. 8′.
3. The term *spotting a brick* is the
   a. addition of color to the brick.
   b. arrangement of bricks to form a pattern in the wall.
   c. laying of a brick in mortar to establish the wall line.
   d. spreading of mortar on the brick.
4. A nail no smaller than 10d Common should be used for attaching the line because
   a. it helps to prevent the line from becoming dislodged.
   b. it matches the size of the mortar joint.
   c. it does not leave an excessively big hole in the mortar joint.
   d. it is easier to tie the line to a nail this size.
5. The main reason for tugging on the line before calling to the mason on the other end to tighten it is
   a. to check that the line is at the correct height.
   b. to be sure that the line is securely in position and will not pull loose.
   c. to take the initial stretch out of the line.
6. When setting a trig brick, the line should be sighted periodically from corner to corner to assure that
   a. the wall is plumb.
   b. the bond pattern is correct.
   c. the trig brick is in proper alignment with the wall.
   d. the trig brick is the same size as the other bricks.
7. The correct distance to position bricks back from the line is
   a. 1/8″.
   b. 1/4″.
   c. 1/16″.
   d. 3/16″.

8. The last brick to be laid in the center of the wall is called the
   a. filler brick.
   b. closure brick.
   c. key brick.
   d. header.

9. Double jointing the last brick laid in the wall is important because it
   a. forms a stronger wall.
   b. allows for expansion in the wall.
   c. forms a more waterproof joint.
   d. divides the joints more evenly.

PROJECT 6: USING THE CORNER POLE

OBJECTIVE

- The masonry student will be able to lay a 4" brick wall in the running bond, using the corner pole as a guide. The corner pole can be used as a replacement for the corner on almost any job.

EQUIPMENT, TOOLS, AND SUPPLIES

2 mortar pans or boards
Mixing tools
Mason's trowel
Brick hammer
Plumb rule
2 corner poles and braces
Chalk box
Ball of line
Convex sled runner
2' square
Mason's modular rule
1 pair line blocks
Brush
Approximately 133 standard-sized bricks
1 1/2 bags cement or lime*
27 shovels of sand
Supply of clean water
1, 8" x 8" x 16" block to use as holder for level

*Lime may take the place of cement in a training situation, since it can be reused.

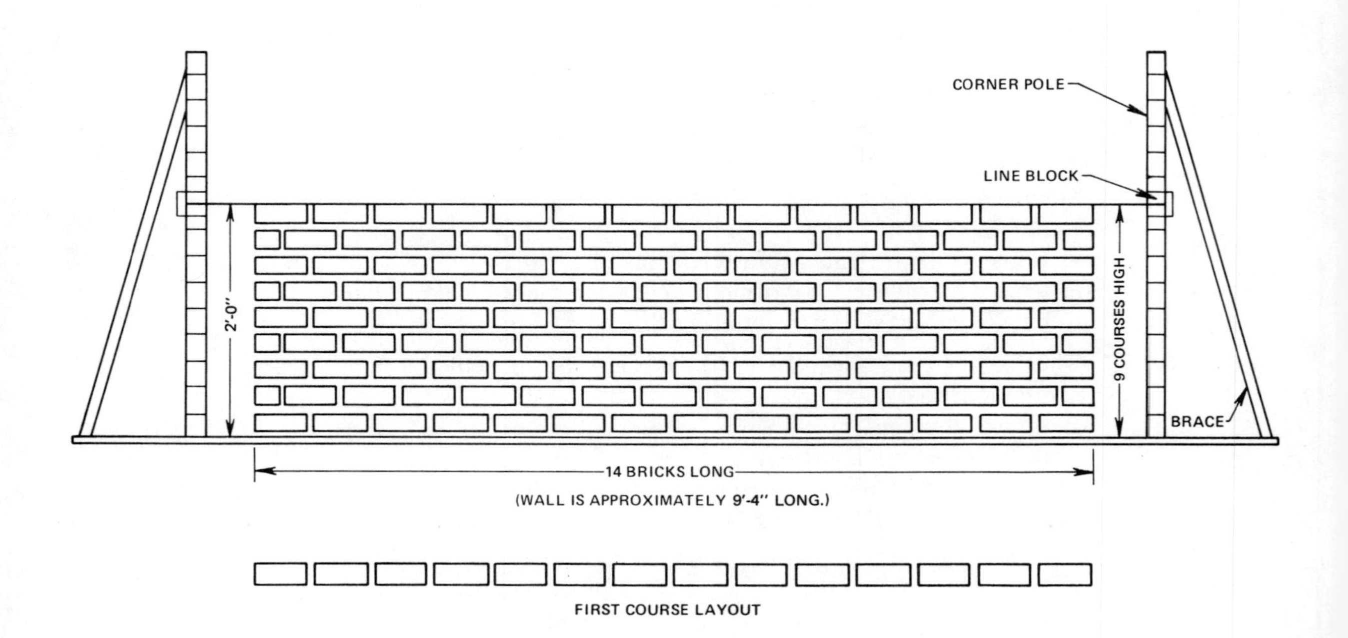

SUGGESTIONS

- Position materials and mortar pans approximately 2′ from the wall to allow for working room.
- Be sure the line is on the correct coursing mark on the corner pole.
- Spread the mortar as uniformly as possible.
- Use solid head joints on all work.
- Remove the excess mortar from the back of the wall at the completion of each course.
- Strike the joints when mortar is thumbprint hard.
- Keep the work area neat and free from debris.

PROCEDURE

1. Mix the mortar.
2. Stock materials in the work area and load the mortar pans.
3. Strike a chalk line approximately 11′ long on the concrete floor to act as the wall line.
4. Set up the corner poles at each end of the chalk line. Plumb and brace them into position with the wall line.

   Note: If corner poles are not available, a stack of block set up plumb or an angle-iron brace also work as guides.
5. Dry bond the first course which is 14 bricks in length.
6. Attach the line and blocks to the corner pole at the height of the first course.
7. Lay the first course in the mortar.
8. Move the line up 1 course.
9. Cut bats for the second course and finish laying the course as shown on the plan.
10. Build the wall to the required height as shown on the plan, plumbing *jambs* (ends of wall) on every course.
11. Strike the mortar joints with a convex sled runner jointer and brush.
12. Recheck the wall with a level (plumb rule) at the completion of the work.
13. Always follow good safety practices.

Unit 16
Building the Brick Corner

OBJECTIVES

After studying this unit, the student will be able to

- lay out the first course of a wall with the correct number of stretchers in preparation for building a corner.
- construct a rack-back lead.
- construct an outside and inside corner.

Building a corner is one of the more difficult tasks that the mason performs. If the corner is built incorrectly, the wall is also built incorrectly, since the corner serves as a guide for the wall. Building corners (also known as *leads*) on a job requires a great deal of responsibility and care on the part of the mason. Care must especially be taken in leveling and plumbing to assure that the corner is true.

The person building the corner must work far enough ahead of the other masons so that they are not detained from laying bricks to the line. However, rushing the work unnecessarily may result in a shoddy job. Each time a line is run and bricks are stacked for the next course, the mason building the corner should call for the line to be raised another course. It is the responsibility of the mason building the corner to set the proper working pace for all the masons on the line. This assures a steadily moving job and uniform productivity of all masons on the job.

When learning to build corners, the mason should be more concerned with technique and good workmanship than speed. The mason should be certain that a course is satisfactory before proceeding to the next course. Once a corner is out of alignment, it is very difficult to straighten it without breaking the set of the mortar.

THE RACK-BACK LEAD

Very often, it is necessary to build a lead between corners on a long wall. The simplest type of lead or guide for a wall is called a *rack-back lead.* This is usually the type of lead with which the mason first becomes familiar. This type of lead does not have any angles or *returns.* It is merely a number of brick courses laid to a given point by racking back a half brick on each course. The fundamentals of corner building, with the exception of turning the corner, apply to building a rack-back lead.

The first course of a rack-back lead rarely exceeds 6 bricks in length. This is because the plumb rule is 48″ in length, the same length as 6 bricks. After laying the first course, the mason must be sure the bricks are plumb and straight. Each succeeding course is racked back a half brick on each end of the lead until the last brick is laid. Excess mortar should be removed as each course is laid and returned to the mortar pan for reuse. Figures 16-1A through 16-1F show six major steps in the construction of a rack-back lead.

When the lead is completed, check the alignment of the tail end of the lead with the plumb rule. This is done by holding the plumb rule at an angle on the edge of the corner of the bricks at the racking point. This is known as *tailing a lead.* If this is not done, protruding bricks may cause a bulge in the wall when the line is attached to the lead and the wall is built.

The mason should tool the joints on the completed lead with a striking tool and fill any holes. The mason should then brush the head with a soft bristle brush. Finally, the lead should be rechecked to be certain that it is still in proper alignment.

Fig. 16-1A Laying a course in a rack-back lead. Note how the end bricks are racked back a half lap.

Fig. 16-1B Checking height with modular rule. Note how the bricks are laid to number 6 on the rule.

Fig. 16-1C Tooling the mortar joints with a convex sled runner

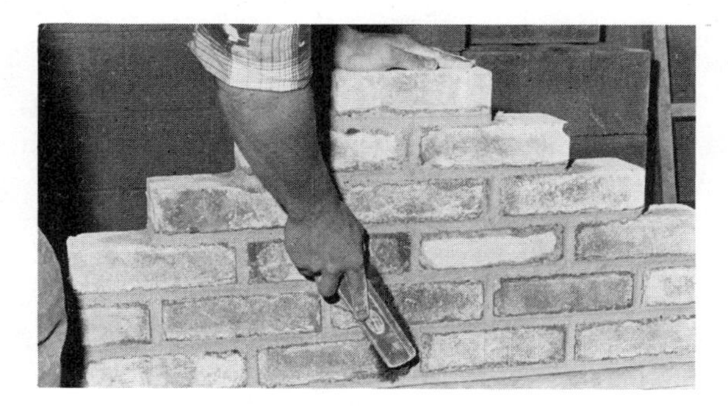

Fig. 16-1D Brushing the lead

Fig. 16-1E Tailing a rack-back lead. Notice that the plumb rule is in line with the corner of each top brick to assure proper alignment.

Fig. 16-1F Completed rack-back lead built on a wooden 2″ x 4″

BUILDING AN OUTSIDE BRICK CORNER IN THE STRETCHER BOND

The number of courses high of a corner should be determined before laying out the first course. Estimating the number of courses is a simple procedure. The sum of the stretchers used on the first course corner layout equals the number of courses high that the completed corner is.

For example, assume that a corner is to be built 9 courses high. To build the corner, 5 bricks must be laid in one direction and 4 bricks in the other direction. A corner that is to be 11 courses high requires 5 bricks on one side and 6 bricks on the other. Corners are usually built the height of a scaffold (4′ to 5′). When that section of the wall is laid, the scaffolding is erected and the corner is built again. This procedure is repeated until the wall reaches the desired height. Extending the lead any further than necessary when building a corner is considered poor practice, since these bricks could be laid more efficiently with the aid of a line. Figures 16-2A through 16-2F shows the steps in laying the first course.

Corners should always be ranged with a line. *Ranging* is the horizontal alignment of masonry units by use of a line tightly drawn between two points. It is done after the first course has been laid. To range corners, fasten one end of the line to the edge of one corner and the other end of the line to the outside edge of the other corner. Adjust any bricks in the line that are not in horizontal alignment until the corners line up perfectly with the range line. The line may then be stored until it is time to lay bricks to the line.

The mortar must be spread as uniformly as possible and should be the thickness of the specified mortar joint. If this is not done properly, the bricks must be removed and laid again. The ability to sense the proper joint thickness is acquired by the mason with practice, as every course is not measured. Many times, an experienced mason building a corner checks for alignment every third course, using a mason's rule or story pole. However, if it is critical that the courses be perfectly aligned, the mason checks every course. The apprentice should check every course on every job until the technique is perfected.

The face of the bricks must be initially laid plumb with the face of the wall without too much shifting of the bricks on the mortar bed. This can be done by constantly keeping the eyes in line with the face of the wall and sighting down the corner. This is especially important on the outermost edge of the corner or wall, Figure 16-3. If done correctly in the beginning, the amount of plumbing to be done later in the job can be decreased greatly and the work made easier. Sighting is a very important skill to the mason, especially when constructing corners. Sighting with efficiency and speed comes only with practice.

All excess mortar should be removed from the ends and inside the corner immediately after the course has been leveled and plumbed. All measurements of height should be done on the edge of the return of the corner. The work is then leveled from that point. Strike the joint in the corner as soon as the mortar has set sufficiently, or is thumbprint hard, Figure 16-4.

To be sure that the corner is plumb, hold the plumb rule against the bottom brick and adjust the other bricks.

Brush the corner carefully after the striking has been completed. This must be done gently so that mortar is not brushed from the joints. Check the corner with the modular rule for the specified height at the completion of the job.

Recheck the corner to be sure that it is plumb and level and is neatly struck. Always recheck the tail end of the lead at the completion of work, Figure 16-5.

With practice, the student will be able to lay the first brick on the corner to the prescribed height and proceed to lay out the first course level, plumb, and on the squared line.

PARGING THE CORNER

Parging is the application of mortar to a masonry wall to prevent water penetration. The mortar, applied with a brick trowel, is known as *back parging.* Parging a brick wall should be done before a backing material is installed.

The mortar coat should be about 3/8″ thick for best results. Mortar that is too thin cracks. Mortar that is too thick cracks and pulls loose from the surface of the wall. A good guide for parging is a piece of wood held on top of the last course and allowed to project over the back of the wall about 3/8″, Figure 16-6.

TOOTHING THE CORNER

Toothing is the construction of a temporary end of a wall by allowing every other end of stretcher bricks to project halfway over the stretcher below. This is done by laying a bat in the mortar on the end of the course previously laid and laying the next brick

Fig. 16-2A Laying out the corner with a steel square

Fig. 16-2B Dry bonding the bricks with the forefinger

Fig. 16-2C Dry bonding the proper number of bricks for a corner 9 courses high. Notice that there are 5 bricks leading in one direction and four in the other direction.

Fig. 16-2D Laying the first course of bricks in mortar. The mason is leveling the corner brick and the end brick before moving the bricks in the center. This is the procedure to follow when working on a base which is not level.

Fig. 16-2E Reversing the procedure and leveling the opposite side. The mason has not yet disturbed any bricks between the leveling points.

Fig. 16-2F Checking the corner on a diagonal to be sure that the first course is level

Fig. 16-3 Sighting down the outermost point of the corner bricks to be certain they are plumb. The corner shown, 5 courses high, is in perfect alignment.

Fig. 16-4 Striking the corner with a V-jointer. Mortar joints should be tooled before they are too stiff.

Fig. 16-5 Checking the corner with the plumb rule to be sure it is properly aligned. This is usually the last step in building a corner.

directly over it, Figure 16-7. After the corner has set a reasonable length of time, the bats are removed very carefully so that the corner is not knocked out of alignment. The bats should not be allowed to set completely in the mortar.

Toothing is used when a corner cannot be extended far enough so that it can be built to the specified height. Toothing may also be done when an opening is needed through which wheelbarrows may pass, or when windows or doors have not been delivered to a job on time.

Although toothing is necessary in specific instances, it is not generally recommended. If a joint is not filled with mortar and finished properly after toothing, there is a much higher chance of water leaking through the joint. In addition, toothing construction requires more time than other brickwork.

When filling a brick toothing with mortar, the mason must make sure the brick that is laid under the overlap is well buttered with mortar. After the brick is laid, the mason points up the mortar joints around the toothing with a trowel and the slicker striking tool. Finally, the mason parges the back of the toothed area with mortar to seal the joints. If a brick wall leaks, it does so most often at a faulty toothing. Following these steps will ensure a leak-proof wall.

Fig. 16-6 Parging the back of the corner. Notice the board at the top of the corner serving as a guide.

Architects and builders permit toothings infrequently, and they are built only under the strictest supervision. Since it is necessary to use toothings in masonry work, however, the apprentice should be familiar with them and be able to install them properly.

INSIDE CORNER

The construction of an inside angle corner is the same as that for an outside corner except that it is reversed. Inside corners are not built as frequently as outside corners, but their use is common in structures that have exposed brickwork in the interior. There may also be inside corners in places where two exterior walls join at an inside angle. If they seem more difficult to build, it is only because they are not built as often as outside corners.

The inside corner should not be attempted until after the construction of a rack-back lead and outside corner have been mastered.

Remember, all types of corners require quality workmanship. Since the corner is the guide for the wall, it must be built level, plumb, and to the specified height.

Fig. 16-7 Toothing of a brick corner. The supporting bats and excess mortar have been removed so that the toothing can receive the wall at a later time. Toothing is not a common practice but is necessary in some cases.

PROGRESS CHECK

At this point, masonry students should have acquired certain basic skills. They must practice these skills as every mason must have them. Students should demonstrate for their instructor each of the following skills. Any skills that are not mastered should be practiced until they can be demonstrated correctly.

- Dry bond a corner.
- Lay out a level course of bricks.
- Plumb a course of bricks.
- Spread mortar to uniform thickness.
- Use a square correctly to square a corner.
- Lay bricks to the line correctly.
- Attach a line pin, nail, and line block to masonry.
- Strike a joint correctly.
- Use a chalk line correctly to lay out a wall.
- Read the modular rule and build to a specified height.

ACHIEVEMENT REVIEW

Select the best answer from the choices offered to complete each statement. List your answer by letter identification.

1. Another name for a lead is a
   a. return.
   b. corner.
   c. bond.
   d. wall.

2. The simplest type of lead to build is the
   a. rack-back lead.
   b. outside corner.
   c. inside angle corner.
   d. Flemish bond lead.

3. The tail of the lead is located at the
   a. point where the lead returns on an angle.
   b. top of the last course laid.
   c. outermost corner of the racking point.
   d. first course of bricks laid on the lead.
4. The number of courses on a corner can be determined by
   a. adding the sum of the stretchers used in the first course layout.
   b. measuring the total length of the layout course.
   c. dividing each course into the total height.
   d. dividing the height by the total number of bricks.
5. A range line is used to
   a. lay the wall.
   b. align the corner when starting the first course.
   c. check the corner to be sure it is plumb.
   d. check the corner for the proper height.
6. Apprentices can learn the skill of plumbing a corner much faster if they master the knack of
   a. spreading mortar.
   b. leveling before the mortar dries.
   c. sighting each course laid.
   d. checking each course for height.
7. The principal reason for parging a corner is to
   a. add beauty to the corner.
   b. strengthen the corner.
   c. create a bond pattern.
   d. make the corner water resistant.
8. Toothing should not be done frequently but is sometimes necessary. The most common fault associated with toothing is
   a. change in the color of the wall.
   b. leakage of the joint where the toothing is.
   c. an unappealing design.
   d. change in the texture of the wall.

PROJECT 7: CONSTRUCTING A 4" RACK-BACK LEAD IN THE RUNNING BOND

OBJECTIVE

- The student will be able to lay out and build a 4" brick rack-back lead in the running bond. This requires the skills involved in leveling, plumbing, and building to a specified height. This is a typical lead that masons build between two outside corners on a long wall.

EQUIPMENT, TOOLS, AND SUPPLIES

Mortar pan or board
Mixing tools
Mason's trowel
Brick hammer
Plumb rule
Modular rule
Chalk box
Pencil
Convex sled runner (medium size)
1 batch of mortar (1 bag lime to 16 shovels sand)
24 standard bricks (3 of these are allowed for waste)
3, 4" x 8" x 16" concrete blocks on which to lay 2" x 4"
1, 8" x 8" x 16" block in which to set the level
Supply of clean water
1, 2" x 4", 6' long board on which to build the project (project could also be built on a concrete floor)

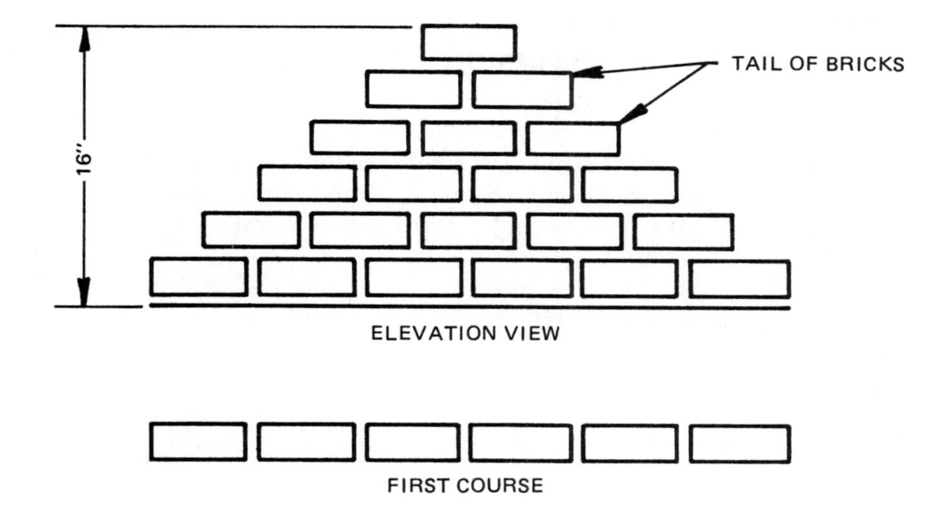

SUGGESTIONS

- Apply a little more mortar than the amount required for a solid joint.
- Apply mortar evenly when forming the head joints.
- Sight each brick which is laid even with the edge of the 2" x 4" or layout line.
- Clean all excess mortar from the project as the work progresses and return it to the pan to be tempered.
- Plumb the lead gently to prevent dislodging the lead.
- When striking mortar joints, hold the hand on top of each course of bricks to hold the lead securely.
- Always use safe working practices.

PROCEDURE

1. Mix the mortar and place it in the mortar pan. The mortar pan should be located approximately 2′ from the immediate work area.
2. Lay out the lead with a chalk line if it is being built on the floor. If using a 2" x 4", the edge of the wood serves as the wall line. Lay the 2" x 4" on 3, 4" blocks.
3. Spread enough mortar on the 2" x 4" or floor for 6 bricks.
4. Lay the first course as shown on the plan.
5. Check the height of the first brick laid, using the number 6 on the modular rule as a reference.
6. Level and plumb the course using the first brick as a guide.
7. Racking back one half brick on each end of the succeeding courses, continue laying the lead until the specified height of 16" (6 courses) is reached. Check each course with the rule. Level, plumb, and align the tail end of the lead.
8. Strike the joints with a jointer as needed. Brush the wall at the completion of the job.
9. Recheck the project with a plumb rule before it is inspected.

PROJECT 8: CONSTRUCTING AN OUTSIDE AND INSIDE BRICK CORNER FOR A 4" WALL IN THE RUNNING BOND

OBJECTIVE

- The student will be able to lay out and build an outside and inside corner 9 courses high in the running bond. These are the two basic corners used in brickwork. Build the outside corner first and the inside corner second. Procedures for both corners are the same.

EQUIPMENT, TOOLS, AND SUPPLIES

Mortar pan or board
Mixing tools
Mason's trowel
Brick hammer
Plumb rule
2" square
Chalk box
Convex sled runner striker
Modular rule
Brush
Pencil
Approximately 45 standard-sized bricks
1/2 batch of mortar (1/2 bag lime to 8 shovels sand)
Clean water
1, 8" x 8" x 16" concrete block in which to set plumb rule

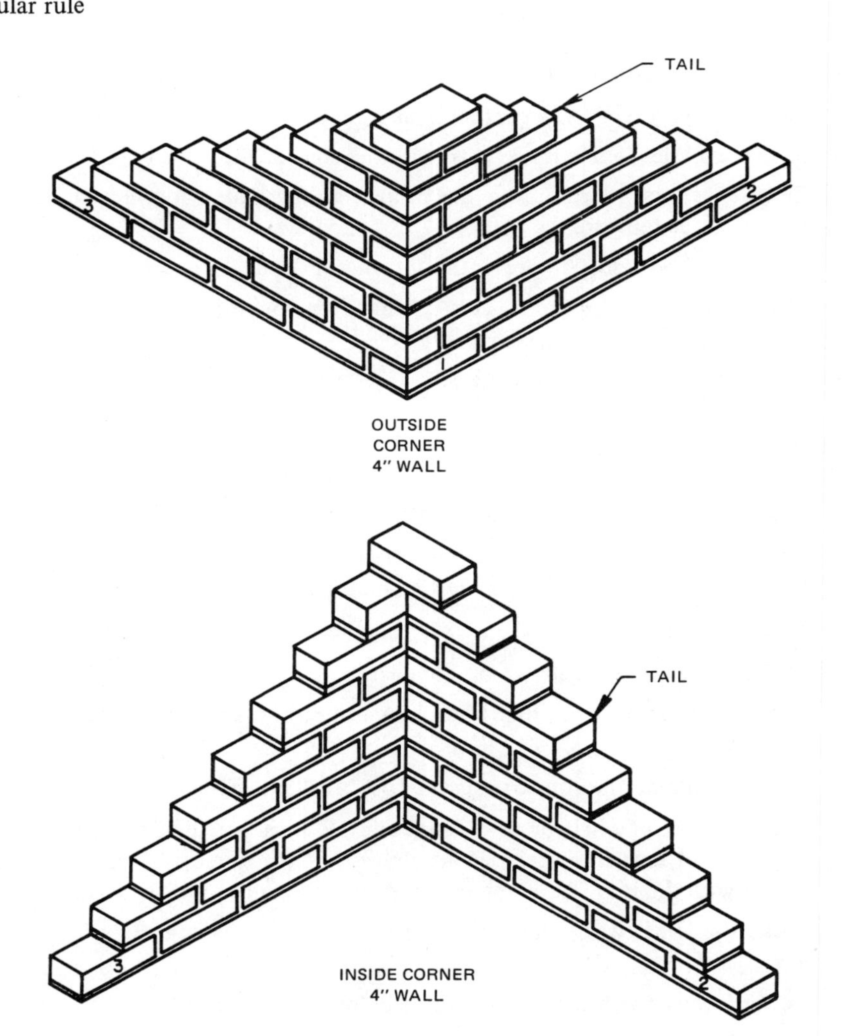

SUGGESTIONS

- Form solid head joints to prevent leakage.
- Select straight, square, unchipped bricks for the corner.
- When building the inside corner, form the head joints in the angle neatly since this is the center of focus. This also ensures against leaking.
- Recheck the corner for squareness after laying the first course of bricks.
- Sight down the corner as each brick is laid.
- Remove the excess mortar as soon as each course is laid.
- Keep all tools and debris away from the immediate work area.
- Observe good safety practices at all times.

PROCEDURE

1. Stock the bricks and place the mortar pan approximately 2′ from the work area.
2. Mix mortar and place into mortar pan.
3. Lay out the corner on the floor using a steel square and pencil. Extend the line a little longer than the actual corner measures with chalk.
4. Lay out the first course dry keeping head joints uniform (3/8″).
5. Lay brick #1 and brick #2 in mortar without moving the bricks in between. (See illustration.) Level and plumb brick #1 with brick #2 and straighten the edge with the plumb rule.
6. Lay brick #3. Level, plumb, and straighten the edge with brick #1. Lay the remaining bricks to complete the course. Level, plumb, and straighten the edge of each corner.
7. Lay the corner brick (# 1) first in each succeeding course and work toward the end of the lead. Level, plumb, and straighten the edge of each course.
8. Check the outermost corner of every course laid with the number 6 on the modular rule.
9. Build the corner 9 courses high. It will measure 2′ in height.
10. At the completion, check the tail of the lead on each side with a plumb rule.
11. Strike the corner with a convex sled runner striker. Brush after striking.
12. Parge the back of the corner with mortar. Be sure that the parging is 3/8″ thick. Remove excess mortar from the corner after parging.
13. Recheck the corner with a plumb rule before it is inspected.

PROJECT 9: LAYING A BRICK CORNER AND BUILDING A WALL IN THE RUNNING BOND WITH A LINE

OBJECTIVE

- The student will be able to build brick leads and a 4" wall in the running bond, maintaining a uniform thickness of all head and bed joints.

  Note: This project, which combines the construction of corners and laying bricks to the line, is typical for a home or small building.

EQUIPMENT, TOOLS, AND SUPPLIES

2 mortar pans or boards
Mixing tools
Mason's trowel
Brick hammer
Plumb rule
2" square
Ball of nylon line
Line pin and nail
Pencil
V-joint sled runner
Brush
Chalk box
Approximately 145 standard-sized bricks (waste included)
1 1/2 bags of lime; 27 shovels of sand
Supply of clean water
1, 8" x 8" x 16" concrete block in which to set level

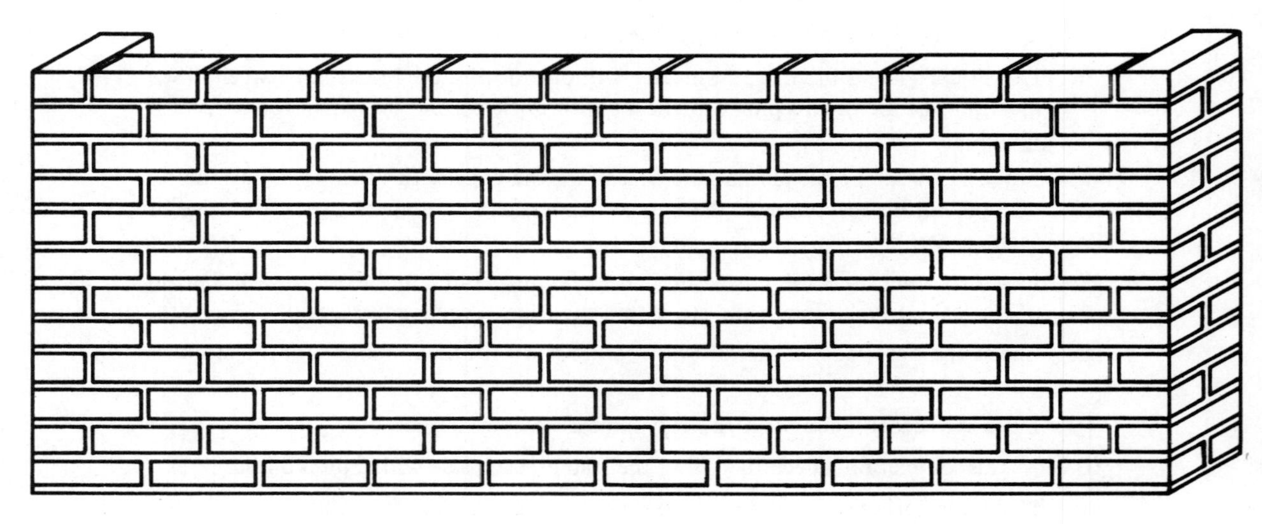

ELEVATION

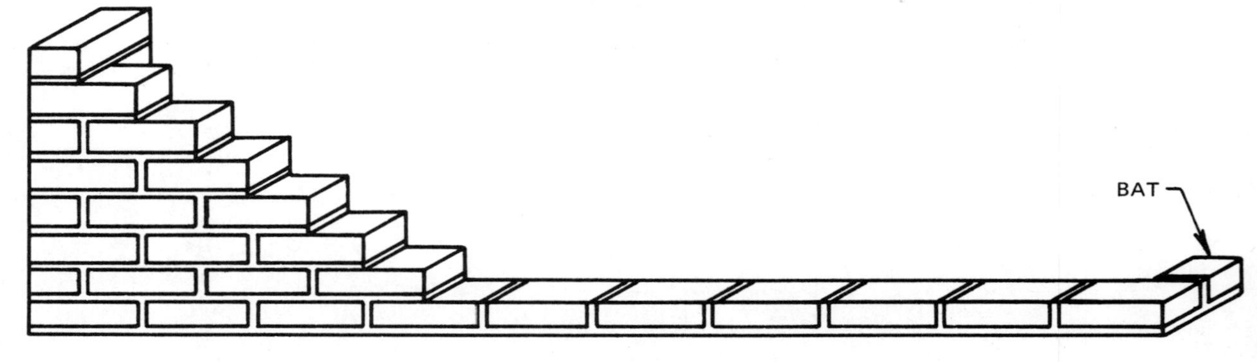

LEAD AND FIRST COURSE

SUGGESTIONS

- Be sure that the base for the construction is clean.
- Keep all materials about 2′ from the wall line.
- Form solid head joints.
- Spread mortar as uniformly as possible.
- Practice pressing bricks into position. Avoid unnecessary tapping.
- Do not spread mortar too far ahead of the brick laying or it will dry before it can be covered.
- Remove the excess mortar from each course before laying the next course.
- Sight each brick laid to be certain the bond pattern is in correct alignment.

PROCEDURE

1. Assemble the bricks in the work area.
2. Mix the mortar and load the mortar pans.
3. Snap a chalk line longer than 8′ on the floor.
4. Mark a line for an 8″ jamb at one end of chalk line with a 2′ square.
5. Dry bond 10 bricks from the point which was squared off to the other end of the line. (The line will be slightly longer than necessary.) This is used as a reference point when laying out the bricks. Use the forefinger for correct spacing.
6. Place a mark at the end of the tenth brick. Use a square to square off this point for the jamb on the other side of the wall.
7. Use a brick of average length as a gauge when laying out the jamb of the lead at each end.
8. Bed the corner bricks at each end of the wall to the correct height (number 6 on the modular rule). Level the two bricks with each other by setting another brick temporarily in the middle as a check point. (A long, straight 2″ x 4″ with a plumb rule laid on top could also be used as a leveling board). Be sure that the bricks are plumb, aligned, and ranged with the layout line.
9. Attach the line by pushing a nail to which the line is attached under the end of the brick on the left end of the wall. Pull the line up over the top of the brick and push the line to the face of the brick. Lay a couple of bricks on top to hold the line in place. Push the line pin under the end of the brick at the right end of the wall. Pull the line tight and wrap it around the pin. Lay a couple of bricks on top of the line to prevent it from dislodging.
10. Pick up the dry bonded bricks as needed and place them in mortar to lay the first course.
11. Cut bats for jambs with the hammer and lay one at each end. Level, plumb, and square the bats with the wall line.
12. Build a lead projecting 3 1/2 bricks at the end of each wall. Check the height of each course with the number 6 on the modular rule. Cut bats as they are needed for jambs. Check jambs periodically to be sure they remain square, plumb, and level.

13. Fill in the space between leads, using the line as a guide.
14. Resume building leads up to the specified 12 courses in height and fill in the wall to the line. Check the work for the correct height (number 6 on the modular rule).
15. Strike the wall with a V-joint sled runner striking tool. Brush the project at completion.
16. Parge the back of the wall and the inside of the jambs with mortar.
17. Recheck the project with a plumb rule before inspection.

Unit 17
Estimating Brick Masonry by Rule of Thumb

OBJECTIVES

After studying this unit, the student will be able to

- determine the number of square feet in a brick wall by the wall area method.
- estimate brick, masonry cement, and sand for small-sized and average-sized jobs.

Estimating by rule of thumb is not intended to be a mathematically perfect method of estimation. It is workable, however, when estimating materials for the construction of small-sized and average-sized jobs. Some examples of these small-sized and average-sized jobs are a home, garage, retaining wall, or chimney. Apprentice masons are not expected to estimate materials for large or complicated jobs. However, they should be able to figure the necessary materials for daily work projects.

THE IMPORTANCE OF ESTIMATING

The rule of thumb method of estimation was developed through years of practical experience on the job. It is designed to allow for some waste on the job. Most small contractors estimate their materials by rule of thumb.

The apprentice should be able to apply basic arithmetic, including addition, subtraction, multiplication, and division, to solve typical estimating problems. The majority of mistakes made in estimating occur in the process of simple arithmetic. The mason should always recheck work before accepting the results as final.

It is important that materials not be grossly overestimated or underestimated. When materials are overestimated, the materials which remain must be moved to another job. This results in wasted labor, time, and money. It may be some time before the materials are used again, especially if they are a special type. Underestimating, on the other hand, is also wasteful of time and energy.

THE WALL AREA METHOD

Most masonry materials used today are based on the modular system. This greatly simplifies estimating for the mason.

For great simplicity and accuracy, the most widely used method of estimation is the *wall area* or *square foot* method. It consists of multiplying the length and height of the wall minus openings.

Rule of thumb estimating can best be explained by using a typical problem as an example. In the discussion of estimating techniques which follows in this unit, a specific problem will be used as an example. Assume that the required joint size is 3/8″.

ESTIMATING SQUARE FEET

A 4″ thick wall is to be built 20′ x 8′. It is built in a running bond, so no extra amounts must be considered for header courses. Estimate the total number of square feet as follows:

```
     20′ (length of wall)
x     8′ (height of wall)
    160  total square feet (sq ft)
```

ESTIMATING BRICK

After finding the total square feet in the wall, the mason may then estimate the number of standard-sized bricks required. There are 6.75 bricks to each square foot of wall area. When estimating by the rule of thumb, round off this figure to 7. This allows for waste and broken bricks.

To estimate the number of bricks, multiply 7 by the total wall area.

```
   160 sq ft of wall
x    7 bricks per sq ft
  1120 total amount of bricks needed
```

ESTIMATING MASONRY CEMENT

Masonry cement can be estimated in two different ways. The first way is by using a base figure of 8 bags per 1000 bricks. The second way is by calculating the number of cubic feet per 100 sq ft of wall area. If the sand is coarse, it may be necessary to increase the amount of masonry cement to improve the workability. Even though the mortar has great strength, if it does not handle well on the trowel, the cost for the job greatly increases since the mason's productivity is cut. This depends on the particular situation.

When using portland cement and lime-based mortars, a table must be consulted to arrive at the correct proportions. A table must be consulted because the percentage of portland cement and lime differs according to the type of mortar specified. For estimating purposes, only masonry cement will be dealt with in this unit, since it is the most commonly used throughout the masonry business.

To estimate masonry cement for a wall measuring 20′ x 8′, figure 8 bags of masonry cement to every 1000 bricks. Waste is included. By dividing 1000 by 8, it is found that 1 bag of masonry cement is sufficient for 125 bricks. To find the number of bags needed for the wall in question, divide as follows:

```
          8.96 bags of masonry cement
125 ) 1120.00
      1000
       120 0
       112 5
         7 50
         7 50
```

Round off the answer to 9 bags.

ESTIMATING SAND

Sand, one of the most inexpensive materials used in masonry work, is sold by weight. Because it is inexpensive, it is not important to estimate to the exact pound for a job. As a rule, masons order sand by the ton unless the job is very small, and then in hundreds of pounds. The base figure for estimating sand is 1 ton per 1000 bricks, allowing for a normal amount of waste.

There are several factors to consider when estimating sand. If more than 3 tons of sand are needed, 1/2 ton should be allowed for waste. First to consider is the loss of sand when it is piled on the ground. (This is the usual method of storing sand.) Since it is on the ground, some of the sand will mix with dirt and foreign particles.

Also, if the sand is located in a populated area with small children and is unprotected, the mason must assume that a certain amount will be lost. The estimated loss depends on the project.

The wall in the original problem, which required 1120 bricks, would require 1 1/2 tons of sand. This figure allows a sufficient amount for waste. This is as close as the mason would estimate on this particular job.

ESTIMATING MATERIAL COSTS ON THE JOB

The materials needed for a wall 20′ x 8′ would be listed:

```
1120 bricks
   9 bags of masonry cement
1 1/2 tons of sand
```

To estimate the total price of the materials, multiply each by the price per unit. Assume that the bricks are $0.09 each, the masonry cement is $1.60 per bag, and the sand is $7.00 per ton.

```
    11 20   number of bricks
x   $0.09   cost per brick
  $100.80   total cost of bricks

    $1.60   cost per bag of cement
x       9   number of bags
   $14.40   total cost of cement

    $7.00   cost per ton of sand
x     1.5   tons of sand
    3 500
    7 00
  $10.500   total cost of sand
```

Add all of the costs together for the total cost.

```
  $100.80   for bricks
    14.40   for masonry cement
    10.50   for sand
  $125.70   total cost of materials
```

ESTIMATING LABOR COSTS FOR BRICK MASONRY WORK

After materials have been estimated, labor costs must be added before a final price for the job can be determined. Many factors affect costs for labor, including weather conditions, accessibility of materials, and other hidden costs. The contractor must expect some hidden costs for such things as equipment repair, insurance, and depreciation of equipment. In addition, the contractor must estimate how much work the masons and laborers will produce. Finally, the contractor must realize that costs will vary depending on geographical location and the availability of skilled masons.

Labor Wage Rates

Factors like unionization, geographical location, and the amount of work available determine wage rates. Rates vary and change constantly in response to these and other factors.

In estimating labor costs, it also helps the contractor to know how many bricks a mason can lay per day. This amount depends on the type of bond being used. However, according to the International Union of Bricklayers and Allied Craftsmen, a national average for masons using running bond is about 675 bricks per 8-hour day.

In addition, the contractor must estimate the number of laborers needed. Generally, this figure is one laborer for every four masons, but it may have to be adjusted for certain job conditions.

After determining costs for labor and materials, the contractor adds them together to reach a final bid or price for the job. As you can see, estimating can be a complicated process, and estimating large jobs should be left to a professional estimator or contractor. However, whether the job is large or small, careful attention to mathematics will eliminate many estimating mistakes that often appear.

ACHIEVEMENT REVIEW

Solve each of the following problems concerning the estimating of materials and cost and determining total area.

1. a. A brick wall 40′ long x 8′ high is to be built in the running bond. What is the total area in square feet?
   b. Determine the number of bricks needed to build this wall.
   c. Find the number of bags of masonry cement needed for this wall.
   d. Determine the amount of sand needed for the wall.
2. An 8″ brick retaining wall is being built. Figure the amount of materials needed to build the wall if it measures 100′ long x 4′ high.
3. A 4″ brick veneer wall is to be built across the front of a store. The wall is 50′ long x 5′ high. Estimate the total amount of materials needed for the total cost of the wall if the bricks cost $80.00 per thousand, masonry cement costs $1.70 per bag, and sand costs $8.00 per ton.
4. An 8″ brick garden shed 10′ wide, 10′ long, and 8′ high is to be built. There is one door which measures 3′ x 7′. Estimate the amount and cost of materials for the job, deducting the door area from the total wall area. Use the unit prices given in question 3.
5. A job calls for 7500 bricks. How much sand and masonry cement is needed for the job?
6. How many bricks could be laid with 25 bags of masonry cement?
7. If a mason lays 675 bricks a day, how many bricks would the mason lay in a 20-day period?
8. A large brick masonry job employs 44 masons. If one laborer is required for every four masons, how many laborers are needed on this job?

Unit 18
Laying Concrete Block to the Line

OBJECTIVES

After studying this unit, the student will be able to

- make bed joints and head joints for concrete block.
- lay concrete block to the line.
- cut concrete block with a brick set and hammer.
- describe and lay insulated concrete block.

The process involved in laying concrete block to the line is very similar to that of laying bricks to the line. There are some important differences, however, due to the fact that concrete block is much larger and heavier.

The placement of materials in the work area is of prime importance. Concrete block must be stacked on the pile with the bottom side down. The top of the block has a larger shell and web which makes it easy to distinguish. The units must be handled carefully, as they easily chip and crack.

Concrete block should be kept dry at all times. When highly absorbent bricks become very dry, they are wet with water. This is not the case with concrete block. Moisture causes block to expand in size. If they are built into the wall when they are wet, they will later dry and shrink, causing cracks between the mortar and the block. Plastic covering is an inexpensive way to protect block. The block should be stored on pallets off the ground to prevent dampness from being absorbed.

SPREADING MORTAR

As a rule, mortar is bedded only on the outside edges of concrete block. Very seldom is a solid bed joint used. Bedding on the outside webs of block is known as *face shell bedding.* It is swiped on with the edge of the trowel.

All joints and holes should be filled when the block is laid so that all of the mortar adheres and dries at the same rate of speed. The mortar must be of the proper consistency. Mortar which is too runny or soft causes the block to sink below the line, which requires relaying of the block. Mortar which is too stiff also presents a problem, since the block requires a great amount of pressure to lay it in the mortar. This could cause damage to the block. Block should never be moved after they have been laid. Movement after the initial set may cause leaky walls and loss of bond strength.

The mortar on bed joints should never be dusted with dry cement to stiffen the mortar. The cement dust prevents the formation of a strong bond between the mortar and the block. It is also a time-consuming and expensive method of stiffening mortar. Nothing but mortar should be placed in bed joints, except wire reinforcement if specified.

A well-built block wall can result with good planning, proper care and spacing of materials, and accurate use of tools.

LAYING THE FIRST COURSE

The apprentice should learn to lay block to the line before attempting to build a corner. As is the case in brickwork, block is laid to the line, 1/16" away. The line is attached either to a corner pole or to a prebuilt concrete block corner.

Block should always be laid in the wall with the wider web of the block facing up, Figure 18-1. Block laid with the wider web down will not be any less

Fig. 18-1 The two major types of concrete block, the 2-celled (left) and the 3-celled (right). Both are 8 inches in length. The 2-celled block is shown with the thicker web facing up, the position in which it is laid in the wall. The 3-celled block is shown with the bottom side facing up. Notice the difference in web thickness of the 2 blocks.

Fig. 18-2 Forming a head joint on a concrete block. The mortar is applied to the ears, or ends, of the block.

strong. However, it is easier to apply mortar on the wider web and less mortar is wasted if it drops into the cell of the block.

The first course should be laid with great care, since a level and plumb wall depends to a great extent on the first course. Sweep or brush the base before spreading the mortar. Spread a solid bed of mortar on the base. Do not furrow the mortar on the first course. A solid joint ensures a watertight wall and proper bonding with the base.

APPLYING THE HEAD JOINT

Full head joints should be formed on both *ears* (end edges) of the block to be laid. With the trowel, pick up enough mortar from the mortar board to form the head joints. Do not fill the entire trowel with mortar, since this much mortar is not needed for head joints. Stand the block on its end in a vertical position and apply head joints on both ears of the block with a downward swiping motion of the trowel. The mortar should then be pressed down on the inside of the ears of the block so that it will not fall off when lifted up and placed in the wall, Figure 18-2. If a head joint becomes dislodged from the block as it is picked up, reapply fresh mortar before laying.

Lift the block firmly by grabbing the web at each end of the block and lay it on the mortar bed joint. Do not move the block with jerking motions. The mortar should stay intact. The trowel should remain in the mason's hand when laying block to save time. The hand should curl over the handle of the trowel and grasp the web of the block at the same time. This may be difficult in the beginning but will come with practice.

> **Caution:** Do not lay 2 blocks together so that the two square ends meet unless it cannot be avoided. If the blocks are laid and shoved together in this position, the mason's thumb could be crushed.

POSITIONING BLOCK

Practicing the techniques involved in laying block will determine the easiest methods for the individual mason. By tipping the block a little toward the body and looking down the face side of the block, the mason can position the block in relation to the top edge of the block in the course below. The block should then be rolled back slightly so that the top of the block is in correct alignment with the line. At the same time, the block is being pressed back toward the last block laid so that the proper amount of mortar oozes from the head joint.

Concrete block should be laid so that the top of the outside edge is level with the top of the line and located 1/16" back from the line, Figure 18-3. The

Fig. 18-3 Laying concrete block to the line. The mason is aligning the block with the line so that it is flush with the course underneath.

bottom outside edge of the block is then in line with the top edge of the course below.

The method in which the block is set on the bed joint is very important. The block should be laid gently in the mortar so that it does not sink too far. The block should not be released too quickly, or it will have to be relaid. By slightly delaying the release of the hand from the block, the block absorbs the moisture from the mortar and takes an initial set. After the block is set, mortar should ooze out of the head and bed joints. If it does not, there was not enough mortar used in the joints. This presents problems in tooling joints and may also prevent a watertight joint from being formed.

To assure a good bond, do not spread mortar too far ahead of the actual laying of the block, since the mortar stiffens and loses its plasticity and strength. Positioning of concrete block must be done before the mortar stiffens or the bond strength is destroyed and cracks result. As each block is laid, cut off the excess mortar with the trowel on a slight angle, Figure 18-4. If the work is progressing at an efficient pace, this mortar may be used for the next head joint. Never allow mortar which has been cut off to fall to the ground, since this mortar cannot be used again.

Fig. 18-4 The trowel should be held at the proper angle when cutting off excess mortar.

Fig. 18-5 Tapping a concrete block into place. The trowel must remain centered on the block to avoid chipping the face of the block.

ADJUSTING BLOCKS

If a block is set unevenly on the mortar, first check to see if a pebble or other foreign matter has become lodged between the mortar and the block. If this is the case, remove the block from the mortar bed and lay it in fresh mortar. If the block simply requires readjustment in the mortar, it is permissible to tap it into place, Figure 18-5. The block should not be tapped with the trowel extended over its face, as the face of the block could be chipped or mortar on the trowel could be splattered over the wall. Although it is not recommended to tap the block with the handle of the trowel, it is permitted if done sparingly. If it is done frequently, the handle becomes rough and worn, which may be bothersome to the mason. The mason's hammer should be used if adjustment cannot be easily made with the trowel.

INSTALLING THE CLOSURE BLOCK

The closure block, the last block laid in the wall, is most prone of all the block to leakage. It is imperative that the mortar joints are strong and that the spacing of the block is correct. The amount of space allowed for a standard block is the length of the block plus room for 2 mortar head joints. For a standard concrete block, this measurement is a total of 16 3/8", 15 5/8" for the length of the block, and 3/8" for each mortar joint.

Fig. 18-6 Laying closure block in wall. It is important that the head joints are on all ends to make full joints.

Before placing the closure block in the wall, apply mortar to form joints on all four edges of the block which is already in place. Next, apply mortar to all four edges of the closure block. Lower the closure block into the space without disturbing the other block, Figure 18-6. Immediately fill any holes that are left in the joint by picking up mortar on the back edge of the trowel and pressing it into the head joints. It is extremely important to apply mortar to all head joints when laying the closure block.

After the closure block has been laid, erect the line for the next course and continue building the wall to the specified height. Some masons stand a number of block on end to apply the head joints rather than applying mortar on one block at a time, Figure 18-7. This practice speeds the process, but the mason must guard against the danger of mortar drying before the block is laid.

REINFORCING BED JOINTS OF CONCRETE BLOCK

When walls require additional strength, wire reinforcements can be laid in the mortar joints, Figure 18-8. The wire is laid on the wall with the mortar spread on top of and fully surrounding it. In places where two pieces of wire meet, there should be a

Fig. 18-7 Applying mortar to form head joints on several blocks at one time can speed up the block-laying process.

Fig. 18-8 Wire reinforcement track laid in mortar joint for additional strength

minimum overlap of 6″. The average distance between vertical courses of wire track is 16″, with 24″ as the maximum allowed. Job specifications should always be consulted before wire reinforcements are installed.

TOOLING JOINTS

Joints on concrete blocks are usually concave joints, with the V-joint ranking second in popularity. A jointer slightly larger than the mortar joint should be used to form the joint.

The sled runner jointer, preferably 16″ long, should always be used when striking joints on concrete block work because it strikes a straighter joint, Figure 18-9.

Strike the head joint first and the bed joints last. Enough force should be used to press the mortar tightly against each side of the joint so that the surface is smooth, water resistant, and free of cracks. When striking, fill any cracks or holes with fresh mortar. Never pick up old or dead mortar to use for jointing. At the completion of the striking process, allow sufficient time for the mortar to dry before brushing the wall. If it is brushed too soon, the mortar joint may be destroyed.

CUTTING CONCRETE BLOCK

Concrete block can be cut with the use of various tools. The masonry saw provides the most precise cut. However, the apprentice first learns to cut block with the hammer and blocking chisel (or *brick set*) since a masonry saw may not always be available. Proper eye protection must be worn when cutting block.

Lay the block on its side and mark where the cut is to be made with a pencil and ruler or straightedge.

Fig. 18-9 Striking a joint on a block wall with a sled runner jointer

Fig. 18-10 Cutting a concrete block with the brick set and hammer

Score one side of the block lightly with the blocking chisel. Turn the block over carefully and score the other side. Set the block with the bed side down and with the holes in a horizontal position. With the brick hammer, lightly tap the top of the block above the point at which the scoring was done, Figure 18-10. The block piece should break evenly. If it does not, rescore the block again. Even when it is cut correctly, the block may not always be perfectly straight. If it is not straight enough, grasp the web of the block nearest the cut end and even the edge with a hammer. Cut downward in a vertical position so that the shell of the block does not break. When it is not necessary that the block be cut accurately, the brick hammer may be used alone.

Do not cut concrete block on a hard surface such as a concrete slab. The hammer or brick set combined with the hard surface may cause the block to shatter or break irregularly.

SAFETY PRACTICES ON THE JOB

Masons should be especially careful when lifting concrete block because of the heavy weight. When lifting the block, keep the feet close together and the block close to the body. If uneven ground or ditches are present in the work area, the chance of a shift in body weight is greater than under normal conditions. A fall or slip with a concrete block in hand could cause a severe injury to the mason.

Cutting concrete block should always be done with consideration for the safety of other workers. Flying pieces of block can cause serious injury to the eyes. Cut the block downward and away from yourself and others. Holding the hand high on the chisel will result in fewer injuries to the hands. Burred chisels should be ground off immediately, as flying pieces of metal can cause injury. Eye protection should be worn when doing any cutting. Cutting concrete block calls for special attention to safety rules.

REPOINTING THE WALL

If patching or repointing must be done after the mortar has hardened, the mortar must be chiseled out with a joint chisel to a depth of at least 1/2". All loose mortar should be removed with a brush. The joint should be wet with water. To do this, use a brush or splash water directly on the cut portion. The joint is then repointed with fresh mortar using a steel sucker tool. Wetting the joint delays the setting time and, therefore, produces a better bond. Line pin and nail holes should be repointed immediately after the pin or nail is removed from the wall. Be especially careful that the adjoining units are not chipped or cracked by the cutting operation when cutting out joints.

CARE AND PROTECTION OF THE WORK

Walls which are constructed above grade line (or ground level) are subject to wind pressure. At times, the pressure can amount to more than 20 pounds per square foot (lb/sq ft) of wall area. While the walls are being built, they should be protected from the wind and given some type of temporary support. This is especially important in the early stages of construction before the mortar has hardened to full strength. As a rule, walls higher than 8′ to 10′ should be braced. This is done by *shoring* (or propping up) timber or framing lumber about every 8′ to 10′ of wall length on both sides of the wall, Figure 18-11. After the *joists,* or floor supports are laid in place, the bracing is removed. Bracing similar to this is also used to prevent walls from caving in when a bulldozer backfills a foundation.

Fig. 18-11 Concrete block wall held in position with wooden braces. After the joints are constructed, the braces are removed.

When the work is complete for the day, it should be protected from the elements. Walls which are rain-soaked may take months to dry and efflorescence may develop. Sheets of plastic or tarpaulin should be used to cover the top of the walls to prevent moisture from entering the core of the blocks. The covering should cover and hang over the wall at least 2′. Loose boards or planks laid on top of the walls are not sufficient protection.

INSULATED CONCRETE BLOCK

A new energy saver, the insulated concrete block, is appearing in commercial and residential masonry construction. A 3"-thick polystyrene insert placed in the cores of the block provides the insulation, Figure 18-12. Although insulated concrete block costs more

Fig. 18-12 Insulated 8 x 8 x 16-inch concrete block. Note how the polystyrene fits in the cell of the block.

than noninsulated block, the initial costs are offset by savings in energy costs.

Insulating properties of building materials are usually stated in terms of an R value. Ordinary block produced from a mixture weighing 95 pounds per cubic foot (lb/cu ft) has an energy rating of R-2.86. In comparison, insulated block weighs 80 lb/cu ft and has an energy rating of R-9.52, which is within 1.50 points of the recommended R-11 value for walls. Insulated concrete blocks can be built in single walls or in double walls with an air cavity between for additional insulation.

Building Insulated Block Walls

In an insulated concrete block wall, a center bed joint must be applied to seal off the joint. This joint is applied directly over the insulation. Wire joint reinforcement is then laid in the mortar. The wire in Figure 18-13 protrudes to tie in the other cavity wall, which will be built later.

The insulated block is then laid and pressed into position to the line. Take care that the joint reinforcement is pressed into the mortar joint, Figure 18-14.

High energy costs are making insulated block very popular with builders. The apprentice mason should be familiar not only with the basic block-laying process, but with the concepts behind the demand for energy-efficient masonry walls as well.

Fig. 18-13 Mortar bed joint and wire reinforcement in place on insulated block wall

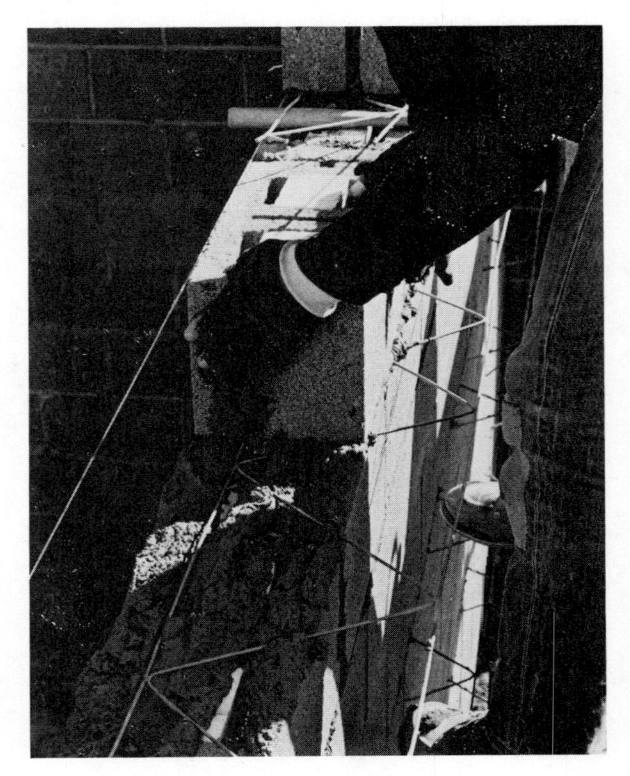

Fig. 18-14 Laying insulated block to the line. All excess mortar is cut off with the trowel.

ACHIEVEMENT REVIEW

The column on the left contains questions on concrete block and their production. The column on the right lists terms. Select the correct term from the right-hand list to answer each question on the left.

1. What is the term for spreading mortar only on the outside web of a concrete block?
2. Concrete blocks have a top and bottom edge. Which thickness of web should be facing up when they are laid in a wall?
3. A solid joint should be spread on the base when starting the first course. What is the main reason for this?
4. What are ends of stretcher blocks where the mortar head joints are applied known as?
5. If mortar is spread too far ahead on a wall, it dries too quickly. What type of wall strength does this affect?
6. When adjusting a block to the line, where on the block should the mason tap?
7. What is the last block that is laid in the wall called?
8. What is the standard height at which wire reinforcement should be installed in a bed joint?
9. What is the most common joint finish on concrete block construction?
10. When an accurate cut is needed and a masonry saw is not available, what is the best tool to use?

a. Brick set
b. Bat
c. Closure block
d. Ears
e. Concave
f. Header
g. Wider web
h. Rake out
i. 24″
j. 16″
k. 8″
l. Face shell bedding
m. Center
n. Jointing
o. Bond
p. To waterproof
q. Jamb
r. Butt edge
s. Brick hammer
t. Furrowing
u. For design
v. Narrower web

PROJECT 10: LAYING A CONCRETE BLOCK WALL

OBJECTIVE

- The masonry student will be able to lay an 8′ x 8′ x 16″ concrete block wall in the running bond with the use of a corner pole.

  Note: This concrete block wall is typical of a block wall which surrounds a stairway in an apartment building.

EQUIPMENT, TOOLS, AND SUPPLIES

2 mortar pans or boards
Trowel
Plumb rule
Chalk line
Line blocks
Modular rule
Convex sled runner
Brick set chisel
Ball of nylon line
Brick hammer
Brush
2 corner poles
Mixing tools
60, 8″ x 8″ x 16″ concrete blocks (4 are allowed for breakage)
2 bags of mason's lime, 32 shovels of sand
Supply of clean water

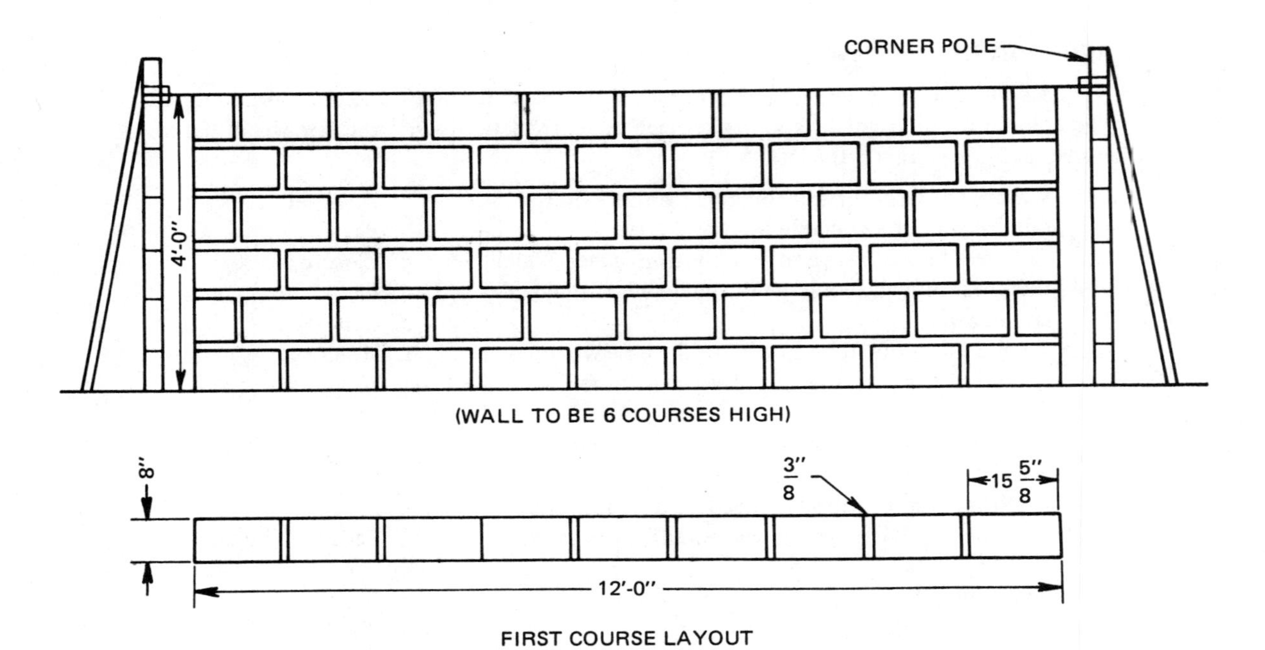

SUGGESTIONS

- Use only face shell bedding for the bed joints.
- Spread the mortar on the concrete block with the front of the trowel blade.
- Check the jambs every course to be sure they are plumb.
- Return all mortar which is removed from the wall for retempering in the pan.
- Be sure that the wider web of the block faces up on each course.
- Do not splash the wall with mortar.
- Be certain that each block laid is spaced the correct distance from the line (1/16").
- Use eye protection when cutting the blocks.
- Strike joints when the mortar is thumbprint hard.
- Follow proper safety rules at all times.

PROCEDURE

1. Stock the work area with concrete block and space the mortar pans the correct distance from the work to allow for sufficient working room.
2. Mix the mortar to the desired consistency.
3. Lay out the first course using a rule to check the bond.
4. Set up the corner poles on the wall line and erect the line for the first course.
5. Mark the corner poles in 8" divisions.
6. Cut the necessary half blocks as shown on the plan.
7. Move the line as needed and erect the wall to a height of 4' - 0" or 6 courses.
8. Strike the mortar joints with a convex jointer as required.
9. Brush and recheck the wall with the plumb rule before inspection.

Unit 19
Laying the Block Corner

OBJECTIVES

After studying this unit, the student will be able to

- lay out the first course of block on a corner.
- install wire reinforcement in bed joints.
- build a concrete block corner to a specified height.

BUILDING WITH BLOCK VERSUS BUILDING WITH BRICKS

A concrete block corner and a brick corner are constructed in similar ways. The main differences are that concrete block are larger and heavier than brick, and mortar is applied in a slightly different manner. Because block are heavy, hollow, and somewhat awkward to handle, care must be taken to prevent them from chipping (also called *spalling*). In addition, since they are larger and heavier than bricks, block must be laid gently in the mortar bed to prevent sinking. Sinking would require relaying the unit.

Because concrete block are greater in length and height than brick, they may be more difficult to keep level and plumb. Adjustment of the block after they are laid may also be more difficult. For this reason, it is very important that the first course be the correct height. Excessive bedding or squeezing of mortar joints is not an acceptable method of adjusting block.

Another important consideration is the length of time needed to lay the units in the wall. Laying a corner using concrete block requires approximately 40 minutes. Laying a corner with bricks requires about 3 hours. Both of these figures are based on normal working conditions.

The volume of bricks and block also relates to efficiency in construction. For example, to build a corner to scaffold height with concrete block requires 7 courses. To construct the same corner with bricks requires 21 courses. When comparing volume, 12 bricks are required to replace 1, 8″ concrete block in a wall. It is not possible for any mason to lay 12 bricks as rapidly as another mason could lay 1, 8″ concrete block. The time and cost of materials and labor saved through the use of concrete block rather than bricks is the principal reason for the popularity of concrete block walls.

CONSTRUCTING THE CORNER

The corner is the key to a straight, plumb wall. It is, therefore, important for the beginning mason to perfect a step-by-step procedure in corner construction. Figures 19-1A through 19-1G show the steps involved in the construction of a concrete block corner.

Preparing the First Course

As is the case in preparing to lay out a brick wall, it is extremely important that a good bond between the first course and the base be established. The corner should never be laid out on mud or loose dirt. It should be laid out on a clean concrete base or footing. In warm, dry months it is good practice to dampen the footing with water to prevent too rapid absorption of moisture before the mortar is spread. Use a brush and a bucket of water to dampen the area where the blocks will be laid. The footing should not be saturated with water. Never soak the block, as this causes cracks from shrinkage in the mortar joints. In the wintertime, it is just as important not to spread mortar over ice or snow.

Fig. 19-1A Leveling the block. Block should always be leveled by length and width.

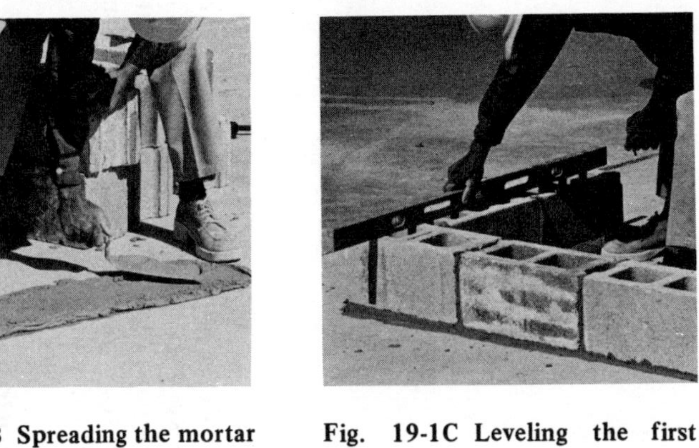

Fig. 19-1B Spreading the mortar bed for the first course

Fig. 19-1C Leveling the first course

Fig. 19-1D Checking the first course on the corner to be sure that it is plumb

Fig. 19-1E Checking the first course along the top of the block to be sure that it is straight

Fig. 19-1F Swiping mortar on the block with the trowel. Notice that this technique differs from the one used in brick construction.

Fig. 19-1G Measuring the corner for the proper height. The modular rule should read number 2 at the top of the corner.

Locate the point of the corner and strike a chalk line across the wall to the opposite corner. Repeat the procedure on the other side of the corner. If it is not convenient to strike a chalk line, a ball of nylon line can be unrolled along the side where the corner is to be built. This line aligns the corner which is to be built with the opposite corner. This practice ensures the construction of a true, straight wall. Check the accuracy of the chalk line with a steel square before laying any block on the corner.

Laying the First Course

The first course must be laid with care to ensure that it is level, plumb, and properly aligned. It is good practice to lay the plumb rule on the footing or base before any mortar is spread to determine if the base is level. Spread a solid bed of mortar with the trowel. Do not *furrow* (or groove) the mortar as this may detract from the strength of the bond. A solid joint produces more waterproofing qualities.

It is essential that the first course be perfectly level to the specified height. This is usually determined by measuring from an established mark located close to the corner with a rule. The mason should attempt to correct any differences in height on the first course. The block may require reducing or bedding with split-block or brick to correct a faulty footing. Footings are not always poured perfectly level. Any variance of 3/4" or under usually can be corrected in the mortar joints without any noticeable effects.

Lay the corner block first and align it with the wall line. Pull the line tight on the corner block to the opposite corner and lay out the correct number of block needed to reach the specified height. (The scaffold height is usually about 4′ or 5′.) As in the construction of brick corners, the sum of the block in the first course equals the number of courses in the height of the corner.

It is a good practice to level the block lengthwise and crosswise on the first course. As concrete block are very square, it can be assumed that if the block is level in these two ways, it will also be plumb on the face. Do not remove excess mortar that has been squeezed from the block immediately, as it could cause the block to settle unevenly. This is usually done after the second course has been laid.

When a block corner consisting of 1 unit in thickness is being built, check only one side of the block to be sure the corner is plumb. Walls consisting of 2 units laid back to back must be plumbed on both sides and aligned. This is done because although concrete block are reasonably square and straight, they are not always exactly the same width. Select the side that is the most important as far as appearance is concerned (usually the side facing out). Plumb the point at which every block corner returns. Then plumb the end of the lead and check the complete course with a plumb rule. The line should be attached to the side which was plumbed. If necessary, both sides can be struck or tooled.

BUILDING THE CORNER TO THE SPECIFIED HEIGHT

Each course should be aligned before proceeding to the next course. Since the corner is racked back one-half block on each course, stop spreading mortar one-half block from the end of the course.

Each block laid on the corner should be checked for the proper height and leveled from the point. This prevents the possibility of having to re-lay the course in fresh mortar.

After completing each course, cut off excess mortar with the trowel and return it to the mortar board or pan. Temper the mortar as needed to retain its plasticity and workability.

INSTALLING WIRE REINFORCEMENT IN JOINTS

If joint reinforcement is called for in the building specifications, install it at the proper height. When turning a corner, cut one side of the wire and bend the wire on a 90° angle rather than cutting the wire in half, Figure 19-2A. A special prefabricated wire for corners is also available, Figure 19-2B.

Be sure that the wire is bedded solidly in the mortar joint. Lap the wire at least 6" over the next piece if more than 1 length of wire is needed. This lends adequate strength to the joint if lateral pressure is encountered.

STRIKING JOINTS

Several types of joints may be used with concrete block, Figure 19-3. As previously mentioned, the concave and V-joints are most commonly used. There are times, however, when a flush joint is required. In a flush joint, the joint is even with the face of the concrete block. This is usually done when a wall is to be painted and the mortar joints should not be evident.

There are two different ways of finishing a flush joint. The first involves carefully cutting off mortar and pointing any holes. This is the simpler method. The second and most often used method is to cut off the mortar and rub the joints with a piece of flat

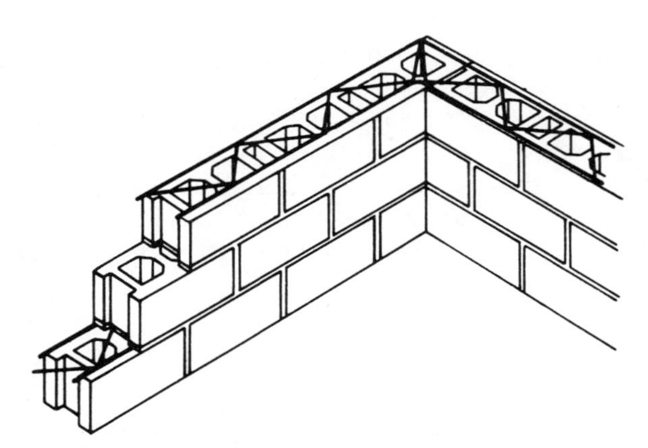

Fig. 19-2A Reinforced wire returning around corner

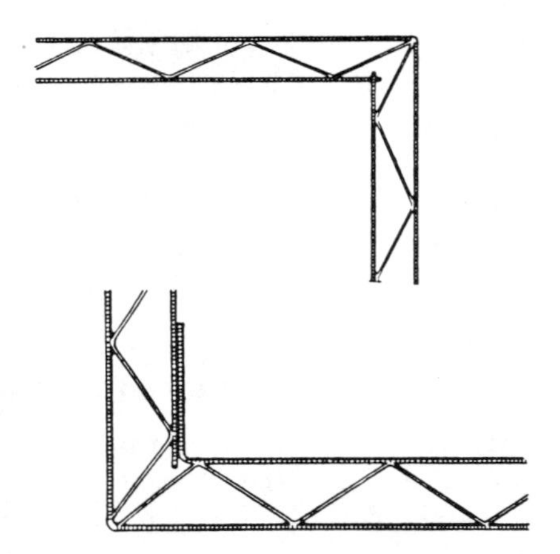

Fig. 19-2B Prefabricated reinforced wire corners for concrete block

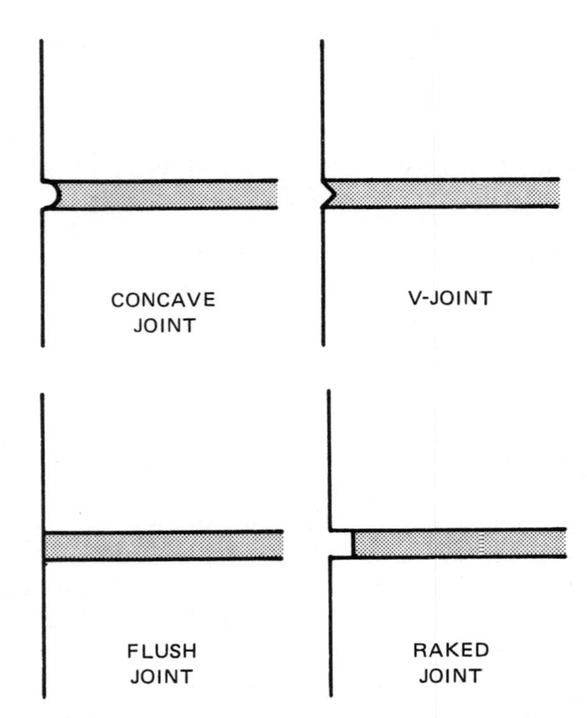

Fig. 19-3 Various types of joint finishes used on concrete block

Fig. 19-4 Finishing a flush joint by rubbing with a piece of block

material to seal all holes or voids, Figure 19-4. Any flat piece of material, such as a broken block, scrap of wood, rubber heel, or rubber ball split in half may be used. All masons working on the same wall should use the same type of rubbing material so that the overall appearance is the same. Whatever material is used, it is important that all holes are filled and that the mortar is rubbed flat to meet the surface of the block. Guard against excessive rubbing and smearing of the block, since paint does not completely hide the joints but only covers them.

The raked joint can be used to achieve a special effect but is usually not recommended for concrete block. If a raked joint is to be used, do not rake deeper than 3/8", since most block is laid with face shell bedding, and the joint will be weakened and subject to water penetration. Excessive raking out of the mortar weakens the wall. Outside concrete block walls should not be raked out, as there is a greater possibility of the mortar joints leaking.

When the corner is completed, brush it off as soon as the mortar has sufficiently set. Lightly restriking the joints after brushing sharpens the edges of joints and removes any small particles of mortar which remain. This is required practice on a job where the work must have a very clean appearance. Be sure to check the end of the lead before leaving the corner, Figure 19-5.

Fig. 19-5 The last step in the construction of a block corner is to check to be certain that the ends of the block are in line.

PROGRESS CHECK

At this point, masonry students should have acquired certain basic skills. They must practice these skills, as every mason must have them. Students should demonstrate for their instructor each of the following skills. Any skills not mastered should be practiced until they can be demonstrated correctly.

- Spread mortar on a concrete block without excessive waste.
- Level a concrete block with a plumb rule.
- Plumb a concrete block with a plumb rule.
- Cut a concrete block with a hammer or brick set.
- Lay concrete block correctly to the line.
- Install a closure block correctly while retaining solid joints.
- Strike or tool mortar joints using the joints described in this section.
- Build a concrete block corner to a specified height using the modular rule as a guide.

ACHIEVEMENT REVIEW

Select the best answer from the choices offered to complete the statement. List your choice by letter identification.

1. Concrete block must be handled very carefully to prevent spalling. *Spalling* is
   a. shrinkage of the masonry unit.
   b. chipping or flaking of the unit.
   c. cracking of the unit.
   d. discoloring of the unit.

2. Concrete block should never be soaked with water before it is laid because
   a. it discolors the unit.
   b. the strength of the block is affected.
   c. shrinkage cracks may occur in the joints.
   d. a good bond cannot be established with the mortar.

3. Furrowing a mortar joint on the first course is not recommended because
   a. the compressive strength of the wall is affected.
   b. the joint would allow water penetration.
   c. good suction between the block and the mortar is not necessary.
   d. only face shell bedding is used.

4. A range line is used for
   a. the layout of the wall.
   b. checking the elevation of the finished grade.
   c. lining up one corner with the opposite corner.
   d. checking the corner for the finished height.

5. When more than one piece of wire reinforcement is to be used, it is best to lap each piece over the next piece at least
   a. 12″. c. 8″.
   b. 16″. d. 6″.

6. If a concrete wall is to be painted and it is important that the mortar joints not be noticeable, the best mortar joint finish to use is the
   a. concave joint. c. V-joint.
   b. flush and rubbed joint. d. raked joint.

7. A joint finish which should not be used on outside concrete block walls that are exposed to weather is the
   a. concave joint. c. V-joint.
   b. raked joint. d. flush joint.

8. The correct number on the modular rule when marking off the height for concrete block is
   a. 6. c. 2.
   b. 3. d. 5.

9. If a concrete block corner is being built to a height of 56″, the number of courses in height is
   a. 4. c. 6.
   b. 5. d. 7.

10. Raked joints should not have a depth greater than
   a. 1/4″. c. 6″.
   b. 3/8″. d. 7″.

PROJECT 11: LAYING AN 8″ X 8″ X 16″ CONCRETE BLOCK CORNER IN THE RUNNING BOND

OBJECTIVE

- The masonry student will be able to build a concrete block corner 7 courses high in the running bond.

  Note: A block corner such as this is used for interior partitions and corridors.

EQUIPMENT, TOOLS, AND SUPPLIES

Mortar pan or board
Mixing tools
Mason's trowel
Brick hammer
Plumb rule
Chalk box
2′ framing square
Convex sled-runner jointer
Mason's modular rule
Brush
28, 8″ x 8″ x 16″ concrete blocks (7 of these must be corner block)
1 bag of lime or masonry cement (Lime may take the place of cement in a training situation as it may be reused.)
18 shovels of sand
Supply of water

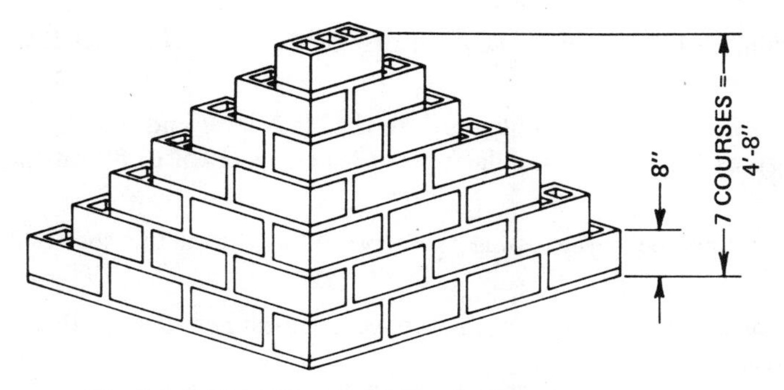

8" x 8" x 16" CONCRETE BLOCK CORNER
IN THE RUNNING BOND

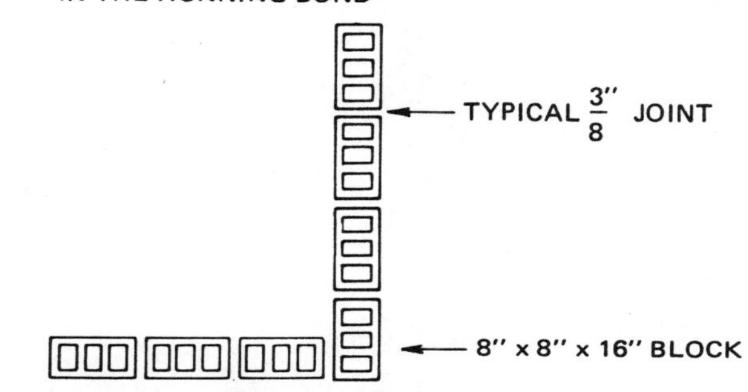

PLAN VIEW OF FIRST COURSE LAYOUT

SUGGESTIONS

- Space the mortar pan or board approximately 2′ away from the corner to allow sufficient working room.
- Snap a chalk line longer than the one actually needed to serve as a reference point.
- Tie the block together on the stock pile by reversing every other course.
- Mark the distance the block will be laid on the base to prevent spreading the mortar too far.
- Measure the height of every course laid.
- Remove excess mortar from the corner and return it to the mortar pan. Temper it as needed.
- Strike the mortar joints when they are thumbprint hard. Do not wait until the entire corner is built.
- Observe safe working practices at all times.

PROCEDURE

1. Mix the mortar.
2. Prepare the work area by stocking the specified amount of concrete block and placing the mortar in the mortar pan.
3. Brush off the area where the corner is to be built.
4. Square a corner with the framing square.
5. With the plumb rule and chalk box, extend the squared line far enough to build the corner.

6. Spread a solid bed of mortar. Lay the first block, being sure that it is level and plumb.
7. Lay the rest of the first course, being sure that it is level and plumb and that the height measures 8″ on the modular rule. (Scale 2 is equal to 8″ on the modular rule.)
8. Continue building the corner using these procedures until the specified height is reached.
9. Strike the corner on both sides with a convex sled-runner jointer. Brush the work when the striking is finished.
10. Recheck the end of the lead and the corner for the correct height with the plumb rule.

Unit 20
Estimating Concrete Block by Rule of Thumb

OBJECTIVES

After studying this unit, the student will be able to

- estimate concrete block by using the rule of thumb method.
- estimate mortar and sand for jobs which require concrete block.

As is the case in estimating bricks for masonry, a measuring system which uses modular concrete masonry units can be used to simplify estimating. For accuracy, the square foot wall area method is considered the best method for large jobs. When using the wall area method, determine the amount of square feet in the wall and deduct all openings from this figure. A percentage of the figure is included to account for waste and breakage. The total figure is calculated by use of a table giving the number of block in each square foot. Additional tables state the correct amount of sand and masonry cement needed to lay the concrete block.

Rule of thumb estimating is usually adequate for a small-sized or average-sized job. When it is necessary to estimate large quantities of materials, tables should be consulted. There are also slide rules that speed the estimating process. Tables and slide rules should be used by masons only after they learn the rule of thumb method.

DETERMINING THE NUMBER OF UNITS

This method involves adding all of the wall lengths around the *perimeter* (outside) of the building and figuring the total lineal feet. The number of concrete block needed to lay 1 complete course around the building is determined. The plans are studied to determine the height of the wall, and the number of courses necessary to build the wall to the specified height are figured. The number of block needed to build 1 course is multiplied by the number of courses in the height of the wall. The result is the total number of concrete block needed to build the wall or project. Deduct the number of block in all the openings in the building.

ESTIMATING MATERIALS FOR A FOUNDATION BY RULE OF THUMB

Estimating Block

Assume that the foundation is of average size, measuring 50′ (length) x 32′ (width) x 8′ (height). It is to be constructed of block measuring 8″ x 8″ x 16″.

The first step is to determine the total lineal feet around the outside of the foundation.

Step 1. Determine the length and width of the foundation.

50′ (length of foundation wall)
32′ (width of foundation wall)

Step 2. Double these 2 figures since there are 2 walls on each side of the foundation.

50′ x 2 = 100′ (total length of foundation)
32′ x 2 = 64′ (total width of foundation)

Step 3. Add the total length and width together.

100′ + 64′ = 164′ (total lineal feet in foundation)

Three concrete blocks are required to lay 4′ of length in a concrete block wall. The ratio of blocks to feet of length is 3/4 to 1, or 0.75 to 1.00.

Step 4. Multiply the total lineal feet in the foundation by 0.75 to find the number of concrete block required to lay 1 course around the foundation.

Note: Be sure to place the decimal point in the correct position.

```
    1 64 (total lineal feet of foundation wall)
x   0.75
    8 20
 114 8
123.00 (Total number of concrete block for
          1 course)
```

Step 5. Find the number of courses in the height of the wall. Since 1 concrete block measures 8″ in height including the mortar joint, divide 8 into the height expressed in inches to find the number of block of which the height consists.

12″ (number of inches in 1 foot)
x 8 (number of feet in foundation)
= 96″ (height in inches)

96″ ÷ 8 = 12 (courses high)

Step 6. Find the total number of concrete block in the wall by multiplying the number of block in each course by the number of courses in the height of the wall.

```
Multiply:    123
           x  12
             246
            123
            1476 (total number of block re-
                   quired for foundation)
```

Note: There are no openings, so no deductions are made.

Allowing for Openings

When openings are specified in a concrete block structure, a certain amount of block should be deducted before estimating the number of block for the job.

A concrete block building which will have 2 windows and 2 doors is to be constructed. The doors measure 32″ x 6′-8″. The windows measure 2′ x 4′. The number of concrete block to be deducted is estimated as follows:

Step 1. Add the width of the 2 doors.

32″ + 32″ = 64″ (total width of doors)

Step 2. Determine the number of concrete block that would be included in that space. Divide 64″ by the length of 1 block (16″).

64″ ÷ 16″ = 4 (number of blocks per course)

Step 3. Express the height of the doors in inches.

6′ = 72″
72″ + 8″ = 80″

Step 4. Determine the number of courses by dividing the height of the doors by 8″.

80″ ÷ 8″ = 10 (total number of courses for both doors)

Step 5. Multiply the number of courses of block in the doors by the number of block on each course.

10 x 4 = 40 (concrete block to deduct for doors)

The space for windors is estimated in the same manner.

Step 1. Add the width of both windows.

24″ + 24″ = 48″ (total width of windows)

Step 2. Divide 48″ by the length of 1 concrete block (16″).

48″ ÷ 16″ = 3 (number of block per course)

Step 3. Each window is 4′, or 48″ in height.

Step 4. Divide 48″ by 8″ (height of 1 concrete block).

48″ ÷ 8″ = 6 (courses of block in window space)

Step 5. Multiply the number of courses by the number of block per course.

6 x 3 = 18 (total number of block to deduct for windows)

To find the total number of block to be deducted, add the block for door space to the block allowed for window space.

40 + 18 = 58 (total block to be deducted)

To find the total number of block needed, subtract the number of block deducted for openings from the total number of blocks needed for the wall.

1476 - 58 = 1418 (total number of block needed)

Estimating Masonry Cement

To estimate masonry cement by the rule of thumb method, assume that 1 bag of masonry cement is used to cover a certain number of units. Face shell bedding is used for most block work. Therefore, no extra allowance must be made for block of different widths. The exception to this rule is the 6" concrete block, since the web is thinner and more mortar is forced into the holes between the webs of the block.

One bag of masonry cement is sufficient for 30 concrete block. This figure applies to the standard 70 lb bag. If portland cement and lime are used, a table must be consulted.

To estimate the masonry cement for the foundation in question, divide 1418 by 30 (the number of block that can be laid per bag). The answer gives the number of bags of masonry cement needed to build the foundation.

1418 ÷ 30 = 48 (bags of masonry cement)

Estimating Sand

One ton of sand is needed for each 8 bags of masonry cement. If more than 3 tons of sand are needed, 1/2 ton should be allowed for waste.

Step 1. Multiply 30 (concrete block) x 8 (bags of masonry cement per ton) to find the number of blocks for each ton of sand.

30 x 8 = 240 blocks per ton of sand

Step 2. Divide 1418 by 240 to find the necessary sand.

1418 ÷ 240 = 6 tons of sand

Step 3. Add 6 tons and the 1/2 ton for waste to find the sand needed.

6 tons + 1/2 ton = 6 1/2 tons of sand

TOTALING MATERIALS

The materials needed for the job in question would be listed as such:

1418 concrete block
48 bags of masonry cement
6 1/2 tons of sand

These materials are adequate to build the foundation and allow a reasonable amount for waste.

ESTIMATING LABOR COSTS

After materials have been estimated, labor costs are estimated. Labor costs for block work are affected by many of the same factors that affect labor costs for brickwork. These factors can be reviewed in Unit 17.

To estimate labor costs, the contractor will want to know the average production rate for masons laying concrete block. The International Union of Bricklayers and Allied Craftsmen supplies these figures for the amounts of block that a journeyman mason can lay in an 8-hour day:

- 225, 8" x 8" x 16" blocks
- 210, 6" x 8" x 16" blocks
- 225, 4" x 8" x 16" blocks
- 200, 10" or 12" x 8" x 16" blocks

As with bricklaying, these figures will vary depending on specific job conditions.

POINTS TO REMEMBER

Rule of thumb estimating is the most commonly used method for figuring materials for the average or small job. Figure 20-1 details the important points in the method.

Remember, the skills involved in estimating by rule of thumb are learned only through experience and on-the-job training.

- **Multiply the lineal feet of the wall by 0.75 to determine the number of concrete block for one course.**
- **Divide the height of the wall expressed in inches by 8 inches to determine the number of courses in the height of the wall.**
- **One bag of masonry cement is the recommended amount for thirty concrete block. (For 6-inch block, estimate 25 block per bag of masonry cement.)**
- **One ton of sand is needed for every eight bags of masonry cement. Waste is about 1/2 ton if the quantity required exceeds 3 tons. Be liberal when estimating sand, as it is usually the least expensive material used by the mason.**
- **Water does not require estimating as a material. As a rule, 5 gallons of water are needed for each bag of masonry cement. Note: The allowance for water depends on the moisture content of the sand being used.**

Fig. 20-1 Estimating materials by rule of thumb method

ACHIEVEMENT REVIEW

Solve each of the following problems using the information presented in this unit. Show all of your work.

1. A foundation measures 70′ x 40′.
   a. What is the number of lineal feet in a foundation?
   b. Estimate the number of 8″ x 8″ x 16″ concrete block needed to lay 1 course completely around the foundation.
   c. The foundation is to be built 8′ high. Estimate the number of courses for this height.
   d. Estimate the total number of concrete block needed to build the foundation. There will be no windows, doors, or other openings to consider.
   e. Estimate the amount of masonry cement needed to build the foundation.
   f. Estimate the amount of sand needed to build the foundation.
   g. If masonry cement costs $1.70 per bag, what is the total cost of masonry cement for the job?
   h. If concrete block measuring 8″ x 8″ x 16″ cost $0.38 each, what is the total cost of the concrete block for the job?
   i. If sand costs $8.00 a ton, what is the total cost of sand for the job?
   j. What is the total cost of materials for the job?

2. A basement without any openings is to be built from 8″ x 8″ x 16″ concrete block. The outside dimensions are 60′ x 32′. How many block are needed for 1 course around the basement?

3. A wall is to be built of concrete block measuring 12″ x 8″ x 16″. How many courses are needed if the wall is 10′ high?

4. A concrete block garage is to be built of 8″ x 8″ x 16″ block. The outside dimensions are 20′ x 24′. The total height of the wall is 10′ - 8″ from the footings to the top of the wall. There are 2 large garage door openings, each measuring 8′ x 8′, and 1 standard door opening measuring 32″ x 6′ - 8″. Estimate the number of concrete block needed to build the garage. Be sure to deduct for all openings.

5. A motor repair shop 20′ x 40′ is to be built of 10″ x 8″ x 16″ concrete block. The walls are to be 10′ high. There are 6 windows measuring 40″ x 48″, and 1 large door measuring 8′ x 8′. Estimate concrete block, masonry cement, and sand needed to build the shop. Estimate the total cost of materials using the following prices.

10″ x 8″ x 16″ concrete block	$0.45 each
Sand	$8.00 per ton
Masonry cement	$1.60 per bag

6. Based on the rate of 225, 8″ x 8″ x 16″ blocks laid per 8-hour day, how many days are required for one mason to lay 4500 blocks?

Unit 21
Building a Composite Wall with Brick and Concrete Block

OBJECTIVES

After studying this unit, the student will be able to

- describe the different methods of bonding walls with brick headers and metal wall ties.
- install these ties in a composite wall project.

A *composite wall* is formed when two separate *tiers* or *wythes* (single vertical thicknesses of walls) of masonry units are combined to form a single wall. This requires that each individual tier be bonded together vertically at a specified height so that the wall retains its strength. The generally accepted height at which the tiers are attached is 16″.

There are two different methods of bonding walls together. One method is by the use of a brick header laid across the wall. The second method is by installing metal wall ties. The use of the brick header is an older, more established method. Brick headers provide more strength than other methods. This method is often used in the construction of solid masonry walls. Metal ties and heavy-gauge reinforcement wire have been greatly improved in recent years and are acceptable for most jobs. This is the most common method of tying tiers. Either type of wall tie can be used to effectively join bricks and block together.

TYING TIERS TOGETHER WITH HEADERS

Problems Presented by Headers

The installation of brick headers in walls may present special problems if they are not laid correctly. Extreme care must be used when applying the *cross joint* (a mortar joint which is applied across the full length of a brick) on the header. If extreme care is not taken, moisture penetration may cause leakage through the wall, Figure 21-1. Single head joints applied only to the ends of the header are not acceptable.

Another problem presented by headers is the difficulty in being sure that the set of the mortar is not broken when tying bricks and heavy block, Figure 21-2. Sometimes, the concrete back-up wall is laid ahead of the brick wall. This only occurs when a mason is working on the outside of the brick wall. In this case, brick headers are laid across the block and bricks when the face brick wall reaches the proper height. This is done so that the tiers may be bonded together. This height is usually 16″. Im-

Fig. 21-1 Mason applying a cross joint to a header brick. Note the solid joint which guards against moisture penetration.

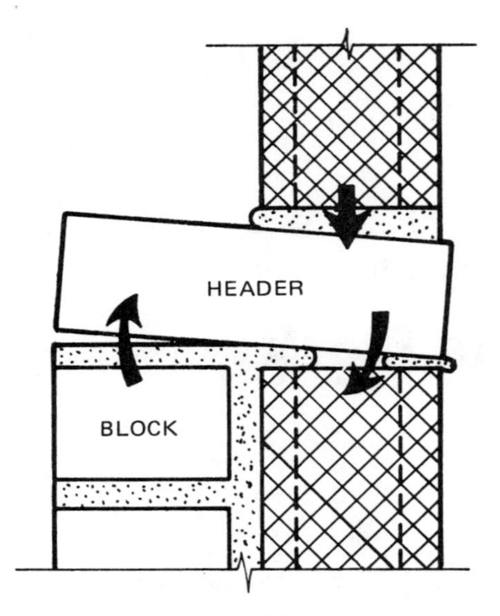

Fig. 21-2 Effect of a heavy back-up block on brick header. Notice that the set of the mortar has been broken.

mediately following the laying of the brick headers, a heavy concrete block course is laid in back of the header course. This block course rests on the back half of the header.

With most types of concrete block used as backup, the mortar is spread by the face shell method or by spreading it only on the webs of the block. This is approximately a 1 1/2" strip of mortar. If the block is laid before the mortar under the header stiffens, the header brick settles unevenly and a crack may develop in the mortar joint under the header. This presents the possibility of moisture penetration. It also detracts from the beauty of the building, since the header lies twisted in the face of the wall.

Expansion and contraction caused by moisture and heat also create problems since the rate of expansion and contraction differs significantly for each material. Movement of the block due to shrinkage creates pressure on the brick header. This may result in hairline cracks developing in the mortar joints surrounding the header course. As masonry units become more complex and new materials are developed, some flexibility in wall ties is essential if materials with different properties are to be used. This is one of the prime reasons the metal tie has become popular in recent years. It lends more flexibility to the wall.

There is also the matter of economy to consider. A header course takes a great deal of time to install as compared with the fairly simple task of laying metal ties or wire joint reinforcement in walls. The materials used to lay header courses are also more expensive than metal reinforcements, as a full header course requires twice as many bricks as a stretcher course.

Laying Headers in the Wall

Remember that it is important to build the brickwork to the height of the header course before laying any back-up block. Block work can be *humored* (gradually adjusted so that a difference in height is not evident) to remain in line with the brickwork. If back parging of the brickwork is specified, it is best to apply it to the bricks before the concrete block is laid since they would interfere with the work. The block must also be laid exactly level in height to match the outside brickwork or the header will not lay level across the two tiers.

Building a 12" Wall with Headers. When building a 12" wall using brick headers, there are two different methods of tying the header to the back-up tier. The first involves using a special concrete block called a *header block* (or *shoe block*). Header block are 8" high on one side and the height of 2 courses of standard brick, including the mortar joint, on the other side. The side of the block which receives the header course is approximately 4" in width, leaving a remainder of about 4" on the opposite side. The brick wall is laid 5 courses high. The concrete block backing wall is laid 2 courses high with the header block laid on the second course. The header course is then laid across the header block. This method of installing header block is used in cases where the brick header should not be seen from the inside of the structure, Figure 21-3. A regular 8" block now can be laid on top of the header block which will bond the brick and block together. The building of the wall to the next header level is now resumed.

The brick headers may also be installed on the sixth course by turning the header block upside down. Either method is acceptable and depends on the architect's specification of the height of the headers. Consider an 8" wall consisting of 4" of brick and 4" of concrete block is built and a header is used to tie the tiers together. Unless the mason uses snap headers, there is no way to avoid having the header course exposed on the back side of the wall, Figure 21-4.

Standard bricks do not actually measure 4" in width (they measure 3 3/4") and an 8" concrete block is approximately 7 5/8" wide. Therefore, a 12" wall will always have a 5/8" space between the two tiers or wythes of units. This space provides the mason with finger room when laying the back-up units. This space between the two tiers

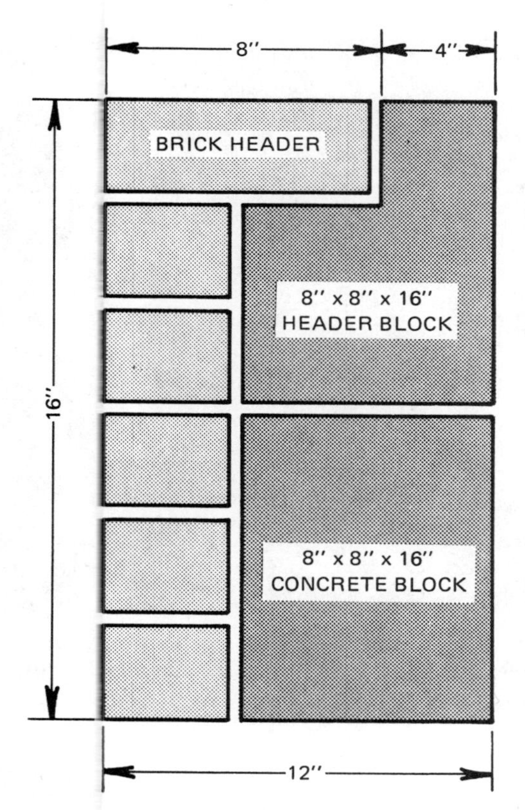

Fig. 21-3 Wall section showing header used in conjunction with header block

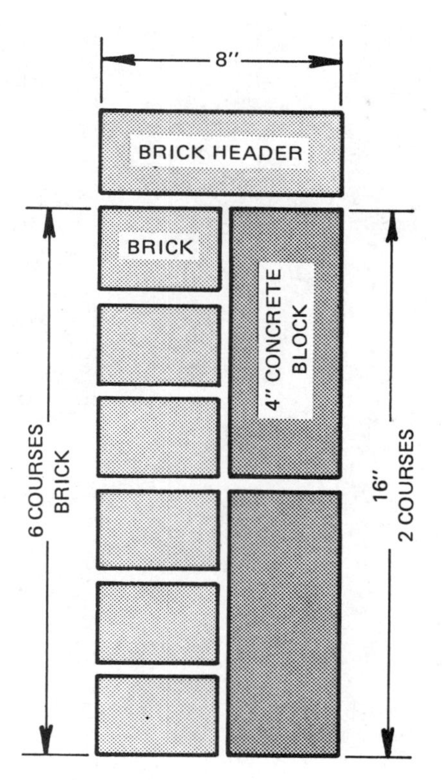

Fig. 21-4 Wall section showing use of brick header on 8-inch wall

of masonry units is called a *collar joint.* After the back-up units have been laid and are level with the front tier of the masonry units, the collar joint may be slushed solid with mortar or left hollow. *Slushing* is the trade term for filling in the collar joint in a masonry wall with mortar. If slushing of the wall is done too vigorously, it may cause the wall to be pushed out of line. Be sure that the masonry work has taken an initial set before slushing. Specifications of the job dictate when to slush walls.

If the inside face of the concrete block is to be plastered or is otherwise not to be exposed when completed, the header block can be replaced by a course of stretcher bricks located behind the brick headers. This allows for the full 12" thickness of the wall. It is more costly to do this, however, as it requires the use of an additional course of bricks.

USING METAL TIES

Prefabricated Metal Joint Reinforcement

The most popular type of metal tie used today is the *prefabricated metal joint reinforcement* or rod joint reinforcement (also called *wire track*). The truss and ladder type design is designed to be installed in the bed joints, Figure 21-5. The metal wire tie consists of two parallel wire rods connected to a single diagonal cross rod which is welded to the outside parallel rods wherever they touch. The truss design which is formed provides great strength. Wire reinforcement ties the masonry work together. It also minimizes shrinkage in the mortar joints and helps prevent cracks due to settlement. For a composite wall such as one constructed of bricks and block, the wire should usually be installed no more than 16" on center (vertically). This is an approved method for a brick header providing the collar joints are filled completely with mortar.

Reinforced wire is usually manufactured in 10' lengths. It should overlap at least 6" when two pieces are used together. This ensures that the maximum tensile strength will not be interrupted throughout the length of the bed joint. It is extremely important that wire track be embedded and covered completely with mortar before any masonry unit is laid on top of it, Figure 21-6.

Fig. 21-5 Brick and block tiers tied together with reinforced wire track

Fig. 21-6 Mason spreading mortar over joint reinforcement. The wire must be completely covered with mortar to be effective.

When using reinforced wire to tie bricks to concrete block, be certain that the two separate walls are perfectly level with one another since wire reinforcement does not bend easily. Never allow the wire to extend past the ends of the wall farther than necessary, Figure 21-7. Extended wire is not only a safety hazard but is also difficult to work around.

The Z Tie

The term *Z tie* comes from the appearance of the tie when it is laid in the wall, Figure 21-8. It is made from copper-coated or zinc-coated steel which is 3/16" in diameter. One tie is usually installed every 4 1/2 sq ft of wall surface. The Z tie is available with a drip crimp in the center so that water does not cross the tie (used mainly in cavity walls). It is also available without the drip crimp (used in solid masonry walls).

The Rectangular Tie

The *rectangular tie* is rectangular in shape. It is constructed from the same materials and is the same thickness as the Z tie. It is sometimes called a *box anchor.* The rectangular tie is recommended for use in walls where both masonry units used are hollow. It can be obtained with or without a drip crimp in the center. For best results, one tie should be installed every 4 1/2 sq ft of wall surface.

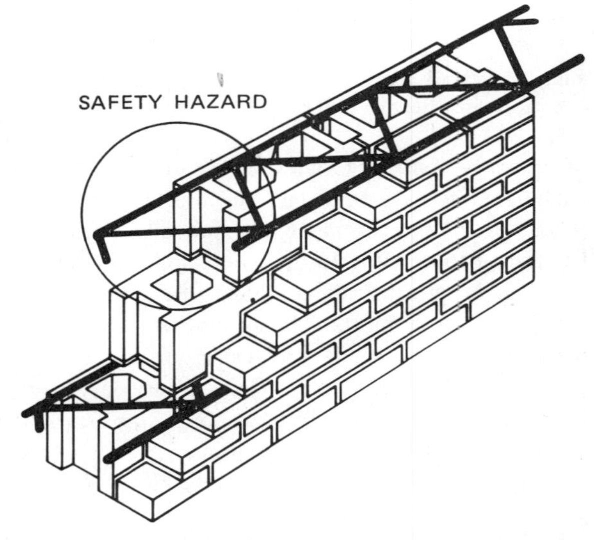

Fig. 21-7 Prefabricated metal joint reinforcement extending past walls could cause serious injury.

The Dovetail Anchor

Building a masonry wall next to a concrete wall or column requires a special tie. The *dovetail anchor,* used for this purpose, takes its name from the resem-

Fig. 21-8 Z tie used in cavity wall construction. Notice the drip crimp in the center of the tie which allows drippage of water before it crosses the tie.

blance to the tail of a dove, Figure 21-9. The anchor fits into a slot in the concrete work. The slot is metal with a fiberglass filler. This prevents the slot from being filled with concrete when the wall is poured. The mason removes the filler material and inserts the tapered end of the dovetail anchor into the slot. The anchor is then turned over much the same as a key is turned in a lock, until the flat side is facing up. The slot in the concrete wall, tapered to match the anchor, holds it in place. Minimum dimensions of a dovetail anchor are 7/8" in width with a 16-gauge thickness.

The Veneer Tie

Veneer ties are constructed of a corrugated metal, Figure 21-10. The wrinkle in the metal aids in bonding the mortar to the steel, which increases the holding power of the tie.

Veneer ties are usually used to tie masonry veneer to wood frame construction. They are sometimes used to tie two masonry walls together. These ties are not recommended for use in load-bearing walls.

The ties used to attach masonry veneer and wood frame construction should be made of galvanized steel not less than 7/8" wide with a 22-gauge thickness. If the tie is to bond two masonry walls, be certain that a tie of a much heavier gauge is used.

Regardless of the type of metal tie used, it should never be placed closer than 1" from the outside surface of the wall, Figure 21-11. If the tie were placed any closer, the joint would not be of the proper thickness and would have a tendency to pop out, exposing the tie. Each method of tying masonry walls has certain advantages over the other. The choice of tie depends upon the requirements of the job. Metal ties are being used more frequently in masonry because they provide adequate strength, economy, better control of shrinkage and expansion in mortar joints, and are faster for the mason to install. All methods discussed have been widely tested and accepted throughout the masonry industry.

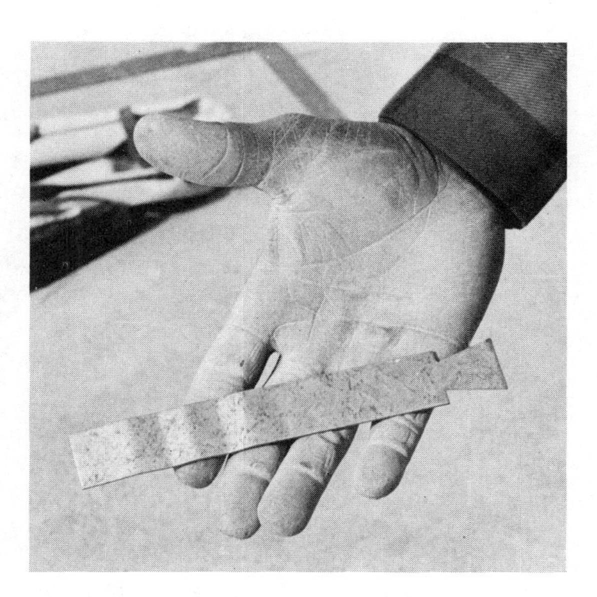

Fig. 21-9 Dovetail anchors are used to tie masonry work to concrete walls or columns.

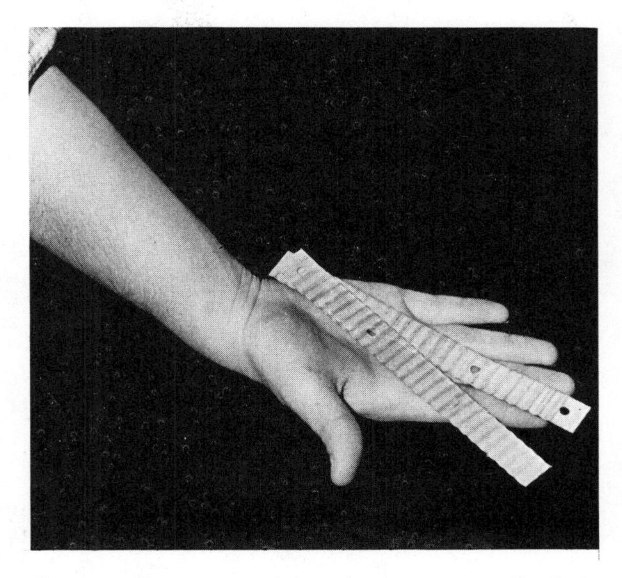

Fig. 21-10 Corrugated metal veneer ties. The holes provide spaces for nails when the ties are used with wood construction.

Fig. 21-11 Mason laying a brick and block concrete wall. Reinforcement bonding wall lies more than 1 inch from the outside of wall.

ACHIEVEMENT REVIEW

Select the best answer from the choices offered to complete the statement. List your choice by letter identification.

1. When two different types of masonry units are laid together to form a single wall, the structure is called a
   a. tier.
   b. wythe.
   c. composite wall.
   d. veneer wall.

2. The principal reason for leakage in composite walls is the use of
   a. metal ties.
   b. collar joints.
   c. hollow blocks.
   d. poorly formed cross joints.

3. The space in the center of a composite wall which is filled with mortar is called a
   a. head joint.
   b. collar joint.
   c. expansion joint.
   d. cross joint.

4. Slushing collar joints should not be done to a great extent because
   a. it may cause the wall to be pushed out of line.
   b. it may cause the mortar joints to lighten in color.
   c. it decreases the strength of the wall.
   d. it is a very wasteful practice.

5. The most popular type of metal tie used today is the
   a. Z tie.
   b. rectangular or box tie.
   c. dovetail anchor.
   d. veneer tie.
   e. wire reinforcement.

6. The principal reason for using metal ties is because they
   a. are very strong.
   b. reduce cracks due to shrinkage.
   c. are easier to install.
   d. are very inexpensive.

7. For best results, wall ties should be installed vertically every
   a. 12″.
   b. 16″.
   c. 24″.
   d. 32″.

8. The dovetail anchor is designed mainly for tying
   a. brick to block.
   b. brick to tile.
   c. masonry units to concrete.
   d. brick to stone.
9. Veneer ties are not recommended solely for tying two tiers of masonry because
   a. they do not bend easily.
   b. they are not as strong as other wall ties.
   c. they do not bond well to the mortar joint.
   d. they are too costly to use on composite walls.

PROJECT 12: LAYING A BRICK CORNER FOR A 12" COMPOSITE WALL IN THE COMMON BOND BACKED WITH 8" X 8" X 16" CONCRETE BLOCK

OBJECTIVE

- The student will be able to lay out and build a brick corner 8 courses high in the common bond by tying the work together with a full header on the first and seventh courses to form a composite wall. This project will give the student practice in combining basic materials to form a single wall and practice in laying header courses.

  Note: Corners such as these are used in buildings where load-bearing solid walls are required, such as warehouses, office buildings, and factories.

EQUIPMENT, TOOLS, AND SUPPLIES

Mortar pan or board
Mixing tools
Mason's trowel
Brick hammer
Plumb rule
2" square
V-sled runner jointer
Modular rule
Brush
Pencil
Clean water
1, 8" x 8" x 16" concrete block in which to set plumb rule

Note: The student will estimate the number of bricks and/or block needed by consulting the plans. Based on the number of bricks and block, the student will estimate the needed mortar (lime and sand) by the rule of thumb method.

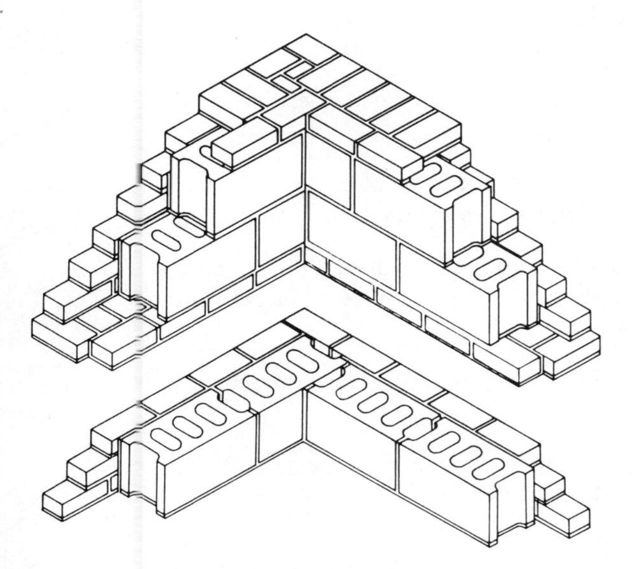

Plan of first course of block

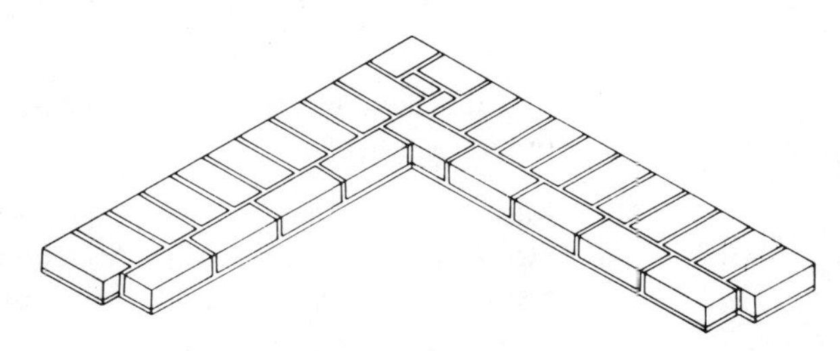

Plan of first course

SUGGESTIONS

- When establishing a reference line for the corner, be sure to extend the square line longer than is actually necessary.
- Select straight, unchipped bricks for the corner point.
- Sight down the corner as each brick is laid.
- Form solid head joints to prevent leakage.
- Cut off all excess mortar from the back of the bricks with the trowel and return it to the mortar pan before attempting to build the backup with concrete block.
- Do not lay concrete block until the brickwork has been constructed to header height.
- Do not lift concrete block over the brickwork; lay them from the inside of the wall.
- Level and plumb the corner on both sides.
- Keep all scraps and tools which are not being used away from the work area.
- Use good safety practices at all times.

PROCEDURE

1. Stock materials approximately 2′ from the work area and set up the mortar pan.
2. Mix the mortar and load it in the pan.
3. Lay out the corner on the floor using a steel square and pencil. Strike a chalk line longer than the actual corner length will be over the top of the pencil line.
4. Measure back from the outside corner line 12″ on each end of the corner and strike the inside line of the corner.
5. The first course to be laid is the header course. Cut 2, 6″ pieces to be used as starter pieces for the corner. Lay out the first course as shown on the plan.
6. Lay the inside stretcher course, being sure that when completed, the width of the wall measures 12″.
7. Recheck the corner with the steel square. Lay the 6 courses of bricks as shown on the plan. Use the number 6 on the modular rule as a reference for course spacing.
8. Back up the corner with 8″ concrete block as shown on the plan. Be sure that the top of each course of block is level with the top of each 3 courses of bricks. Level and plumb all work.
9. Lay the next header course as shown on the plan, using all solid cross joints between each header.
10. Recheck the corner for the proper height, using the number 6 on the modular rule. Strike the brickwork with a V-jointer. Brush the corner. Do not strike the concrete block work since it does not show.
11. Tail the ends of the corner before having the work inspected.

PROJECT 13: LAYING A 12″ BRICK AND CONCRETE BLOCK COMPOSITE PANEL WALL BONDED WITH MASONRY WIRE REINFORCEMENT

OBJECTIVE

- The student will be able to lay out and build a 12″ brick and concrete block panel wall using metal wire reinforcement to tie the wall every 16″ in height.

  Note: Many of the composite masonry walls built today use masonry wire reinforcement. This project is typical of a panel built between a series of windows on an apartment building.

EQUIPMENT, TOOLS, AND SUPPLIES

Mortar pan or board
Mixing tools
Mason's trowel
Brick hammer
Plumb rule
Brick set chisel
Small square
Chalk box
Convex sled-runner striking tool
Modular rule
Pencil
Brush
Heavy-duty wire or bolt cutters
Supply of clean water
1 length of masonry wire reinforcement (minimum 4′)

Note: The student will estimate bricks, concrete block, and mortar from the plan.

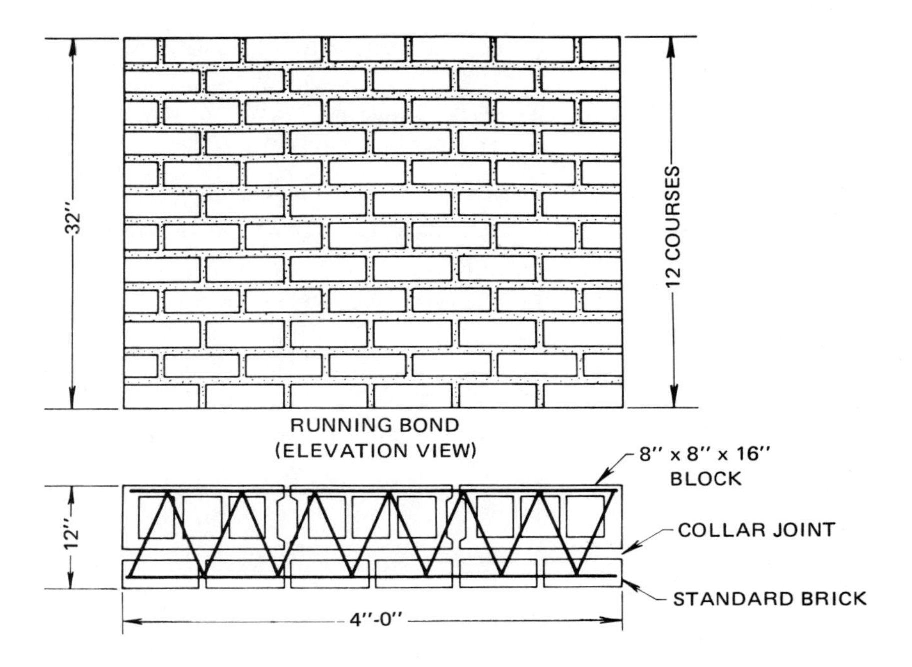

SUGGESTIONS

- Select straight, unchipped bricks for the end of the jambs.
- Cut all bats needed with a brick set before starting the brickwork. (The same holds true for concrete block halves that are needed unless factory halves are available.
- Be sure to clean all excess mortar from back of the brickwork before laying the backing block. Mortar which is not cleaned may push the wall out of line.
- The mortar may be tempered so that it is thinner for the brickwork than for laying the block.
- Do not lift the concrete block over the brick wall; lay them from the other side of the wall.
- Keep all scraps and tools away from the work area to prevent injuries.
- Be especially careful when handling wire reinforcement. Contact with the sharp edges could cause injury.

PROCEDURE

1. Stock materials approximately 2' from the work area.
2. Mix the mortar and load the mortar pan.
3. Strike a chalk line for the project.
4. Dry bond the bricks as shown on the plan.
5. Lay the first course in mortar, leveling and plumbing the course. Square the ends of the wall with a small square.
6. Check the height of the brickwork as it is laid with a modular rule (number 6 on rule).
7. Lay 6 courses of brickwork, which will be the height to which reinforcement wire is installed.
8. Back up the brickwork with concrete block in the running bond to tie height.
9. Install wire track in the wall, being certain that it is 3/8" to 1/2" from the outside of all face work.
10. Continue laying the brickwork to the height specified on the plans. Strike the work as needed with the convex jointer.
11. Back up the brickwork with the remaining concrete block. Strike the block with the convex jointer.
12. Brush the work and recheck the wall with a plumb rule before having it inspected.

Unit 22
Cavity and Reinforced Masonry Walls

OBJECTIVES

After studying this unit, the student will be able to

- describe techniques in building a cavity wall.
- discuss construction details of a masonry reinforced wall.
- build a cavity wall from given plans.

The durability of masonry is dependent for the most part on its resistance to the penetration of moisture. The source of the moisture may be weather or conditions that exist inside the wall. Differences in humidity inside and outside the building cause moisture to condense and collect on the inside.

There are two basic methods used to prevent penetration of moisture into the wall from the outside. One is to provide cavities between the outside and inside wythes of the masonry. The structure formed is called a *cavity wall.* The moisture that collects in the wall is removed from the cavity by weep holes. *Weep holes* are small holes in the bed joints at the bottom of head joints of a cavity wall which permit the leakage of excess moisture.

A second method is to provide a barrier in back of the exterior wythe. Barrier walls may be metal-tied, reinforced masonry, or bonded masonry. Reinforced masonry has steel rods between the two wythes, and sometimes in the cores of the wall. Concrete or *grout* (mortar or concrete which is mixed so that it is liquid enough to flow into the cavity of the wall, completely filling the void around the reinforcing steel) is poured in the cavity and puddled or vibrated. *Puddling* is the process of forcing an object into the grout so that it consolidates and is compact around the reinforcing steel. It is a process similar to vibrating concrete in a form. Both cavity and reinforced walls, when constructed correctly, have a high resistance to rain penetration and will remain reasonably dry. The design of any wall should be based on the degree of moisture and the elements to which it will be exposed, and the climate of the area in which it is built.

The degree of exposure varies greatly throughout the United States, from severe on the Atlantic seaboard and Gulf Coast, to moderate in the midwest, and slight in the more arid sections of the west.

CAVITY WALLS

The BIA recommends the construction of cavity walls where there is severe weather exposure, or where a maximum resistance to rain penetration is desired, Figure 22-1. Cavity walls have been built for many years in England and, during the past 40 years, have become popular in the United States.

The cavity wall consists of two tiers or wythes of masonry separated by a continuous air space not less than 2″ wide. The air space acts as an insulator. The cavity wall may be composed of any two types of masonry units but as a rule is constructed of bricks and concrete block. Metal ties are used to connect the two wythes of masonry. Cavity walls can also be masonry bonded if the tying unit does not completely close off the cavity.

For exterior walls, the facing tier is usually of the minimum thickness permitted by local building codes. Most codes define this as a 4″ wall, making the smallest cavity wall 10″ thick. When constructed of brick masonry, 2, 3″ units in addition to the air space are acceptable, resulting in an 8″ cavity wall. This practice is limited to 1 story construction.

Fig. 22-1 Typical cavity wall constructed of brick and concrete block. Note the Z ties which bond the block and bricks together.

The features of the cavity wall include excellent resistance to rain penetration, good insulation due to the air space located in the center of the wall, and ease of construction. Remember, however, that a certain amount of moisture is always present in cavity walls.

Since the primary function of the air space is to serve as a barrier against moisture, anything that would reduce the effectiveness of the moisture barrier should be avoided. The selection of the insulating material should in no way impair the cavity. In view of the continued depletion of natural energy resources, moistureproof insulation in cavity walls is a great asset. The new rigid insulation boards and plastic-formed insulation are especially effective. When cavity walls are insulated, the insulation must permit the moisture to drain from the wall so that it is not carried across the cavity to affect the inside wall.

Flashing in Cavity Walls

Flashing is installed in masonry walls to divert moisture, which may enter the masonry at floorlines, rooflines, where chimneys come through the roof, and above windows or doors, Figure 22-2. Flashing should extend through the outer face of the wall and be turned down so that moisture drips off it. Install flashing under brick windowsills as well, since water tends to collect on the sills before finally running off. The dryer a masonry wall is, the dryer the interior wall will be.

Copper is the most popular flashing material for masonry construction, although zinc, aluminum, lead, and plastic flashings are also available.

Masons must provide *weep holes* above the flashing to allow water to drain from the wall. Weep holes are located in the head joints of the first course which lays on the flashing, Figure 22-3. Weep holes should be provided at intervals of 16" to a maximum of 24". Spacing depends on the severity of the moisture problem as determined by the architect. The most common material used to form weep holes is a piece of sash cord or rubber tube about 3/8" in diameter. These materials should be soaked in oil to aid in removal from the wall. After the wall is built scaffold high and before it completely hardens, the mason pulls the weep rope or tube from the wall, leaving a clean hole. Sometimes metal screening or fibrous glass is placed in the open weep hole and left in place to act as a wick which draws moisture from the cavity.

When weep holes are installed, they should never be located where they will be covered with ground

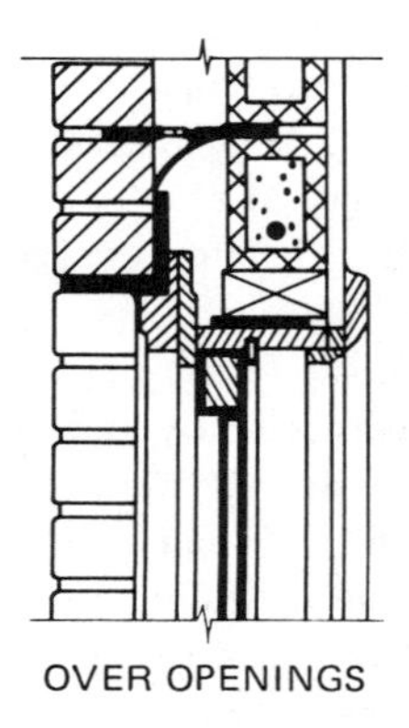

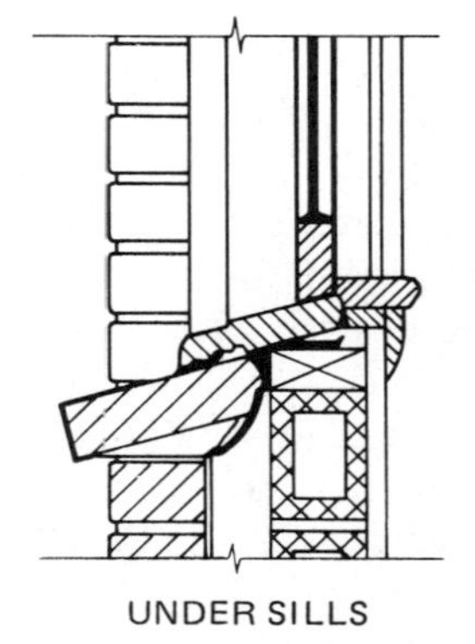

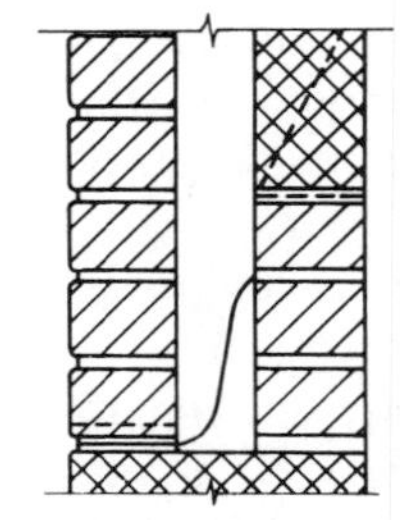

Fig. 22-2 Installation of flashing

when the building is completed. For this reason, weep holes are usually installed on top of the third course of bricks above ground level.

Keeping the Cavity Free of Mortar

If mortar falls into the cavity of the wall, it may form *bridges* upon which moisture may cross or it may drop to the flashing, blocking drainage of the weep holes. To prevent this, the mason should use a trowel to flatten any mortar projections on the inner face of the cavity.

After the first set of wall ties has been installed, a drip strip may be laid on the ties to catch any mortar that may drop, Figure 22-4. A *furring strip* (wood strip) is an effective drip strip. A heavy wire or string can be attached to the wood strip to make it easy to remove the drip strip from the cavity. After the wall is built to the next level of ties, the wood strip is carefully removed so that no mortar falls into the cavity.

Another popular method of removing the mortar droppings is to lay every third brick dry (without mortar) when laying out the first course. After the wall is completed, the mason removes the dry brick and cleans out the mortar droppings from the wall. The dry bricks are then relaid in fresh mortar, sealing the wall, and the weep holes added. The spaces made when the dry bricks are removed are known as *clean-outs* in cavity or reinforced wall construction.

Tie Placement

In a properly constructed cavity wall, both wythes of materials must be adequately and properly tied together. The ties must be firmly embedded in and bonded to the mortar. To achieve this, the two wythes surrounding the cavity must be laid with a completely filled mortar bed, and the ties must be placed firmly and solidly in the bed. Extensive tests prove that wall ties have excellent tying capacity if they are well bedded in the mortar joints.

Most building codes require at least 1, 3/16" diameter steel wall tie or equivalent in every 4 1/2 sq ft of wall area. The most common type of wall tie for brick construction is the Z tie. In addition, there are rectangular and U-shaped ties for use with hollow back-up units with vertical cores.

The most important factors to consider when using wall ties in cavity walls are the following:

- Corrosion resistance
- Proper spacing
- Placing the wall tie firmly in the mortar bed joint to assure a good bond
- Using cavity wall ties with preformed crimps in the center. The crimps collect moisture which then drops into the cavity.

Prefabricated horizontal joint reinforcement is acceptable for tying cavity walls when used with rectangular ties. The rectangular tie ties the wall together and the joint reinforcement gives added stability. Horizontal joint reinforcement is not usually required in brick masonry walls since they are not subject to shrinkage stresses. However, using them makes it more likely that the mason will remember to lay the cavity wall ties.

Fig. 22-3 Location of weep holes in masonry wall

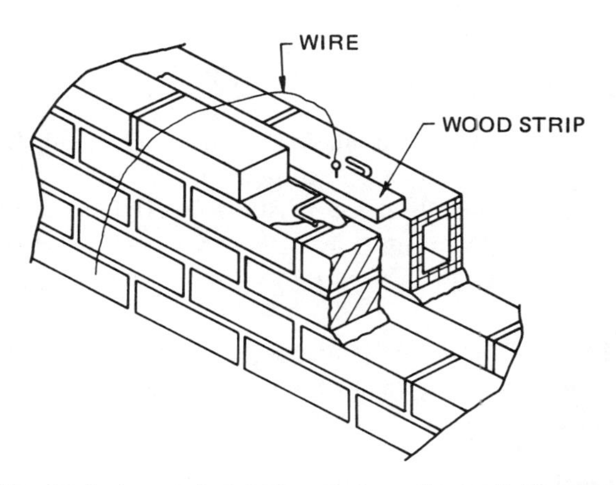

Fig. 22-4 A wood strip inserted on the wall ties prevents mortar from dropping into the cavity of the wall.

The Importance of Good Mortar in Cavity Wall Construction

Mortar has an important bearing on the strength of cavity walls so it is critical that masons follow the recommendations of the architect for the proper mortar mix.

Tests by the BIA indicate that Type M or Type S portland cement-lime mortars, under ASTM C270 or BIA Designation M1-72, provide maximum bond between masonry units and mortar. When wind velocities in excess of 80 miles per hour (mph) are expected, Type S mortar is recommended for cavity wall construction. Under average conditions, Type M may be used. In any case, the designer selects the lowest strength mortar that is still compatible with the structural requirements of the building.

Insulation of the Cavity Wall

In addition to their ability to protect interior walls from moisture, cavity walls also serve as sound and temperature insulators. These features have made cavity walls even more popular.

While the air space between the two wythes of masonry acts as an insulating layer, placing insulating materials inside the cavity increases insulating capacity considerably.

Some important requirements for cavity wall insulation are the following.

- Insulation must permit moisture drainage within the cavity.
- Insulation must not hold moisture so that it loses its thermal insulating efficiency.
- Insulation must support its own weight without settling.
- Insulation must be resistant to rot, fire, and vermin.
- Granular fill insulation is usually poured directly from the bag into the cavity or from a hopper placed on top of the wall. The insulation should be protected from moisture during installation.

Types of Insulation. Suitable types of insulation are granular fills and rigid boards. Two types of granular fill insulation have been tested by the BIA and found to meet the above requirements. They are water-repellent vermiculite masonry fill and silicone-treated perlite loose fill.

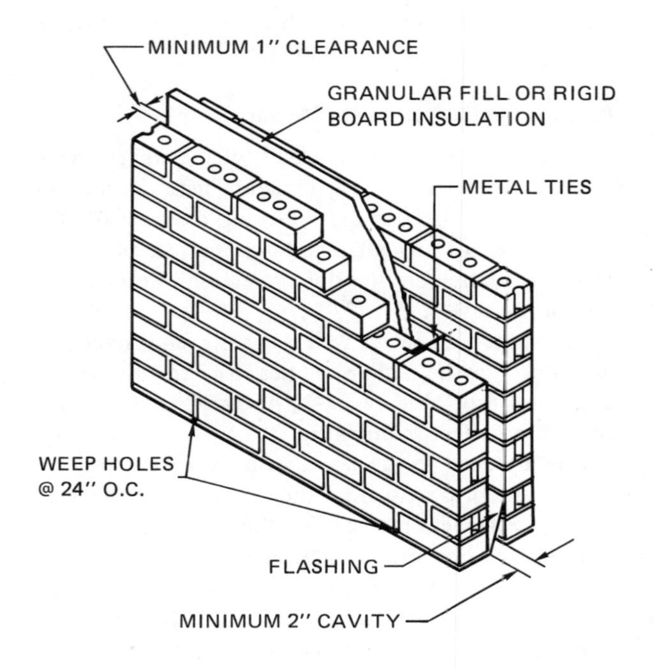

Fig. 22-5 Typical, insulated brick cavity wall

Rigid board insulation can be molded polystyrene, expanded polyurethane, rigid urethane, cellular glass, preformed fiberglass, or perlite board. The rigid board insulation fits flush against the cavity face of the internal wythe. There should be at least 1" of air space between the cavity face of the external wythe and the insulation board, Figure 22-5. The 1" space supplies room for drainage of the cavity to weep holes at the base of the wall. All boards must fit between ties and abut against the ties. An adhesive holds the insulation in place.

Insulated cavity walls, whether filled with granular fill or rigid board insulation, are still popular. They offer resistance to moisture and, equally important in light of high energy costs, they can greatly decrease the amount of energy needed to heat and cool buildings.

REINFORCED MASONRY WALLS

A reinforced masonry wall is built like a cavity wall with an air space between the wythes. However, this air space is later filled with reinforcing steel rods and grout. For even greater strength, steel rods can also be placed in the hollow cores of the masonry units. Grout is poured into the cavity and puddled or vibrated to consolidate it around the reinforcement steel. Reinforced masonry walls are designed to supply greater strength and to withstand greater than normal stresses.

Steel Reinforcement Rods

The steel rods in reinforced masonry are the same type as rods used in reinforced concrete. Projections on the outside surface, called *deformations,* help the grout adhere to the rods. When placing rods in a reinforced wall, masons usually first place short dowel rods into the concrete footing before it hardens. Two wythes of masonry are then built, and the long reinforcements are wired securely to the short dowel rods to form continuous steel rods inside the cavity. As the wall increases in height, the vertical rods are periodically wired to horizontal rods for extra strength, Figure 22-6. The stress the wall will bear determines the size of the rods used.

Mortar Joints in Reinforced Masonry

It is extremely important to use solid mortar joints in reinforced masonry walls. A nonfurrowed bed joint is preferred, with the inside of the mortar joint angled with the edge of the trowel blade, Figure 22-7. This taper helps to prevent mortar from protruding beyond the bricks but still allows full bed joints. If mortar does extend into the cavity, carefully cut off the excess mortar. Be careful not to drop any in the space where the reinforcement is to be placed. If mortar does drop into this space, the grout will not fill in solidly around the reinforcing steel.

Fig. 22-6 Vertical steel reinforcement rods are tied to the horizontal rods to attain greater lateral strength in the wall. Grout will eventually be poured between the rods.

The head joints between each masonry unit must be filled completely with mortar. This is done by completely buttering the ends of each unit and shoving them forcefully into place. The joint formed is known as a *shove joint.* When this method is used by a competent mason, it produces a solid section of wall with no voids.

Cleanout holes are provided during construction by leaving out some masonry units along the base of one side of the wall. These holes can be used not only to clean out the cavity but also to position and tie the steel rods to the dowels in the footings. Mortar that projects more than 1/2" into the cavity should be removed so that the grout fully fills the cavity. Many contractors place a layer of polyethylene film or a layer of sand on the bottom of the cavity. Mortar dropped on such layers is easily removed.

Use of Grout in Reinforced Masonry

Grout may be fine or coarse. Fine grout is used when the grout space does not exceed 2 1/2" in thickness. Coarse grout is used for spaces which measure more than 2 1/2". The aggregate used in coarse grout should be about 3/8" in size. The thickness of the grout between the masonry units and the reinforcement rods should be at least the diameter of the bars. This ensures that grout will surround each rod.

The grout must contain sufficient water so that it flows easily between the walls. One means of measuring the consistency of the grout is by using a 12" high slump cone. (A *slump cone* is a metal truncated cone. A *truncated cone* is a cone with the top cut off.) The cone is placed on a flat surface with the

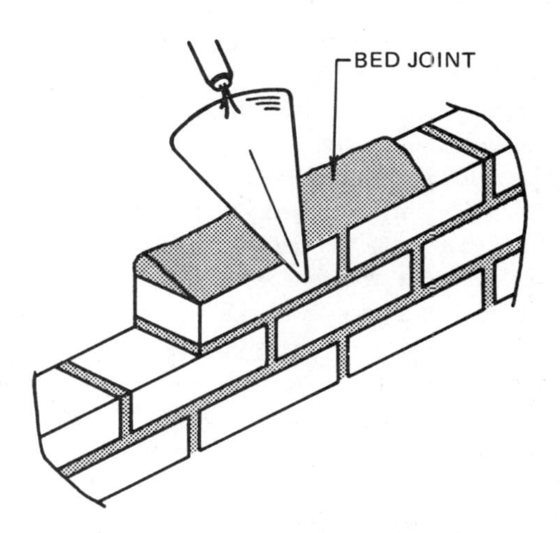

Fig. 22-7 Beveling the bed joint with the blade of the trowel

large end facing down and is filled with grout. The cone is then removed. The distance the grout falls below the top of the cone, known as the slump, shows the consistency of the grout. Grout for use in reinforced brick masonry should have a slump of 9″, using a 12″ slump cone.

Grouting Methods

There are two accepted methods of placing grout in the wall, the low lift method and the high lift method. In low lift grouting, one wythe of wall is laid about 12″ high. The other wythe is then laid 1 course higher before any grout is poured into the cavity. Fifteen minutes should be allowed between each pouring of grout. Masons should not work any closer than 10′ to 15′ apart to avoid the possibility of blowouts. In a *blowout,* the grout swells the wall and leaks through to the outside surface. If the wall bows out, or if blowouts occur, the work must be rebuilt since the bond strength of the bed joint is destroyed. Such repairs are costly and time consuming.

In low lift grouting, pour the grout in the wall from a container equipped with a pouring lip so that the grout does not run onto the face of the masonry work. The best procedure is for the masons to work in pairs with one pouring the grout in the wall and the other puddling with a wooden stick to vibrate and pack the grout into the cavity. Puddle sticks cut from 3/4″ to 1″ wood are the most practical size for this purpose. Sticks which are too large are usually clumsy and may cause blowouts.

If operations on the wall are to be suspended for a period of more than 1 hour, it is best to build both wythes of the wall to the same height and pour the grout to within 3/4″ of the top. This gives the next pouring of grout a better chance of bonding to the section of wall underneath.

High lift grouting involves building the wall to the specified height (not greater than 9′) and then pouring the grout into the cavity. The two wythes of masonry are tied together with metal ties similar to those used in a cavity wall. When the wall is built to the full height, the grout is poured with either a concrete bucket or pumping equipment. After the grout stiffens, pouring resumes. Grouting should not be poured more than 4′ high at any one time. Puddling or vibrating should always be done before the plasticity of the grout is lost.

Remember that the cavity should not be filled with grout until the masonry is strong enough to prevent blowouts from the pressure of the grouting procedure. Code requirements often determine the minimum curing time. Under ordinary conditions a cure of at least 3 days is required before any grout can be poured. Only after inspection of the masonry is the grout poured.

Reinforced masonry walls can be constructed to combine beauty with strength, Figure 22-8. Figure 22-9 shows an example of testing the strength of a reinforced masonry wall section. Another advantage of masonry reinforced walls is that they are highly resistant to moisture penetration, penetration of heat and cold, and expansion. The use of reinforced masonry is highly recommended in areas where earthquakes frequently occur. All these features have led to the increased demand in recent years for reinforced masonry construction.

PROGRESS CHECK

At this point, masonry students should have acquired certain basic skills. They must practice these skills as every mason must have them. Students should demonstrate for their instructor each of the following skills. Any skills not mastered should be practiced until they can be demonstrated correctly.

- Proportion and mix mortar ingredients to suit specific job conditions.
- Stock and place materials.
- Fill chalk lines. Hold chalk lines properly when striking a line.
- Cut, spread, and furrow mortar.
- Form mortar joints.
- Dry bond unit.
- Square jambs or returns of structures.
- Read the plumb rule and level structures with it.
- Lay units to the line. Use line blocks, line pins, and nails.
- Strike joints.
- Brush and clean work at the finish of the job.

Fig. 22-8 Reinforced masonry piers arranged on an angle to create an unusual appearance

Fig. 22-9 The strength of a reinforced masonry wall section is tested by stacking in a large amount of bricks on the section. Notice how little the wall is bending under the strain. As a rule, a compressive load is not applied to a wall horizontally.

ACHIEVEMENT REVIEW

Select the best answer from the choices offered to complete the statement. List your choice by letter identification.

1. The wall type that best ensures interior dryness is the
   a. cavity wall.
   b. solid masonry wall.
   c. veneer wall.
   d. reinforced wall.
2. The weep holes in a brick cavity wall should be spaced the same distance apart to be effective. The proper distance is
   a. 36″.
   b. 24″.
   c. 8″.
   d. 4″.
3. The term *grout* is used frequently when discussing masonry reinforced walls. Grout is
   a. caulking used to seal ends of walls.
   b. steel reinforcement used in the cavity of a cavity wall.
   c. special mortar mixed in liquid form.
   d. wall ties used to tie cavity walls together.
4. An 8″ cavity wall is recommended for buildings
   a. 1 story in height.
   b. 2 stories in height.
   c. 3 stories in height.
   d. 4 stories in height.
5. If the cavity wall is to be effective in draining moisture, the weep holes should be installed over the
   a. header course.
   b. metal wire course.
   c. flashing course.
   d. floor level course.

6. The most popular wall tie used to tie cavity walls together is the
   a. veneer tie.
   b. masonry reinforcement wire.
   c. Z tie.
   d. strap anchor.

7. The consistency of grout can be measured accurately by using a
   a. cylinder compression test.
   b. slump cone.
   c. pouring rate scale.
   d. table which lists exact amounts of water to be used in the mix.

8. Steel reinforcement rods have rough projections and designs on the outside surface. These are there for
   a. the appearance of the steel rod.
   b. easier handling by the person installing the rods.
   c. better bonding of the rods to the grout.
   d. easier bending or cutting when joining one or more rods.

9. The principal reason for using a shove joint is
   a. to ensure a more solid head joint.
   b. for greater speed in laying a reinforced wall.
   c. to increase the compressive strength of the joint.
   d. to allow for expansion of the mortar joint.

PROJECT 14: BUILDING A 10″ CAVITY WALL IN THE RUNNING BOND

OBJECTIVE

- The student will be able to lay a 10″ cavity wall in the running bond, and install metal Z ties and weep holes at designated heights.

  Note: Cavity walls such as this are used in the construction of many school buildings since the interior walls remain extremely dry.

EQUIPMENT, TOOLS, AND SUPPLIES

2 mortar pans or boards
Mixing tools
Mason's trowel
Brick hammer
2′ square
Modular rule
8, 2″ ties with drip crimps
3 pieces of 3/8″ rope, each 14″ in length
2 drip sticks 1 3/4″ wide and 3/4″ thick;
one about 76″ long and the other about 16″ long

Ball of nylon line and line pans
Plumb rule
Convex sled-runner striker
Chalk box
Brush
Clean water

Note: The student will estimate the required number of bricks, lime, and sand from the drawing.

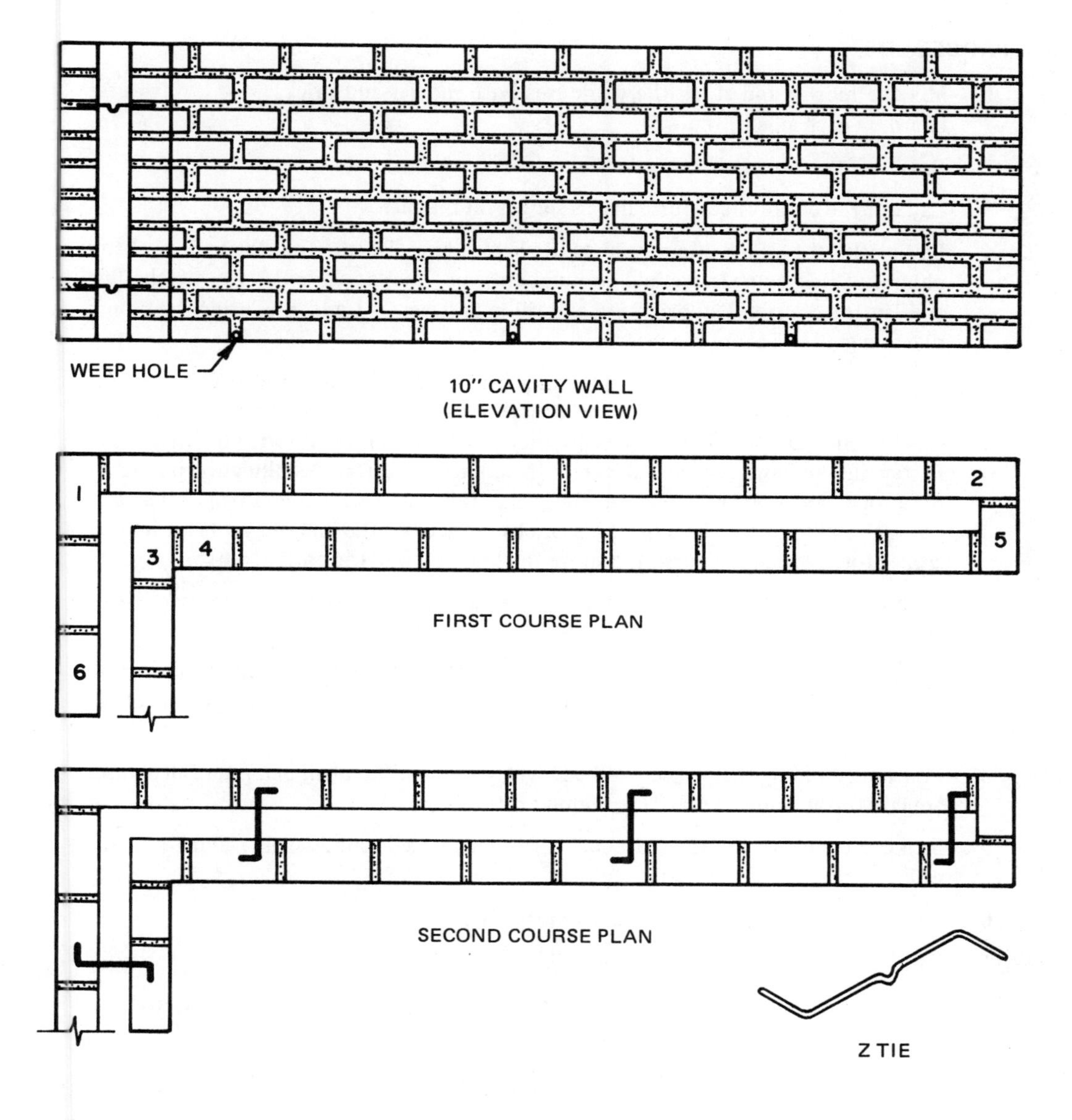

SUGGESTIONS

- Brush off the base carefully to remove any dust or dirt.
- Strike chalk lines longer than are needed.
- Angle the back of the bed joint with a trowel to prevent mortar from falling in the cavity.
- Be sure that the drip stick lies flat on the Z ties to ensure that no mortar falls into the cavity.
- Install the Z ties with the drip facing down.
- Keep all materials and tools away from the immediate work area to avoid accidents.
- Wear eye protection when cutting bricks or mixing mortar.
- Be sure to embed all metal Z ties solidly in mortar.
- Remove the drip stick as soon as you complete the wall.

PROCEDURE

1. Mix the mortar and stock the work area with mortar and bricks. Set up 1 mortar board on each side of the wall.
2. Brush off the base and strike chalk lines.
3. Lay out the entire first course dry as shown on the plan.
4. Lay bricks # 1 and # 2 in mortar to the layout line. Be sure that they are level and plumb. Brick courses are laid to number 6 on the modular rule. Put up the line and lay the course. Lay brick # 6 and fill in between # 1 and # 6, using the plumb rule as a guide. Square the returns.
5. Lay brick # 5 (3/4″) and the adjoining brick. Be sure they are to the line with plumb rule. Measure 12″ from the course already laid and lay bricks #3 and #4 (3/4″) true. Double-check to be sure that all returns are square with the main wall. Fasten the line and lay the balance of the course in mortar, installing weep holes at the proper locations as shown on the plan. Tail out several bricks from brick # 3. Using the plumb rule, be sure these bricks are true to the line. Lay a brick in each corner for the second course and move the line up. Lay the entire course.
6. Install wall ties as shown. Be sure that the front and rear tiers of masonry are level before laying wall ties across the cavity.
7. Lay the drip stick on top of the Z ties and continue laying the project until the wall is 8 courses high. Remove the drip stick and install the second set of Z ties over the eighth course.
8. Continue building the project to the height shown on the plan, being sure all courses are laid to the number 6 on the modular rule.
9. Strike the mortar joints as needed with convex sled-runner striker and brush the wall.
10. Clean as much mortar as possible from the cavity.
11. Pull the weep hole ropes from the wall.
12. Recheck the wall to be sure that it is level and plumb before having it inspected.

SUMMARY, SECTION 5

- Laying bricks and concrete block is more productive with the aid of a line.
- The corner pole eliminates the need for building a corner before erecting a wall and makes the mason more competitive in the building market.
- The masonry corner serves as a guide for the wall and must be built true.
- The mason learns the skills involved in building a corner only by constant practice and strict adherence to basic tool skills.
- When learning to build the corner, do not attempt to work very quickly. Try instead to build the corner as perfectly as possible.
- Pulling and attaching lines to the walls demands safe work practices to avoid injury to the mason and fellow workers.
- Masons spend most of their time laying bricks or concrete block to the line. Using time productively is the key to being a successful mason.

- The mason should be able to estimate the necessary materials for a small brick job by applying the rule of thumb method of estimating. The rule of thumb method of estimating allows for a reasonable amount of waste and is accurate enough for a small job.
- Methods of applying mortar to bricks and concrete block for bed joints are different due to the size of the units.
- Only face shell bedding is necessary when laying concrete block unless otherwise specified.
- Once the masonry unit takes its initial set, it should not be shifted or moved unless fresh mortar is applied and the unit is relaid.
- Mix mortar to the correct consistency for the masonry unit to set properly. As a rule, concrete block require a slightly stiffer mortar than bricks.
- Reinforced wire greatly increases the strength of the bed joint in a masonry wall.
- The best joint finish for bricks or concrete block is the concave joint.
- Always wear eye protection when cutting masonry units.
- When laying concrete block, observe proper lifting practices due to the weight of the units.
- You can erect a structure consisting of concrete block more rapidly than a brick structure since the units are larger.
- Plumb single units of concrete block only on one side.
- When a block wall is to be painted and the mortar joints are not to be evident, use a flat, rubbed joint.
- Rule of thumb estimating for concrete block is done by calculating lineal feet around the structure and converting that figure to block. Multiply the number of block on one course by the number of courses in height for the total number of block needed for the job. Deduct block for all openings.
- When two separate tiers of masonry units are built to form a single wall, the structure is known as a composite wall.
- Tie a composite wall together with masonry headers or metal wall ties.
- Metal ties of different types are now being used more than brick headers to tie walls since they allow for more expansion and contraction in the mortar joints.
- Regardless of the type of metal tie selected for a job, make sure to embed the ties solidly in mortar where they rest on the wall.
- To obtain maximum resistance to rain penetration, the best choice of wall type is the cavity wall.
- Weep holes are necessary to allow moisture to drain from cavity walls.
- Cavity walls may be insulated for greater efficiency.
- Keep cavity walls free from mortar droppings so they will drain properly.
- Wall ties for cavity walls must have a drip crimp in the center so that moisture is not carried to the walls.
- Reinforced masonry walls are built when superior strength is required.
- The middle of reinforced walls is constructed of steel rods and mortar or cement grout.

- Use solid mortar joints when building a reinforced masonry wall to avoid blowouts.
- Never attempt grouting of a reinforced wall until the wall has had time to set.
- Reinforced walls offer the strength of concrete and pleasing appearance of masonry work.

SUMMARY ACHIEVEMENT REVIEW, SECTION 5

Complete the following statements referring to material found in Section 5.

1. Most masonry units laid by masons are laid to the ________________.
2. As the mason faces the wall, the left-hand lead where the line is attached is called the ________________. The right-hand lead where the line is pulled and fastened is called the ________________.
3. If a long wall is being built and the center of the wall sags, the sag can be corrected by setting a ________________.
4. Masonry units (brick or concrete block) should always be laid so that the top edge of the unit is even with the ________________ and ________________ inch (inches) from the line.
5. Masonry corners are also known as ________________ in the trade.
6. One of the most important points in building a corner is looking down over the work when laying the unit to be sure it is in close alignment. This technique, which reduces the amount of necessary plumbing, is known as ________________.
7. Masonry walls are sometimes plastered on the back or inside of the wall to prevent moisture penetration. This practice is known as ________________.
8. Estimating brick masonry for small jobs is done by the rule of thumb method. To determine the number of bricks in a 4" wall, multiply ________________.
9. When laying concrete block to the line, mortar is applied only on the outside web. This method of applying mortar is called ________________.
10. The most effective and popular joint finish for concrete block is the ______________.
11. The major difference in laying bricks and concrete block is because of the ________________ and ________________ of the units.
12. When cutting bricks or concrete block, the mason should always wear __________ for safety.
13. Additional strength can be added to mortar bed joints by installing _____________.
14. When a masonry wall is built of 2 different tiers of masonry units and bonded together by some type of tie, it is known as a ________________.
15. The composite wall may be tied in two different manners, by use of a __________ or ________________.
16. Concrete block are estimated by the rule of thumb method by determining the number of lineal feet around the total structure and multiplying by ________________.
17. The number of concrete block required to lay 1 course around a structure is multiplied by ________________ to determine the total number of block needed to build the structure.

18. Before figuring the total amount of materials for a job and their costs, all ____________ must be deducted.

19. A masonry wall that consists of 2 tiers of masonry with an air space in the center is called a ______________.

20. A masonry wall that has 2 tiers of masonry with steel reinforcement and grout in the center is known as a ______________.

21. Masonry walls should be built to suit specific climates or weather conditions. In a very wet, humid area, the best wall type to use is a ______________.

22. Grout used in reinforced masonry walls must be more ______________ than regular mortar to fill all voids and holes.

23. For wall ties in a cavity wall to remain free of moisture, they must have a ______________.

24. Aside from the architectural beauty of reinforced masonry, the most important feature is ______________.

25. Under ordinary conditions, before placing grout in a reinforced masonry wall, the period of time that must pass is ______________.

Section 6
Masonry Practices and Details of Construction

Unit 23
Masonry Supports, Chases, and Bearings

OBJECTIVES

After studying this unit, the student will be able to

- describe how lintels are installed in a masonry wall.
- construct a pier or pilaster using given plans.
- improve tool skills by building various masonry projects.

After studying and learning the fundamentals of basic masonry, the student progresses to the details of laying bricks and concrete block. These processes are not difficult, but demand the application of the information presented thus far and mastering of tool skills to a greater degree.

LINTELS

A *lintel* is a horizontal member or beam support placed over a wall opening to carry the weight of the masonry to be laid over it. The design of the particular building and load to be placed on the lintel are the deciding factors in the type of lintel selected.

Steel Lintels

One of the most common types of lintels in use today is the *structural steel lintel,* or *angle iron.* The steel angle iron, which is shaped like the letter L, should have a thickness not less than 1/4″ and a width not less than 3 1/2″ to support the standard 4″ masonry unit. Angle irons such as this one may be installed back to back when building an 8″ wall. Three angle irons are required to build a wall 12″ in thickness, Figure 23-1.

There are some important items to remember when installing steel lintels over openings. The mason should consult the specifications on the job to determine the *lintel bearing* (the load which the lintel must bear). The lintel should bear a minimum of 4″ on each side of the opening on houses or small commercial buildings.

Most angle irons have been cut from a larger piece of iron with a burning torch. This burning causes small beads of steel to form on the surface of the bottom and top portions of the angle iron. The beads should be removed from the irons with the brick hammer before installing. The masonry units will not lay level over these rough welds. This is especially important if flashing is to be installed over the lintels, as the rough beads of steel may puncture the flashing.

The lintel must be set plumb and level so that the unit which is laid on it is also in the proper position. The steel lintel must be set back 3/8″ to 1/2″ from the face of the wall so that a mortar joint may be formed in front of the steel without cracking, Figure 23-2. Never set the angle iron on the wall until the masonry unit under it has set sufficiently to avoid sinking.

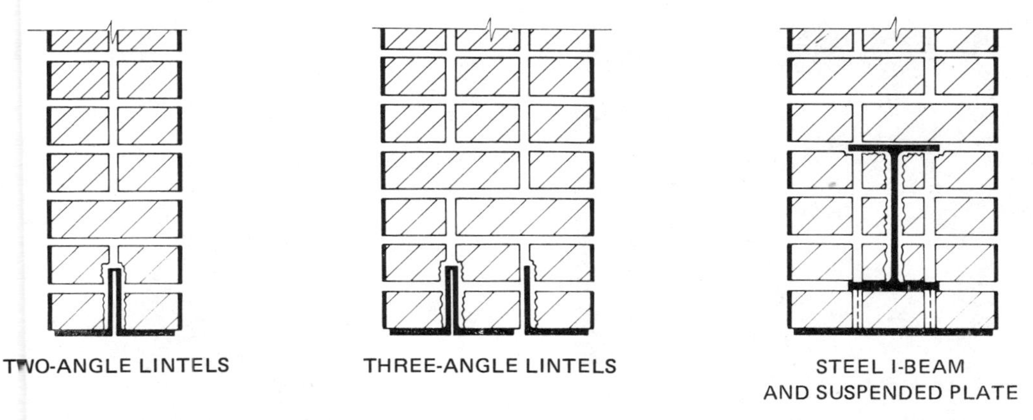

Fig. 23-1 Typical lintel assemblies used in masonry walls. (Left to right) an 8-inch wall requires two-angle irons; a 12-inch wall required three-angle irons; the steel 1-beam and suspended plate combination is used for heavier walls.

Fig. 23-2 Installing a steel angle iron over a window opening. The mason is positioning the lintel about 1/2 inch from the face of the wall to allow room for a mortar joint.

Concrete Block Lintels

Concrete block lintels are made of the same materials and are the same texture as concrete block. These lintels have steel rods running throughout for extra strength. They are available in various thicknesses to suit individual concrete block walls. They are available in lengths to fit most openings. A minimum bearing of 8″ should be provided on both sides of the openings for concrete block lintels.

When installing concrete block lintels, caution should be used since they are bulky and heavy. It is recommended that two masons lift and place the lintels in position. Very heavy lintels should be installed with the aid of power-driven lifting equipment.

Bond Beam Lintels

The *bond beam lintel* is used if the lintel must match the rest of the surface of the wall or if other types of lintels are not readily available. To form a bond beam lintel (also called a *reinforced lintel*), the inside of the brick is sawed out or a special bond beam block is purchased. Steel reinforcement rods are placed in the cavity on a layer of concrete and then the cavity is filled with concrete. The bottom of the reinforced lintel must be braced with lumber until the lintel cures, Figure 23-3.

Architects sometimes specify that a course of reinforced lintel run completely around a building. Many missile sites use this type of construction for extra strength.

PIERS

Piers, Figure 23-4, are vertical columns of masonry that are not bonded to a masonry wall. They can be used as supports for beams, arches, porches, or wherever a free-standing masonry column is needed. The difference between a pier and a column is that columns are usually much higher in proportion to their thickness.

The size and type of construction of a pier is determined by the load the pier will support. Its horizontal length should not exceed four times its thickness. If a pier is hollow, excess mortar must not fall inside the center of the pier since this could cause the pier to swell and be pushed out of position. The cen-

Fig. 23-3 Bond beam lintel. This type of lintel is typical of reinforced masonry walls which cross over an opening.

ters of solid piers should not be filled immediately after the outside course of bricks is laid. Instead, the pier should be laid to a height of 6 courses and then the bricks should be laid in the center.

Materials must be dry if the pier is to be constructed level and plumb. Because piers are relatively small, they are laid more quickly than most masonry projects. If the masonry units are laid wet and the mortar is too fluid, the units do not set properly, making the tesk of leveling and plumbing almost impossible. Piers should also be squared frequently and rechecked to be sure that their positions remain the same throughout the building process.

STEEL BEARING PLATES

Steel bearing plates are placed on top of piers to provide a solid bed for steel beams or girders. They also serve to spread weight over a greater area. Steel plates are available in various thicknesses, but should be thick enough so that they are not bent out of shape under load.

The job foreman should consult individual job specifications for the required size of the plates. The height of the plates are shown on the plans and should

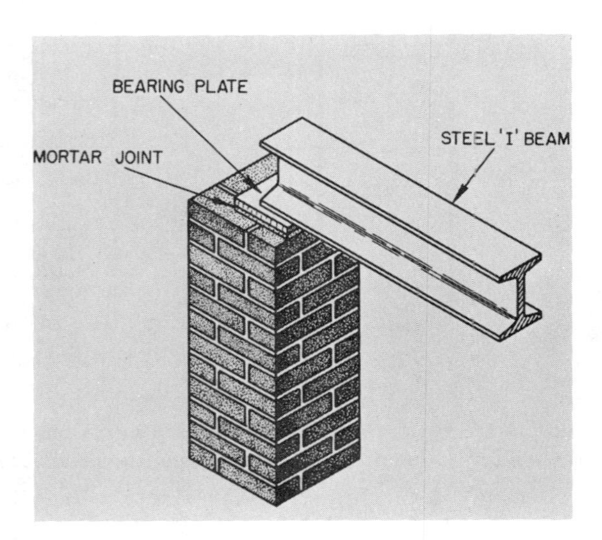

Fig. 23-4 Brick pier with bearing plate and steel I-beam

be strictly followed if the steel beam is to be set at the correct height.

Installing Plates

A full bed of mortar should be spread with a trowel and the plate should be laid on the mortar bed. It is recommended to bed bearing plates in portland cement mortar for greater strength. Always spread more mortar than is actually needed. Then, when the plate is tapped into position with the handle of the hammer, no voids will exist under the plate. Be sure that the bearing plate is level to ensure that the beam sets level on the plate. When steel anchor bolts are installed in piers to draw the plates against them, they must be built into the masonry work beforehand.

PILASTERS

A *pilaster,* Figure 23-5, is much like a pier except that a pilaster is tied into the wall of a structure. Pilasters may be tied to a wall by use of wall ties or by bonding the masonry units into the wall. Some building codes require that pilasters be tied into the masonry wall at least 4″ and extend out from the wall 4″ to 8″. Local regulations will determine this.

Pilasters should be laid from the footing to a predetermined height. If possible, pilasters should never be laid so that units must be cut. For example, a pilaster that extends from a brick wall 6″ would involve cutting 2″ pieces of masonry units to fill the space behind the pilaster.

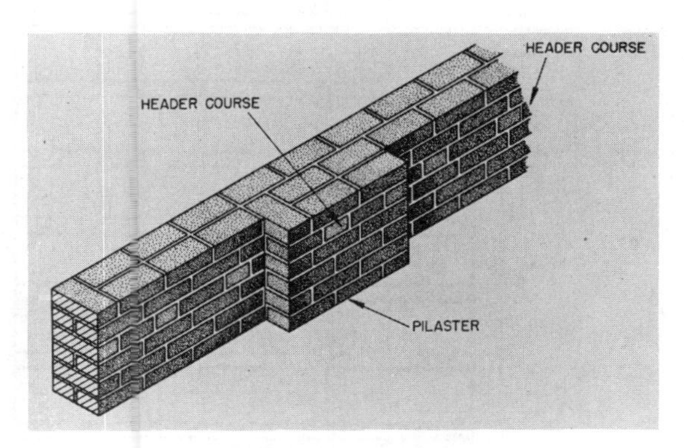

Fig. 23-5 Pilaster extending 4 inches from 8-inch brick wall

Pilasters are used for supporting loads, for ornamental design, and as vertical supports for strengthening masonry walls. When pilasters are used for ornamental purposes or for vertical support, they are also known as *buttresses.* Buttresses are usually finished at the top by a sloping face.

In reinforced concrete masonry structures and large buildings, pilasters are also sometimes used to enclose heating and air-conditioning ducts. Steel columns can be hidden from view and given some protection from fire by constructing pilasters around them. When a pilaster is used for this purpose, the wall is usually only 4" thick since it is acting only as a screen and is not bearing a load.

CHASES AND RECESSES

A *chase,* Figure 23-6, is a vertical or horizontal recess left in a wall to conceal certain items such as plumbing pipes, electrical wires, or heating accessories. Many jobs, especially the commercial type, require that these mechanical items be installed behind the finished surface of the wall. If chases are not left in at the beginning of the construction, the workers must cut them out with a chisel, causing much greater expense and delay in the job.

When buildings are constructed with chases, wall thickness and wall strength are always reduced. This must always be considered when designing a structure. Deep chases should never be built where the wall is to receive a heavy beam or steel bar joists over it.

Some general rules apply to the installation of chases in masonry walls. Do not build chases that are deeper than one-third the thickness of the wall. If the wall is load bearing, install lintels over the chase to sufficiently support the load. Check local building codes if in doubt, but remember, chases that will weaken the wall should never be built.

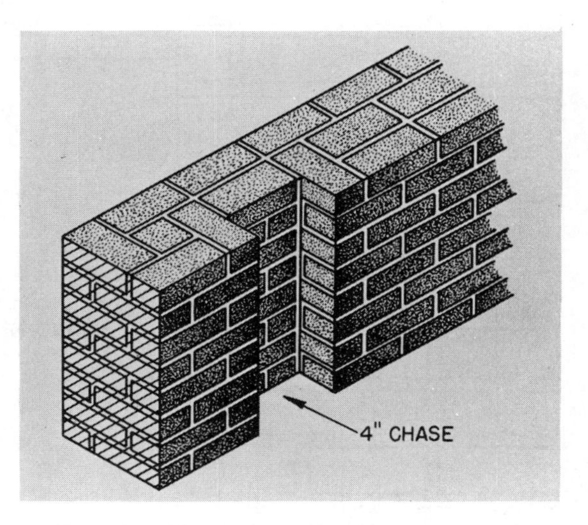

Fig. 23-6 Chase constructed in 12-inch wall

Masons build the chases as they lay the wall. The plans should specify exactly where the chase will be built. If it is not shown on the plan, do not build a chase in a bearing wall without first consulting the architect or the job engineer. Chases are usually on the inside face of the wall and may be 4", 8", or 12" or more wide. The masonry units forming the chase should be bonded into the rest of the wall. Stack joints are not acceptable for bonding chases, since they are weak and may separate when the load of the structure increases. A chase should be plumb with all excess mortar cleanly cut off so that it does not interfere with the installation of pipes or wires.

SOLID MASONRY BEARING WALLS

A basement or foundation bearing wall (a wall which supports joists, concrete slabs, or steel) often must meet special requirements. Many building codes require that the top 8" of such walls be built of either 3 courses of brick or a solid course of concrete blocks, usually 4" or 6" thick. This helps distribute the weight of the structure on the foundation. Holes in the course of the regular concrete block beneath this solid course are filled with mortar. In addition, face shell bedding such as is used on a standard block wall is not acceptable underneath a solid bearing course.

The bearing course, whether it is brick or concrete block, is the last course of the wall which is laid. Joists or concrete slabs rest on the bearing course. It

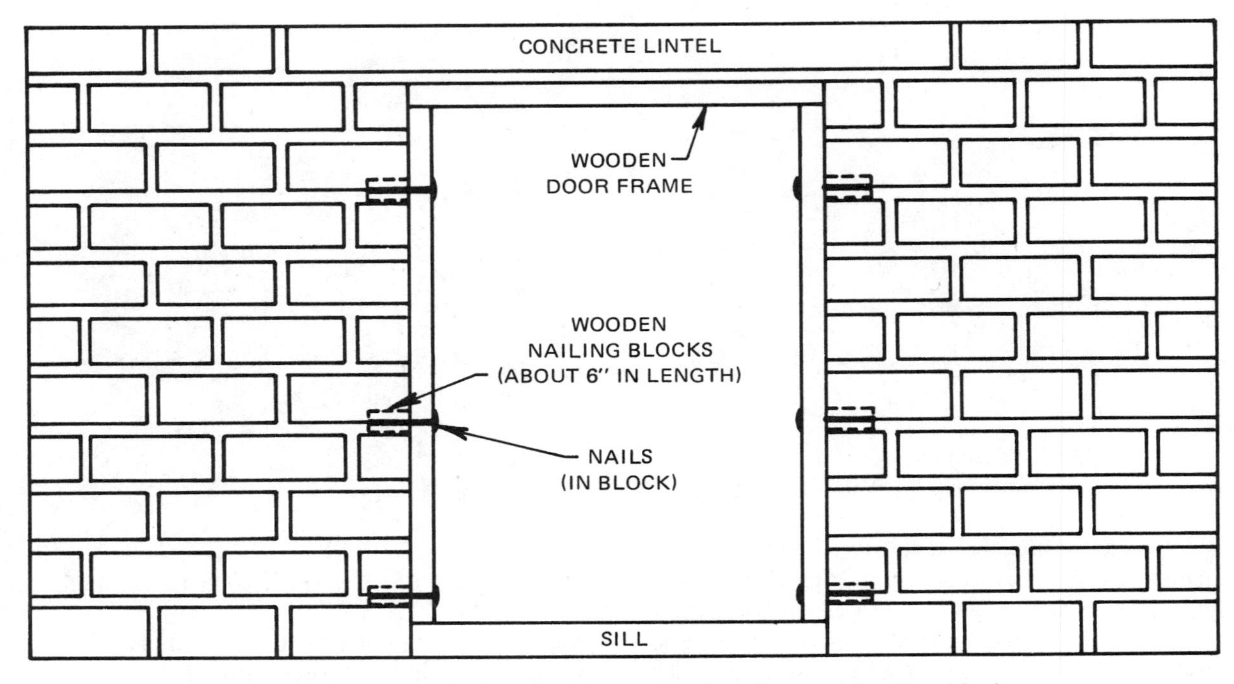

Fig. 23-7 Concrete block wall with wooden door frame and nailing blocks

is, therefore, essential that the bearing course be perfectly level and plumb if the joints are to lay straight on the wall.

LAYING UNITS AROUND DOOR AND WINDOW FRAMES

The mason must often lay masonry units to, against, and around various types of door and window frames. Door frames may be tied to masonry work in different ways. Wooden door frames can be attached to the masonry work by nailing a special tempered nail through the frame and into the unit or by nailing into a wooden block (*nailing block*) that is laid in the mortar against the door frame. When using nailing blocks, 3 blocks are used on each side of the door; one at the bottom, middle, and top of the door, Figure 23-7.

Metal doors are in great demand for commercial construction. Metal ties are used to tie masonry work to metal doors, Figure 23-8. The process involves inserting the tie into the jamb of the door which is filled with mortar to create a solid door jamb. The mason should use caution when filling the door jamb with mortar so that the mortar does not push the door out of alignment. The door should be rechecked to be certain it is plumb after filling it and before leaving for the day. A wooden spreader installed at the bottom of the door helps prevent the door jamb from being pushed out of alignment.

Fig. 23-8 Mason installing metal ties in a metal door frame. Notice that the mortar is filled in solid against the door frame.

Windows are built in walls much the same way as doors are. On brick veneer walls, the bricks are butted flush against the wooden window jambs and the window frame is nailed into the framing of the wall. Metal window frames may be tied with metal ties in the same way as metal door frames. Solid composite walls that have wooden window frames can be secured by nailing a tempered nail through the wooden frame into the masonry unit.

Regardless of how frames are fastened to the masonry wall, the mason must remember that the frame should be positioned the proper distance back from the face of the wall. The position depends on the thickness of the wall; that portion of masonry work which extends from the window frame to the face of the masonry wall is known as the *reveal.* The frame must be kept level on the sill and plumb with the face of the wall. The carpenter has the responsibility of setting the frame. However, the mason should always check the frame to be sure that it is level and plumb before finishing the wall.

ACHIEVEMENT REVIEW

Select the best answer from the choices offered to complete the statement or answer the question. List your choice by letter identification.

1. The primary purpose of a lintel is to
   a. provide a means of expansion for masonry work.
   b. support a masonry unit over an opening.
   c. tie a masonry wall together.
   d. provide an extra measure of strength in a bearing wall.

2. If a 12" brick wall is being constructed and lintels are required for a window, how many lintels should be used if angle irons are specified?
   a. 1 lintel.
   b. 2 lintels.
   c. 3 lintels.
   d. 4 lintels.

3. When a course of masonry units has the inside cut out and steel rods and concrete placed within, the resulting structure is a (an)
   a. angle iron.
   b. hanging plate.
   c. bearing plate.
   d. bond beam.

4. The main difference between a pier and a pilaster is that
   a. the pier is always wider than the pilaster.
   b. piers do not carry any weight and are only built for decorative purposes.
   c. a pilaster is tied to the main wall.
   d. the pilaster is free standing.

5. It is extremely important to have dry bricks when constructing a pier because
   a. the pier is built more quickly than a regular wall and, therefore, may not set with wet bricks.
   b. the wet bricks stain and cause an unsightly appearance.
   c. the mortar joints do not attain full strength.
   d. the mason's fingers become sore with the result that productivity is seriously curtailed.

6. Steel plates are laid on piers in mortar for the purpose of
   a. fireproofing the area where the beam sets on the plate.
   b. distributing the weight of the beam more uniformly.
   c. spot welding the beam to the steel plate.

7. On many jobs involving the installation of pipes and wires in the wall, a recess is built by the mason. This recess is called a
   a. reveal.
   b. pilaster
   c. pier.
   d. chase.

8. Bearing walls which carry joists or concrete slabs should have at least
   a. 6 courses of bricks or 2 courses of solid concrete block at the top of the wall.
   b. 3 courses of bricks or 1 course of solid block.
   c. 9 courses of bricks or 3 courses of solid block.

9. When laying a concrete block wall against a wooden door frame, nailing blocks are installed in the wall. How many nailing blocks are needed for 1 door, including both sides of the door?

a. 6 c. 2
b. 3 d. 8

PROJECT 15: BUILDING A 12" X 16" AND A 20" X 24" HOLLOW BRICK PIER

OBJECTIVE

- The student will be able to lay out, square, build, level, and plumb a hollow brick pier.

  Note: These piers are typical of those built to support a porch or steel beam in a foundation.

EQUIPMENT, TOOLS, AND SUPPLIES

Mortar pan or board
Mixing tools
Mason's trowel
Brick hammer
4′ plumb rule and 2′ level
Convex striking tool
2′ square
Chalk box and pencil
Mason's modular rule
Brush
1, 8" x 8" x 16" concrete block in which to set plumb rule

The student will estimate the amount of standard face bricks, lime, sand, and water needed to build the project based on the rule of thumb method.

12" x 16" HOLLOW BRICK PIER (ELEVATION VIEW)

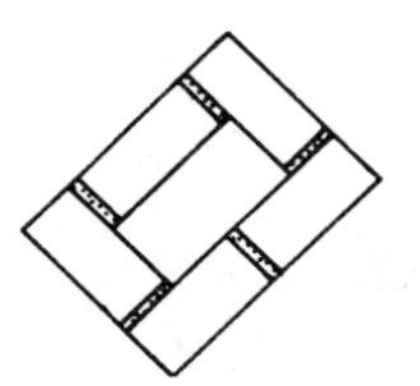

FIRST COURSE PLAN

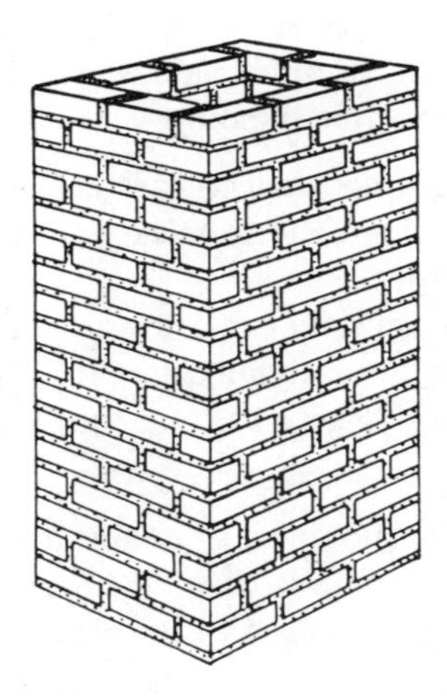

20" x 24" HOLLOW BRICK PIER (ELEVATION VIEW)

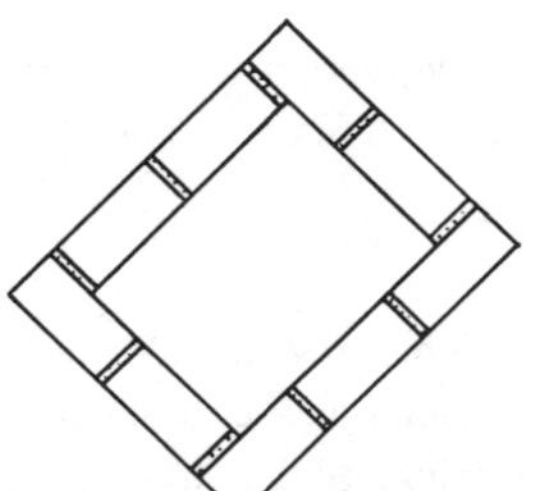

FIRST COURSE PLAN

SUGGESTIONS

- Strike the chalk line longer than the actual size of the pier. This gives a reference line with which to check the work as the project is being built.
- When laying out the first course, select the highest point on the base and start the project there. The rest of the course can then be bedded with mortar to the highest brick.
- Use only single head joints on the immediate corner bricks until the entire course has been leveled and plumbed. This reduces the possibility of the corner brick being pushed out of alignment when the next brick is laid against it. Be certain to fill all joints after the course is completed.
- Use the 2′ level for the first 9 courses since it is easier to handle.
- Keep the levels stored in the block when they are not in use. The 8′ block is not upset as easily as a smaller block is.
- Prevent the mortar from dropping into the center of the pier by cutting it off with the trowel, as an excessive amount may cause the pier to be pushed out of alignment.
- Do not beat on the pier excessively as this may cause the brickwork to settle out of alignment.
- Keep the work area free from materials. Always use safe working practices. Always consider those working nearby.

PROCEDURE

1. Set the mortar pan or board and bricks approximately 2′ from the project.
2. Mix the mortar. Mix only as much mortar as can be used in the time provided. (Generally, never mix more mortar than can be used in a 2-hour period.) Fill the mortar pan with mortar. Clean tools immediately after mixing. Strike a chalk line on one side only.
3. Lay out the first course dry and square it. Mark the layout course with a pencil according to the plan. Remove the layout course and strike a chalk line over the pencil lines, extending the lines past the actual size of the pier for reference lines.
4. Bed the corner brick and check the height with the number 6 on the modular rule. Level and plumb. Spread mortar for the rest of the course and lay the bricks. Be certain that the leveling is done from the original corner brick and completely around the first course. Plumb the course on its corner points and align the pier with a straightedge on all four sides. Do not remove mortar that has been squeezed from under the bricks at this time, as they may settle unevenly. Remove the mortar after 3 courses have been laid.
5. Check the project with the steel square and make any necessary corrections.
6. Repeat the operations involved in laying each course of bricks until the pier has been built the specified number of courses. All brick courses are laid in height to the number 6 on the modular rule.
7. Strike the mortar joints as needed with a convex jointer. Remember to apply the thumbprint test. Brush the work.
8. Recheck the pier to be certain it is level, plumb, square, and the proper size before having it inspected.

PROJECT 16: CONSTRUCTING AN 8", METAL-TIED BRICK WALL WITH 2, 4" X 12" PILASTERS

OBJECTIVE

- The student will be able to lay out and construct an 8" brick wall with 2, 4" x 12" pilasters. The wall will be laid up to number 6 on the modular rule and tied together with metal Z ties.

  Note: The project is typical of garden walls which surround a residence or divide parking lots in an urban area.

  Note: This project can be done by two students working as a team.

EQUIPMENT, TOOLS, AND SUPPLIES

2 mortar pans or boards
Mixing tools
Mason's trowel and small pointing trowel
Brick hammer
4' plumb rule and 2' level
Convex striking tool
2' square
Ball of nylon line
Line pin and nail
Chalk box and a pencil
Mason's modular rule
Brush
Supply of clean water
1, 8" x 8" x 16" concrete block in which to set levels
6 Z ties

The student will estimate bricks, sand, and lime by studying the plan.

SUGGESTIONS

- Strike the chalk line longer than is actually needed to serve as a reference line.
- Use the 2' level to plumb the first 6 courses of the pilasters as it is easier to handle.
- Lay the bricks to the line on the straight section of the wall before laying the bricks on the pilaster. This prevents the pilaster from interfering with the line.
- Square the pilaster as often as is necessary.
- Wear eye protection when cutting bats for the pilaster.
- Cut off excess mortar on the inside of the pilaster and return it to the mortar pan. Temper the mortar for reuse in the wall.
- Be certain that the inside angle joint is pointed solidly and neatly. Use the small pointing trowel.
- Keep the work area free from scraps of materials and follow good safety practices.
- Do not fill in the center of the wall (collar joint) with mortar as this may cause the wall to be pushed out of alignment.

PROCEDURE

1. Mix the mortar and arrange the necessary materials in the work area.
2. Strike a chalk line for the wall. Lay out the project dry as shown on the first course of plan.
3. Bed bricks #1 and #2. Stretch a line from brick #1 to brick # 2 and tighten with the line pin.

4. Lay the rest of the course in mortar. Repeat the procedure for bricks #3 and #4. Lay the pilaster bricks in mortar, being sure to square off of the main wall. Pilaster bricks should project 4 1/2" from the face of the wall, including the head joint.

5. All brick courses are laid to number 6 on the modular rule. On the second course, bed up bricks #1 and #2 being sure they are true with the level. Install the metal Z ties as shown on the plan.

6. Lay up a small 2-brick lead on both ends of the project. Attach the line and lay the courses in mortar. Alternate the line on each side of the wall as the work proceeds. Strike mortar joints as needed with the convex jointer.

7. Repeat this procedure until the wall is 6 courses high from where the last Z ties were installed. Install the second set of Z ties, being sure that the ties are not directly in line with the ties on the second course.

8. After each course of bricks is laid in the wall, lay the bricks for the pilaster level and plumb. Be certain that the pilaster is kept level with the wall. Recheck for squareness.

9. Continue constructing the wall until the number of courses shown on the plan (12 courses) is reached. Strike the joints, fill any holes, and recheck the wall before having it inspected.

PROJECT 17: LAYING A BRICK WALL WITH A PIPE CHASE

OBJECTIVE

- The student will be able to lay out and build a 12" brick wall with a vertical pipe chase in the common bond.

  Note: Pipe chases such as this one are constructed in many commercial buildings to conceal drainpipes descending from roofs.

EQUIPMENT, TOOLS, AND SUPPLIES

1 mortar pan or board
Mixing tools
Mason's trowel
Brick hammer
4' plumb rule and 2' level
Convex jointer or V-jointer
Small square (12" maximum size)
Ball of nylon line
2 line blocks
Chalk box and pencil
Mason's modular rule
Brush
Pointing trowel
Supply of clean water
1, 8" x 8" x 16" concrete block in which to set levels

The student will estimate bricks, lime or cement, and sand from the plan.

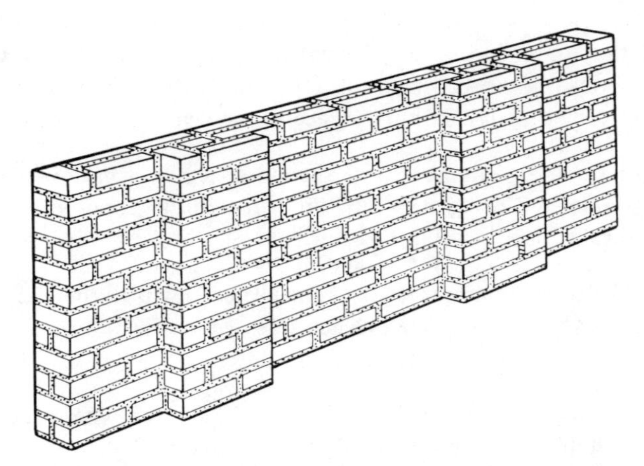

8" METAL-TIED WALL WITH TWO 4" x 12" PILASTERS
(ELEVATION VIEW)

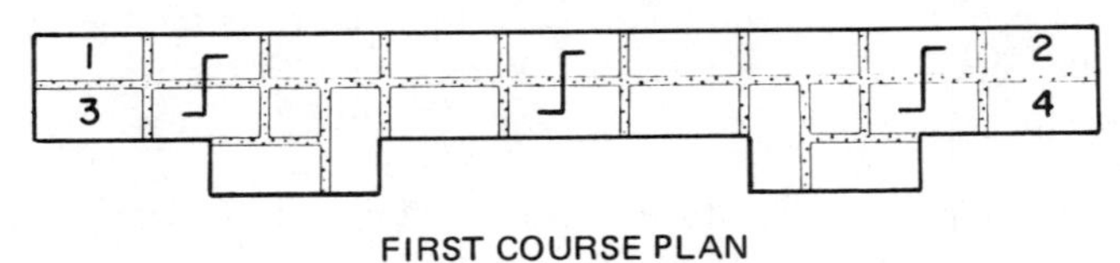

NOTE: TIES ARE SPACED EVERY SIXTH COURSE

FIRST COURSE PLAN

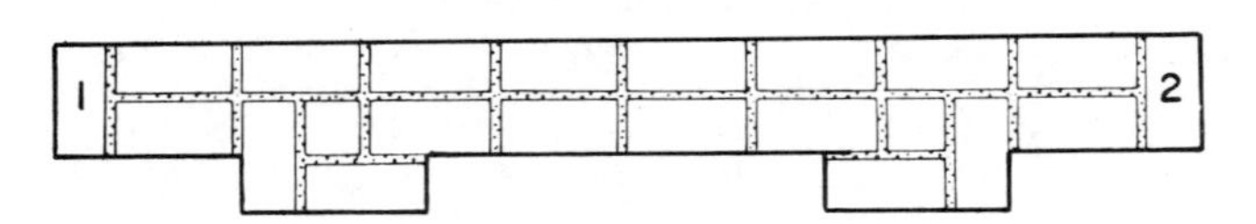

SECOND COURSE PLAN

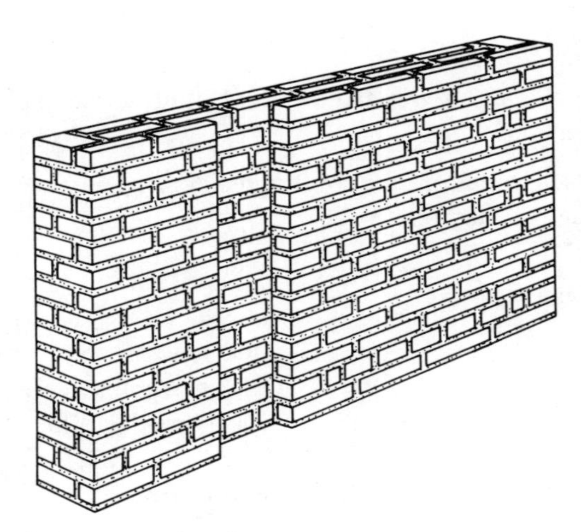

PIPE CHASES
(ELEVATION VIEW)

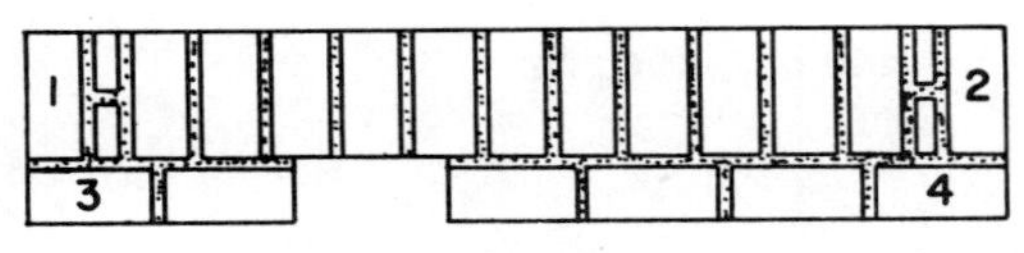

FIRST COURSE PLAN

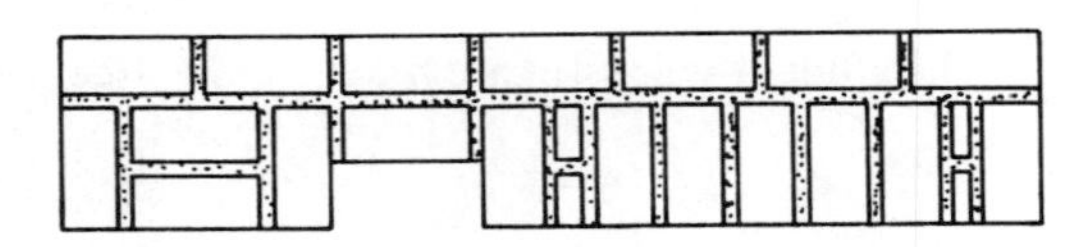

SECOND AND EIGHTH COURSE PLAN

SUGGESTIONS

- Select unchipped bricks for all corners.
- Do not fill in the collar joint with mortar, as it may cause the wall to bulge or be pushed out of alignment.
- Start the header course either with a 2" piece as shown on the plan or with a 6" piece. In this particular project, the 2" starter piece is the stronger tie since 1 extra header could be used in the wall. As a rule, walls with pipe chases do not require striking. The wall in this project, however, requires striking because it is exposed.
- Frequently recheck all corners and chase with the square while the project is being built.
- Practice sighting down the corners as the bricks are laid. This practice helps reduce the amount of plumbing needed.
- Observe safety practices at all times.
- Form solid, well-filled joints between all bricks.
- Cut excess mortar from the backs of the bricks as soon as possible and return it to the mortar pan for reuse. Retemper the mortar as often as needed.

PROCEDURE

1. Prepare the work area and mix the mortar. Stock the materials at least 2' away from the project to allow sufficient working room.
2. Strike a chalk line for the project as shown on the plan.
3. Lay out the first course dry. Lay bricks #1 and #2 in mortar. Erect the line with line blocks and lay the balance of the course. All courses are laid to the number 6 on the modular rule. The first course on the side of the wall that does not have a chase is a header course.
4. Repeat the procedure on the second course. Notice that the header course of the second course is on the same side of the wall as the chase. This is done in 12" walls so that the headers occur over each other on succeeding courses.
5. Build a small lead on each end of the wall (2 or 3 courses high) and resume laying bricks to the line until the next header height is reached (6 courses high).
6. Lay the header course and continue building the wall until the total number of courses shown on the plan is reached.
7. Clean out excess mortar from the chase and point with mortar any remaining holes.
8. Check the project for the correct height with the modular rule. Recheck to be sure it is square, level, and plumb.
9. Strike the project with the convex or V-jointer as needed. Brush work before having it inspected.

Unit 24
Small, One-Flue Chimneys

OBJECTIVES

After studying this unit, the student will be able to

- describe the various components of a chimney.
- explain how a one-flue chimney is built.
- build a one-flue chimney from given plans.
- list general steps in the installation of wood-burning stoves.

Due to the energy crisis and the high cost of conventional types of fuel, the heating of homes throughout North America with wood stoves has made a dramatic return. Although many of the homes that are converting to wood stove heat utilize the fireplace, there is also a great demand for new one-flue chimneys to serve wood stoves.

Local building authorities require permits and inspections of these chimneys while they are being constructed and after the wood stove has been installed in the home. Brick or concrete chimney block are acceptable for building one-flue chimneys for wood stoves. It is important not only that the chimney function correctly but also that there is ample brick laid around all combustible areas to prevent the danger of fire.

The primary function of a chimney is to produce the draft necessary to carry smoke or gases to the outside of the structure without endangering the structure or its occupants. A *draft* is defined as the movement of flue gases or air through the chimney. If a good draft is not present in the chimney, it cannot perform correctly.

A natural draft is influenced by the differences of temperature in the chimney and the outside atmosphere. Air travels down the chimney and returns to form a draft. When the draft is supplied by a fan in the furnace, the draft is said to be forced or induced. Most small chimneys operate on the forced draft principle.

Formerly, chimneys did not have protective linings to prevent the chimney from burning out. Almost all chimneys built today have a fired clay or terra-cotta lining to make them more fireproof and long lasting.

Most areas have building codes which require flue linings to be installed in all chimneys. They are obtainable in various widths and in a standard 2′ - 0″ length. The clay from which the flue lining is made is fireproof and gives many years of service under extreme heat. It is recommended that flue linings be installed in all chimneys.

This unit discusses the one-flue chimney that is normally used in furnace and stovepipe hookups.

THE CHIMNEY BASE

Chimneys should be built on solid concrete footings which have been poured below the *frost line.* Any ground above the frost line freezes when the temperature falls sufficiently low. The depth at which frost lines occur differs in various parts of the country. Local building codes should be consulted to determine the frost line of a particular area. All masonry work should be started on a base which is below the frost line or deterioration of mortar joints may occur.

Footings at the base of chimneys require more concrete than the average footing of a home. This is because of the greater height and weight which is concentrated in a small area. Footings should be a mini-

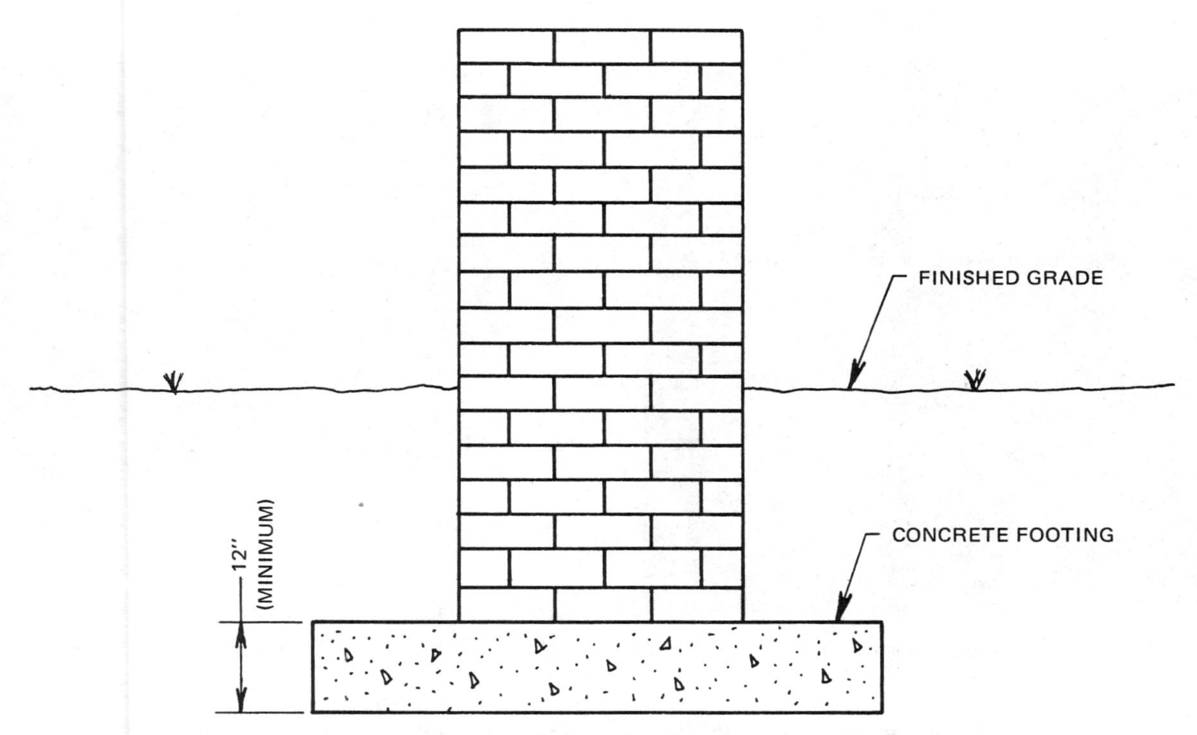

Fig. 24-1 Sectional view of concrete footing and chimney. The base of the chimney should be twice as wide as the chimney itself for proper weight distribution.

mum of 12" deep and twice as wide as the chimney to properly distribute the load over the base area, Figure 24-1.

Chimneys should always be started from footings, never from wood platforms or metal supports. Local building codes should be consulted for proper footing sizes. As a rule, metal reinforcement is used only on large commercial chimneys; it is not necessary for the average-sized or small-sized house chimney.

MORTAR FOR THE CHIMNEY

The same type of mortar which is used to lay the masonry units of a structure can be used to set the flue liners. Standard masonry cement (Type N) is generally accepted for house chimneys. The average house chimney normally does not necessitate the need for a special high-strength mortar. Whether the masonry unit is a brick or concrete block, the same mortar serves equally well. If extra strength is desired, Type M or S (portland cement and lime mortar) can be used. It is important to use solid mortar joints at all times in chimney construction.

UNITS FOR THE CHIMNEY

Any hard, fired masonry unit can be used to build a chimney. However, bricks are selected most of the time. Old or used soft bricks should not be used for the outside face of the chimney or near the hot part of the flue as they may not be able to withstand the dampness or heat.

Concrete masonry units may also be used for chimneys. They may be the solid type or a specially manufactured chimney block with or without a section of flue lining built into the block. If the block does not come with a flue lining, the lining can be inserted as the chimney is built by the same method used on brick chimneys. Generally, the same type of mortar used for building brick chimneys can be used on concrete masonry units.

INSTALLING THE CLEANOUT DOOR

After the chimney has been laid out and construction has started, the mason must determine how to deal with soot from the chimney. All chimneys require periodical cleaning. For easy access to the chimney, cleanout doors are installed near the bottom of the chimney, Figure 24-2.

Cleanout doors should be installed over the third course on a brick structure above grade or the first course on a block structure. They are always located above the level of the finished grade, Figure 24-3. They should never be installed lower than the first

Fig. 24-2 Metal cleanout for soot removal in chimney

Fig. 24-3 Cleanout door properly installed above grade

course on a brick building. When they are this low, there may be a problem since soot builds up rapidly at this point and moisture may prematurely deteriorate the metal door. Cleanout doors should always be installed tightly so that they do not interfere with the draft of the chimney. Inserting a wire or nail in the holes on the flange of the door frame and into the mortar joint helps to hold the door in place while it is being constructed. Cast-iron cleanouts are recommended since they last longer than the more lightweight metals.

THE FLUE LINING

The lining in the chimney serves two purposes. The lining prevents the extreme heat from burning into the masonry work. Also, it adds to the efficiency of the chimney, since its smooth surface prevents soot from clinging to it as readily as to masonry units. Flues may be square or round in design. Round flues are more effective than the rectangular type, as gases tend to travel up the flue in a spiral motion. The corners of rectangular flues contain dead spaces. Therefore, a rectangular flue must be larger to obtain the same results.

Nothing is gained by using a flue lining which is larger than the size the manufacturer of the heating system recommends for the chimney. In fact, when a flue lining of the correct capacity is used, the chimney performs better and is more economical and safe. Using linings larger than necessary is a common mistake in the building trades. The 8" x 12" flue lining is recommended for use with the standard heating unit.

Flue linings which curve in the chimney are as efficient as straight linings provided that there are no abrupt turns or completely flat places. The curvature must be smooth, causing no interference with the passage of air or gases leaving the chimney. Nothing should project within the chimney or block the flue. Any offsets of the flue should not exceed 30° from a vertical position. Mortar projecting from the bed joint between the flues should be scraped off cleanly and all holes filled to prevent the heat from penetrating into the chimney walls.

BEGINNING INSTALLATION

The first section of flue lining should be installed at least 1′ below the point at which the thimble is to be inserted. The first section should be located approximately 2′ from the point at which the joists will rest on the wall. Since the flue liner is 2′ high, only one section is installed below the joist level.

A hole must be cut into the flue lining to receive the terra-cotta thimble, or *flue ring*, in order to make a fireproof connection. The stovepipe or furnace pipe fits into the thimble. This is done by first filling and packing the flue lining with sand to prevent it from cracking while the hole is being cut, Figure 24-4.

A circle is drawn on the flue lining exactly where the hole is to be cut. A small hole is cut in the center with a point chisel. Then a small, light brick hammer or tile hammer is used to gently tap the hole through following the line drawn on the flue. The

Fig. 24-4 Packing flue lining tightly with sand prior to cutting

trick is not to cut too much with any one tap. Figure 24-5 shows a mason cutting the hole in the flue lining.

Care should be taken to see that the flue ring does not project any further inside the chimney than the inside face of the flue lining or the draft may be affected. Installing the thimble requires good workmanship since it must fit snugly in the flue. It is mortared in to prevent air leakage.

Sometimes, flue linings are started by driving a nail into the chimney wall where the flue starts and setting the flue on the nail. This should never be done on any chimney. The nails could become dislodged, causing the flue to settle in the chimney, thereby shearing off the thimble and cracking the joints. The proper method is to project a header course at the height of the bottom of the flue lining flush with the inside of the flue and rest the flue liner on the headers, Figure 24-6.

INSTALLING WOOD FRAMING AROUND CHIMNEYS

All wood or other burnable materials should be kept at least 2″ from chimneys to prevent any chance of fire. This space may be filled with some type of loose fireproof material to act as a fire stop. As a rule, carpenters on such jobs box the opening around the chimney with framing lumber to maintain the 2″ distance.

Fig. 24-5 Cutting hole for thimble with small tile hammer

INSTALLING FLASHING AT THE ROOF

The point at which the chimney passes through the roof must be weatherproofed by the insertion of some type of flashing and counter flashing into the chimney. *Flashing* is a piece of metal such as lead, copper, or aluminum which covers the point at which the chimney and roof meet to prevent the penetration of moisture, Figure 24-7. If the mason does not install the flashing, he or she must be sure to rake out the bed joint sufficiently so that the carpenter can install it at a later time. The recommended method is to mortar the flashing into the chimney rather than to rake out the mortar bed joint, followed by installation of the flashing.

Note: If the joint must be raked out, it should be to a depth of at least 1/2″.

CAPPING THE CHIMNEY

The top of all chimneys should be at least 2′ to 3′ above the ridge or peak of the roof to assure a good draft in the flue. The presence of large hills or taller buildings nearby may necessitate building higher than this to prevent downdrafts.

The top or cap of the chimney may be finished in different ways. Various courses of bricks may be projected and recessed to accomplish different patterns and designs. Precast concrete or stone may also be used for caps.

Years ago, the trend was to make the cap of the chimney one of the important architectural features of the home. Very intricate designs were used to accomplish this. The trend now is to keep the cap as simple as possible. The fewer projections which are on the chimney cap, the less chance of deterioration there is, since snow and water which collect on caps cause the bricks to eventually loosen.

The most frequent method of capping a chimney is to apply a wash of mortar on an angle by approximately 45°, Figure 24-8. The wash is applied from the face of the last course laid to the side of the flue lining with the trowel. A wash coat serves two purposes, to direct air currents upward and to prevent moisture from eroding the chimney top. Regular mortar (Type N) should not be used to apply a wash coat as it is not durable enough for such a project. Instead, a richer portland cement mortar, such as Type M, should be used. The flue lining should project 6″ above the last course of masonry units laid on the chimney.

CHECKING FOR A GOOD DRAFT

When the chimney is completed, the mason should check to make sure that the chimney creates a good draft, Figure 24-9. A simple method of doing this is

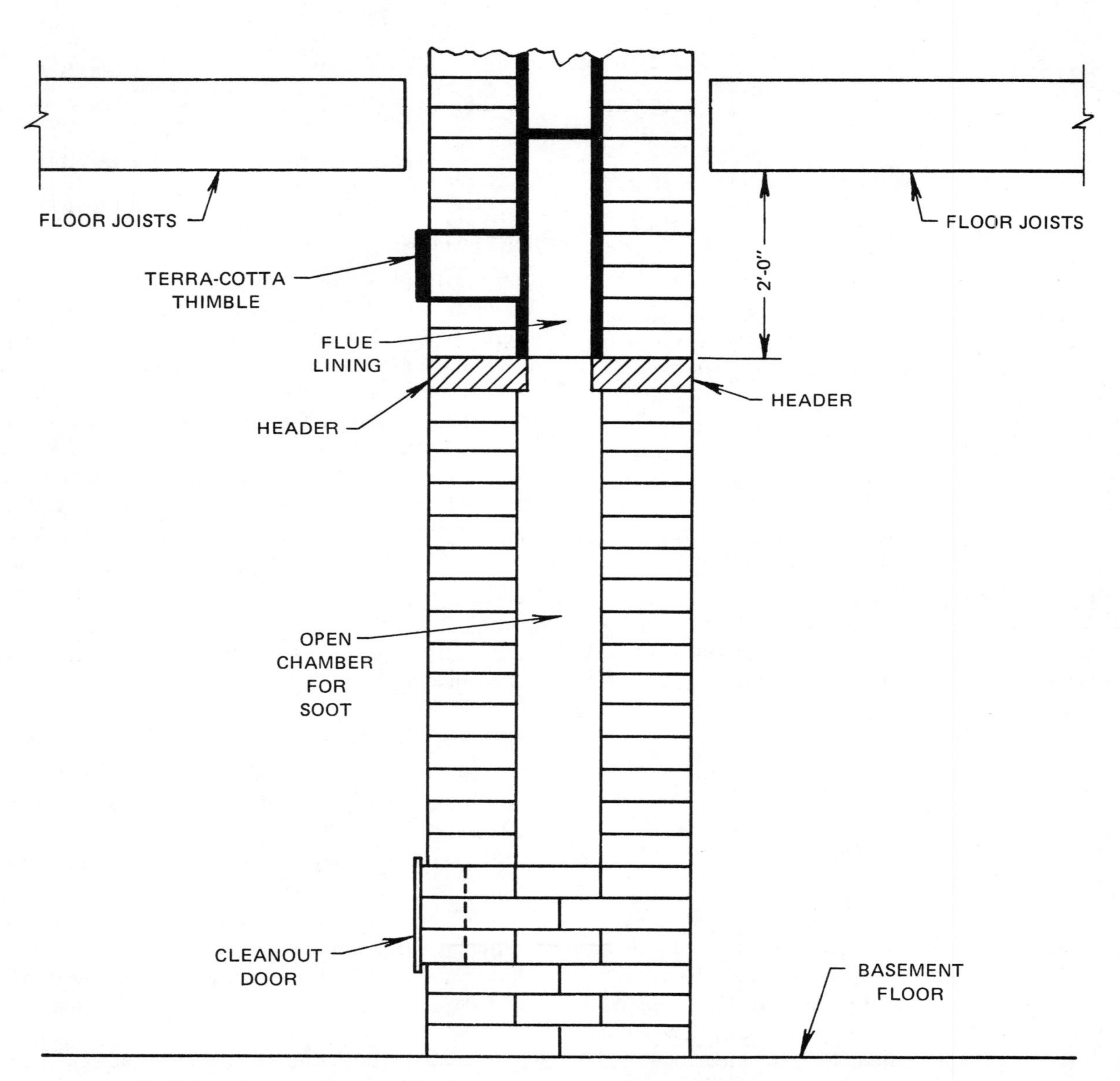

Fig. 24-6 Cross section of chimney showing how the flue lining rests on brick headers

to roll a piece of paper into a long cylindrical shape and light it with a match. Hold the lighted paper near the flue ring in the chimney. The flame should draw into the ring strongly. It may take a short time for this to happen, as the air in the flue will have to warm slightly. If the flame and smoke do not draw into the ring, it will be necessary either to extend the chimney top or to check for an obstruction in the chimney.

THE IMPORTANCE OF GOOD WORKMANSHIP

Care must be taken to lay bricks and block level and plumb on chimneys. Because of the height and width of the chimney, mistakes are especially noticeable. All mortar joints must be solid and uniform in size. Any work built near wood framing of structures must be free of the wood to prevent the possibility of

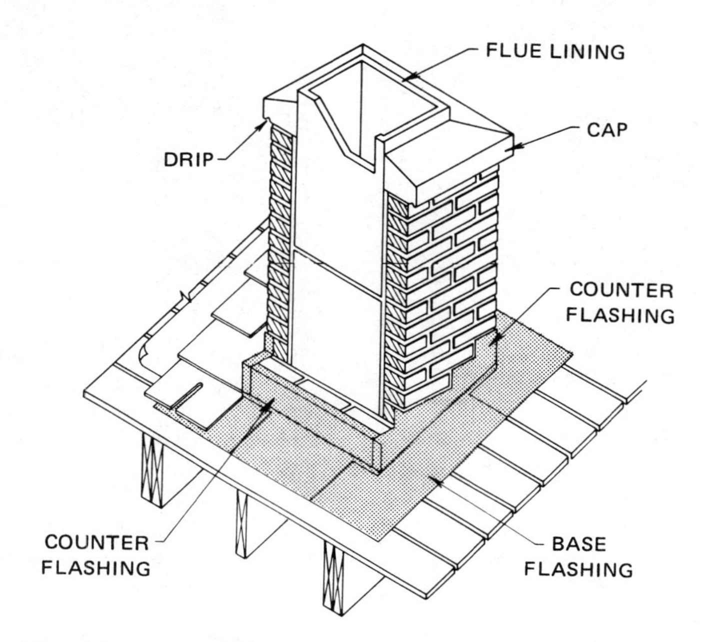

Fig. 24-7 View of chimney showing installation of flashing

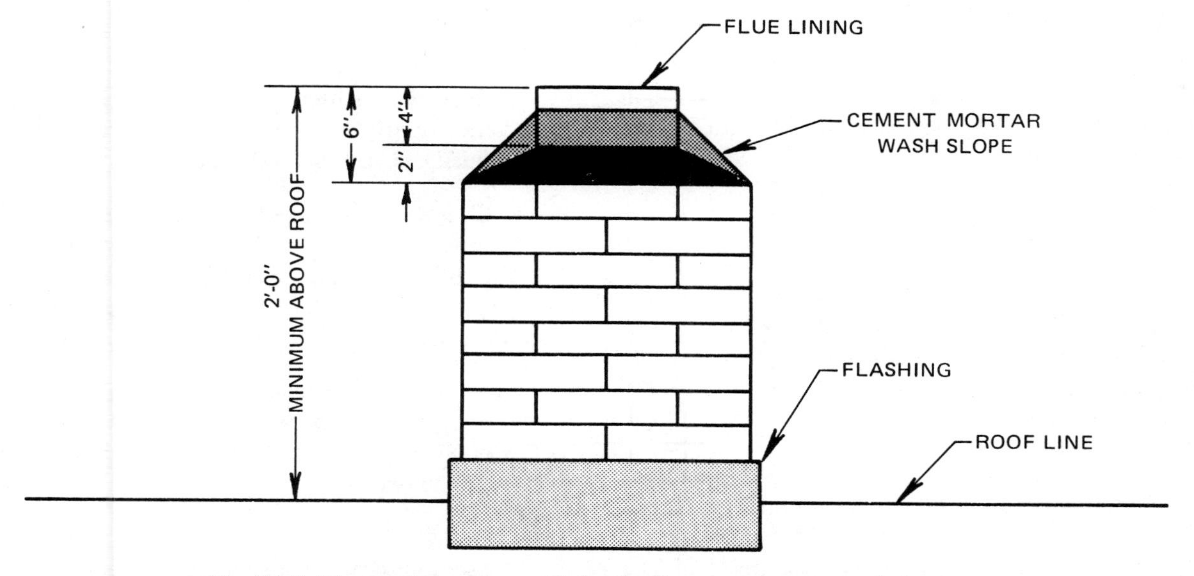

Fig. 24-8 The mortar wash coat helps to direct currents toward the top of the flue and prevents water from entering the chimney top.

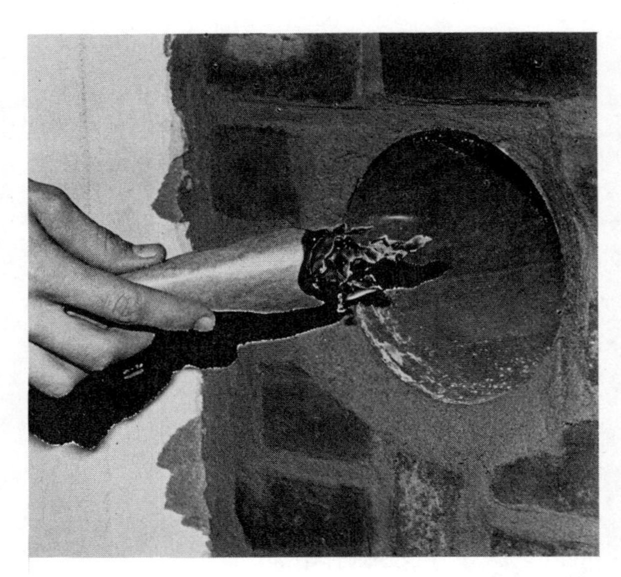

Fig. 24-9 Checking chimney draft with a lighted piece of paper. Flame and smoke should be drawn strongly into the thimble and up the chimney.

Fig. 24-10 A wood stove installed in a one-flue chimney. The brick hearth and wall make this a very safe job. The metal hood causes heat to rise to the floor through the grate.

fire. Tooling and striking of the mortar joints should be done neatly. Excessive smearing of mortar on the face of the work must be avoided.

Flue lining should be cut with a masonry saw, if possible, or packed (reinforced) with sand and cut neatly with a chisel. Broken sections of flue lining should never be used in the chimney.

All materials used in the chimney should be free from dirt and foreign matter. Scaffolding should be removed as soon as possible after the chimney is built to prevent dirt from splashing onto the surface of the chimney.

Note: The chimney is a true test of the mason's skill because the major portion of it is done with the use of the plumb rule.

WOOD STOVE INSTALLATION

The one-flue chimney has been in great demand in the last several years due to the high cost of oil, natural gas, and electric heat. The principal use of one-flue chimneys is to service wood-burning stoves, Figure 24-10. Wood-burning stoves are as safe as any other heating system if sensible safety rules are followed by the homeowner.

Building one-flue chimneys for wood stoves has become a valuable source of work for masons. Therefore, they must be familiar with some basic rules of installation. It is the mason's responsibility to build a fireproof chimney in the home and to make recommendations to the homeowner for safe installation of the stove. Following are some of the more important rules to observe when hooking up a wood stove to a one-flue chimney.

- It is necessary to have a building permit and inspection to build a chimney and install a wood stove. This is governed by local authorities and codes.
- All masonry chimneys with walls less than 8″ thick should have a flue lining. Some areas of the country require a flue lining regardless of thickness. Check local codes.
- Cleanout doors should be provided to remove soot. They should be located approximately 2′ below the inlet (thimble). The door will last longer if it is cast iron. Cleanout doors are available at building trades supply houses.
- An exterior chimney may touch the house siding or sheathing if 8″ of solid masonry is provided on the face between the flue lining and house.
- Masonry chimneys with flue linings built on the exterior of the house must be placed a minimum of 1″ from any combustible materials.

- The size of the thimble (connector) being installed should not be smaller than the stove flue collar as heat may build up at the point where the connector passes through the house wall.
- The terra-cotta thimble, when installed, should be surrounded by no less than 8" of solid brickwork or fireproof materials. Check local codes for the specific measurement in your area.
- The thimble should be installed flush with the inside face of the flue lining and mortared in with fire clay or high strength mortar to prevent burning out.
- The thimble should have an elevated pitch where it meets the chimney. The general rule is that there should be a horizontal position of not less than 1/4" to the linear foot, so that the chimney connection point is higher than the stove connection point.
- Flue linings are available in either modular, nonmodular, or both sizes depending upon the availability in certain areas of the country.

  Modular flue lining is manufactured on multiples of the 4" grid (module) which is the common unit of measurment used in the construction industry. Nonmodular flue lining is not manufactured to the 4" grid and is a carryover from the days when the modular system was not used as a construction industry standard. Today there are still several companies that make the nonmodular-size flues. An example of a nonmodular-size flue lining is 9" x 13", and a modular-size flue lining of approximately the same size is the 8" x 12" flue.

  When determining what size flue lining to use in a chimney for a stove, the cross-sectional area of the flue lining should be approximately 25% more than the cross-sectional area of the thimble (connector). This will promote good draft and will decrease the creosote buildup in the flue. This rule will work for open-face or close-face stoves.

  For example, a stove is being connected to a chimney in which the stove outlet where the stove pipe fits on is 6" in diameter. Multiplying 6" x 6" equals 36". Therefore, an 8" x 8" flue lining is needed because 8" x 8" equals 64". Subtracting 36" from 64" equals a total of 28" which is easily more than the 25% required for good draft. There is nothing to be gained by installing a flue lining that is much more than the 25%. This will only cause most of the heat to go up the chimney and be wasted. However, the closest possible flue lining size that will work out should be selected. Many times it is not possible to match the size perfectly. When this happens, the next larger size should be used. The chimney should also be built higher than the standard 2' above the peak of the roof. It is good practice when building a chimney to look over the area and take note of other houses. If they have a higher than normal chimney or extension, then you will need one also.

  Modular and nonmodular flue linings are available in a wide variety of sizes to fit building needs. Check with local building supply dealers in your area to find out what is available.
- The chimney should be tied to the building with some type of approved wall ties as stated in local building codes.
- The top of the chimney should be a minimum of 2' above the peak of the roof for a good draft.
- Solid masonry walls cannot burn but will conduct heat, so combustible materials should be kept out of contact with masonry walls. It is recommended to set the stove on firebrick or a nonburning masonry base at least 18" from the masonry or noncombustible wall. Stovepipe should always be a minimum of 24" from the ceiling.
- The masonry base or hearth should extend a minimum of 12" on each side of the stove and 16" in front of the stove to protect the floor from hot embers. If the area is carpeted, allow 16" around the stove.
- There is a simple test to see if there is enough clearance for a hot stove. Place your hand on the surface closest to the stove. If you can keep your hand there comfortably, the location probably is acceptable.
- Before hooking up a stove on existing masonry chimneys with flue linings, a visual inspection should be performed to be sure that the chimney is not blocked. A strong light held in the chimney area or a large mirror reflected down the chimney from the top usually provides enough light for a check.

A smoke test indicates the workability of an existing chimney. This is performed by lighting a small, smokey fire or setting off a smoke bomb in a connected stove or fireplace, and then partially closing off the top of the chimney. If any leaks or cracks are in the chimney, the mason will see smoke coming out of them. Repairs can be made based on the results of the test. Stains on masonry or loose mortar joints must also be repaired. An old chimney should be inspected carefully before ever attempting to install a wood stove.

ACHIEVEMENT REVIEW

The column on the left contains a statement associated with the construction of chimneys. The column on the right lists terms. Select the correct term from the right-hand list and match it with the proper statement on the left.

1. Raw material used to make flue linings
2. Passage of air through the chimney
3. Piece of metal such as lead, tin, or copper which is laid in the bed joint of the chimney at the roof height to prevent moisture penetration
4. Mortar applied to the top of a chimney at a 45° angle to prevent moisture from entering the chimney
5. Fireproof terra-cotta flue ring in which furnace pipe fits
6. Object installed in the chimney through which soot is removed
7. Safe depth in the ground at which the footing is poured
8. Distance from house (exterior) to chimney if flue lining is used in chimney
9. Distance from masonry wall to back of wood stove inside house.
10. Height of chimney above peak of roof

a. Cleanout door
b. Wash
c. Frost line
d. Draft
e. Flue lining
f. Flashing
g. Thimble
h. Fire clay
i. 2′
j. Hood
k. 1″
l. 18″

PROJECT 18: BUILDING A ONE-FLUE CHIMNEY

OBJECTIVE

- The student will lay out and build a one-flue chimney of bricks. The student will also install various parts of the chimney, such as cleanout door, thimble, and flue lining.

  Note: Chimneys such as this are built to service furnaces and small heating plants.

EQUIPMENT, TOOLS, AND SUPPLIES

Mortar pan or board
Mixing tools
Mason's trowel
Brick hammer
Plumb rule
Striking tool (convex or V-jointer)
2′ square and pencil
Modular rule
Brush
Point chisel
1 cleanout door
1 terra-cotta thimble
2 sections of 8″ x 8″ x 24″ flue lining (Because of the fragility of terra-cotta flue liners, a wooden model may be used in a shop situation.)

The student will estimate bricks and mortar from the plan. (For training purposes, it is suggested that only lime and sand mortar be used for the work at the top of the chimney, not portland cement.)

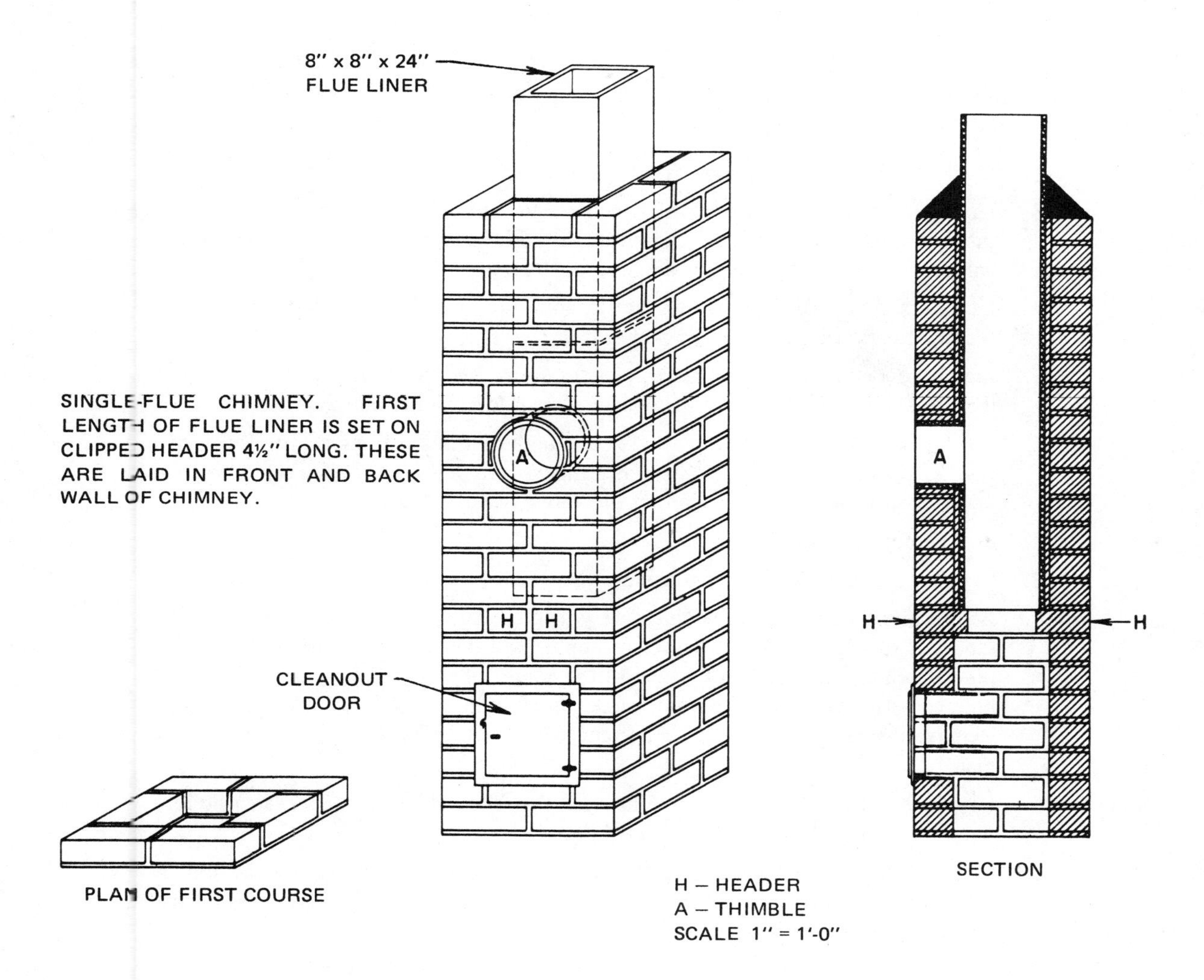

SUGGESTIONS

- Due to variance in bricks and flue liners, it is possible that the flue liner will not fit into the chimney as shown. In this event, increase the size of the chimney by one-half brick in each direction. On some flue linings, the 8" x 8" dimension is taken from the inside rather than the outside. This would require extremely large head joints in the chimney to accommodate the flue liners inside the brickwork. These points must be determined when the chimney is laid out.

- The space between the liners and brickwork should not be filled with mortar or droppings. Only enough mortar should be used to form a good joint and to hold the liners in a level and plumb position. The proper air space allows the flue to expand without cracking the brickwork.

- Use well-filled mortar joints at all times.

- Be careful when handling flue linings as they have sharp edges.
- If flue linings must be cut, do not do the cutting around other workers because the cutting process can result in flying chips. Wear eye protection whenever cutting flue linings.

PROCEDURE

1. Mix the mortar and stock the work area with the needed materials.
2. Dry bond the first course to obtain the correct spacing.
3. Lay the first course in mortar. Level and square all sides. Be sure to check the inside space to determine if flue lining will fit, allowing room for expansion. Bricks will be laid to the number 6 on the modular rule.
4. Lay the second course, installing the cleanout door as shown. Anchor the frame to the brickwork using a metal clip nail or tie.
5. Lay 5 more courses, walling in the cleanout door. Recheck the door to be sure that it is level and plumb. Be sure that each course is level and plumb as it is laid.
6. Cut 4 long bats, each measuring 4 1/2" in length. Lay these in mortar on the eighth course, allowing the rough end to project inside the flue space. The projecting edge provides a bearing on which the first flue lining rests. (See plan.)
7. Strike joints as needed.
8. Lay 2 or 3 additional courses to anchor the header course.
9. Before installing the first flue liner, carry it to the sand pile and pack it tightly with sand. Mark the point at which the thimble will be cut in the flue. With a steel-pointed chisel, carefully cut a hole in the center of the marked area. Enlarge the hole with the head of the brick hammer by chipping until the thimble fits neatly. Remove the sand from the flue liner and install it in the chimney as shown on the plan.
10. Lay the brickwork, installing the thimble at the required height. (See plan.) Solidly point up around the thimble and flue lining.
11. Continue laying brickwork until the top of the lining is reached. Set the second section of the flue liner, being sure it is level and plumb. Cut off any projecting mortar on the inside of the flue lining.
12. Build the chimney to the height shown on the plans. Strike remaining joints and brush the work.
13. Apply a wash coat of mortar on a slope. It should be located on top of the last course against the flue liner. Smooth the work with the trowel.

Unit 25 Expansion Joints, Intersecting Walls, and the Use of Rules

OBJECTIVES

After studying this unit, the student will be able to

- mark courses of various masonry units with the modular rule and the spacing rule.
- explain the term *hog in the wall* and what steps must be taken to correct it.
- describe an expansion joint and explain how it is installed in a masonry wall.
- lay out and construct a concrete block intersecting wall.

THE MASON'S RULES

All masonry units in a wall must be laid to a predetermined, specified height. Close attention must be given to the height of such items as windows, doors, beam pockets, and joist seats. Since mortar joints and unit size may vary, the mason uses two different rules to be certain that mortar joints are divided equally under any circumstances. The two rules with which all masons must acquaint themselves are the *spacing rule* and the *modular rule.*

Building a wall to the same height on both ends of the building, but ending with 1 more course on one end than the other, is a very costly error. With the correct starting point (bench mark) established, proper use of the mason's rules and story pole, and adherence to rules of construction, the mason can be assured that the work will be constructed to the correct level.

The Modular Rule

The modular rule, Figure 25-1, has a standard 72″ marked on one side. The reverse side of the rule has various numbers to match the different sizes of masonry units which are manufactured under the modular manufacturing system. These masonry units may be reviewed by referring to the section in Unit 1 on modular bricks.

It is important to remember that all modular units are bonded together at a 16″ increment. This height may be expressed in terms of a multiple of 16″, such as 32″, 48″, or 64″. Examine the various scales on the rule in Figure 25-1. The last scale, scale 2, accommodates either a concrete block or a masonry unit which equals 2 courses reaching 16″ in height. Concrete block is usually laid off in increments of 8″ instead of a set number on the scale.

Reading the modular rule requires no learning of new techniques. All that is necessary is a basic understanding of how the modular system coordinates with building materials and how masonry units fit into the 4″ modular grid.

The Spacing Rule

The spacing rule was designed before the modular system was adopted. Its purpose at that time was to divide mortar joints evenly in brickwork. It was also used to lay out courses of units since frames and doors at that time were not designed and manufactured in a standard size. The spacing rule is still used today in gauging mortar joints that are not modular.

One side of the spacing rule is marked off in increments of 1/16″ and is read as an ordinary 6′ rule. The other side of the spacing rule is marked off in several groups of numbers from 1 to 0, Figure 25-2. (The 0 actually represents 10.) As the mortar bed joint increases in size, so does the number on the rule representing that course of bricks. The smaller the number, the tighter or thinner the mortar joint is. For example,

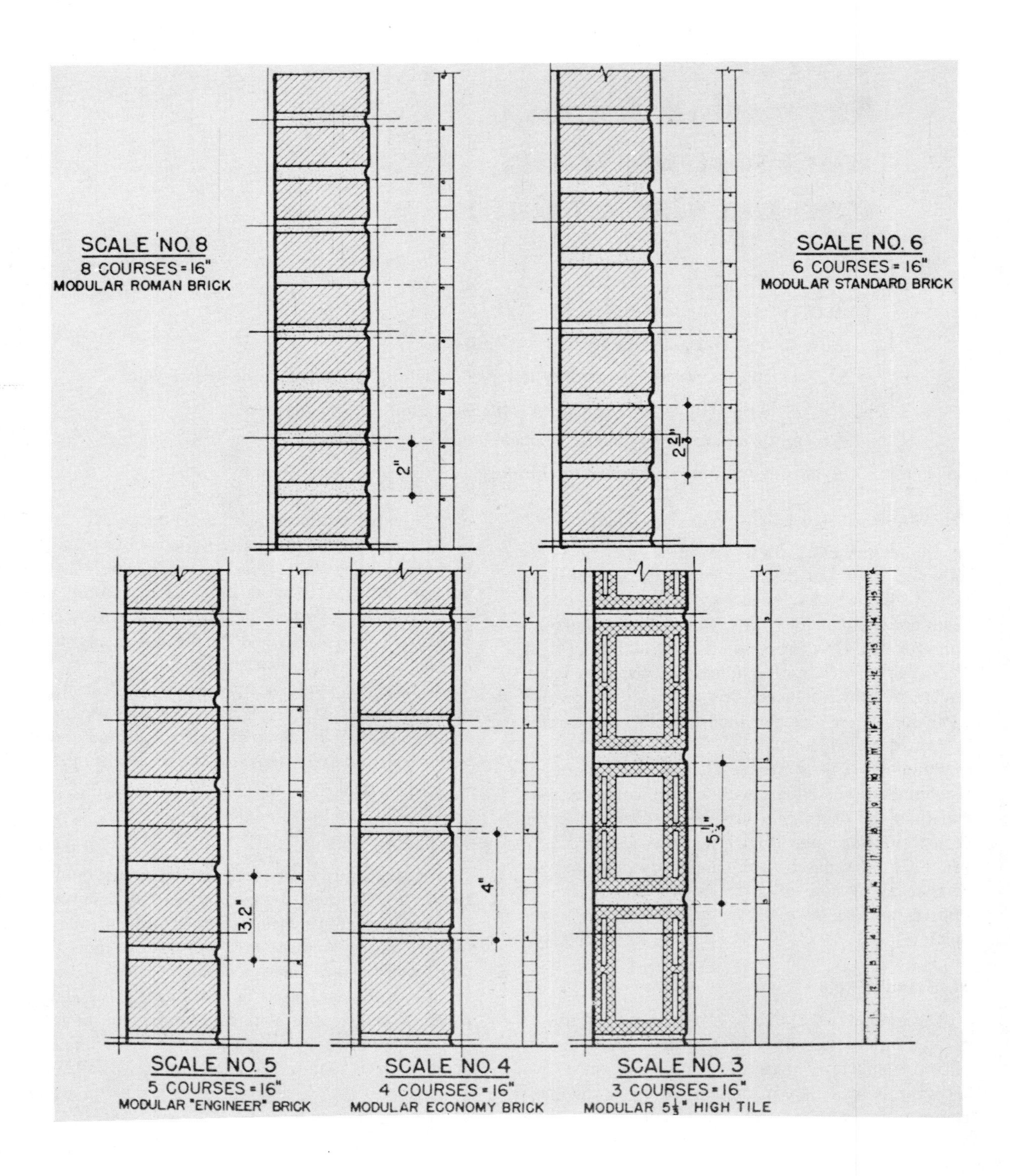

Fig. 25-1 Various scales of the modular rule applied to masonry units based on the modular system of measurement

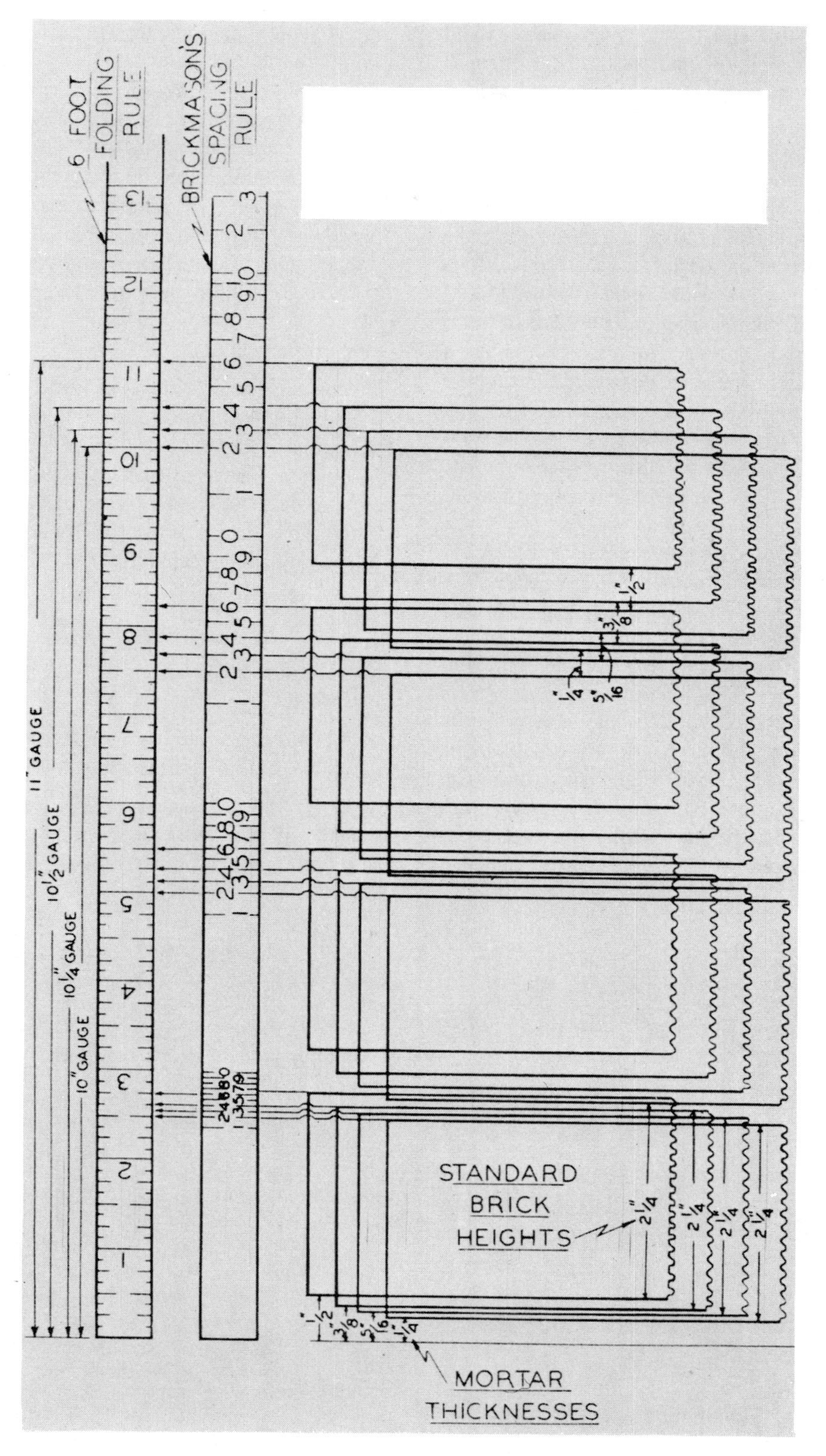

Fig. 25-2 The spacing rule as compared with regular 6-foot rule. The larger the numbers on the spacing rule are, the bigger the mortar joint being formed is.

number 6 on the rule equals 4 courses of bricks and 4 mortar joints which will measure 11" in height. Number 7 on the spacing rule represents 4 courses of bricks laid to 11 1/4" in height. By using various numbers, the mortar joint size can be increased or decreased as needed.

Another important feature of the spacing rule is that the total number of courses shown on the scale side from the bottom of the rule to the top is shown in red. The height of any wall can be checked and, at the same time, the number of courses needed to reach the specific height is known. For example, to lay a standard brick wall to the height of 71 1/2" using the number 6 on the spacing rule, 26 courses of bricks including the mortar joint are required. Do not confuse number 6 on the spacing rule with number 6 on the modular rule, as they represent entirely different measurements.

Compare application of the two rules in this example involving installation of a brick windowsill. The bricks are laid in a rowlock position, but the spacing is figured the same as if the course were in a vertical position. The brick sill is to be laid for a window 36" long.

Checking the distance with the modular rule, the space coincides with the number 4, which will not fit into the modular grid. On the other hand, using the spacing rule, the number 6 is only 1/4" from coinciding perfectly with the 36" mark. Therefore, the number 6 on the spacing rule should be used to gauge the distance, gaining the 1/4" in the mortar joints. Compare the two rules in Figure 25-3, in which 36" is marked.

The mason will seldom use a scale larger that the number 7 on the spacing rule since the thickness of the mortar joint would be too great in most cases. By the same token, the number 4 on the rule generally forms the tightest joint possible. Regardless of which rule is being used, always read from the top of one brick to the top of the next brick.

Hog in the Wall

The term *hog in the wall* is a trade term which indicates that two opposite corners or leads have reached the same height but do not contain the same number of courses, Figure 25-4. This usually occurs when the story pole being used does not have each course numbered from the bottom of the pole to the top. A hog in the wall can be caused by the following:

- A footing which is not level
- A very long wall between doors or windows
- Structures in which the wall is built over a series of concrete footers
- One mason laying a thicker bed joint under the bricks than another mason is using

Correcting this situation is extremely difficult, especially if the wall has been partially constructed. Sometimes it can be corrected by increasing the mor-

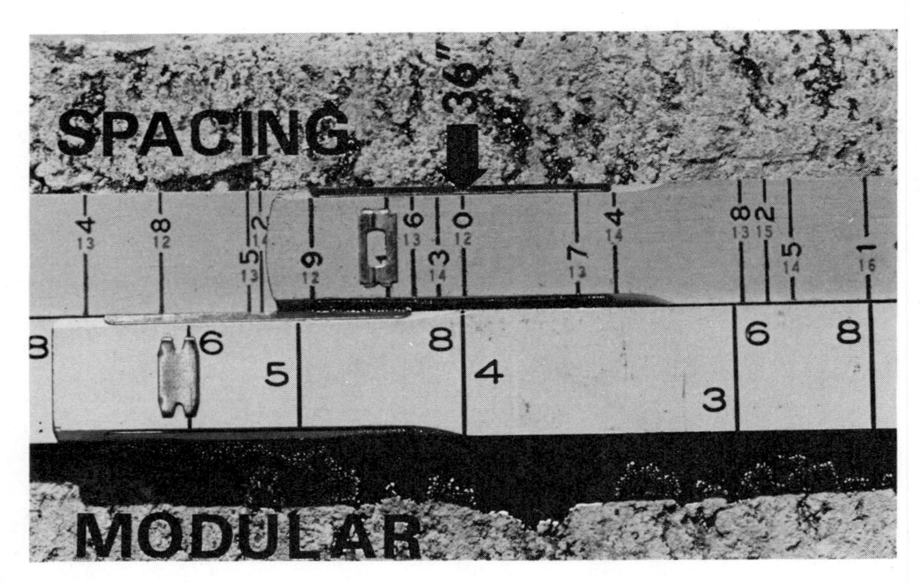

Fig. 25-3 Comparing the spacing and modular rules. The arrow indicates the 36-inch measurement on both rules. Notice the difference in scale of the two rules.

tar joint thickness at one end and decreasing the joint thickness at the opposite end of the wall. This action equalizes the number of courses but usually can be detected in the wall, as there are bigger joints on one end than on the other. This is not considered good building practice and should be avoided. Usually, the best method is to tear down the portion of the wall containing the hog and rebuild it correctly. This is a costly procedure since the work must be torn down and rebuilt, and there are no materials which can be reused. It also delays completion of the job.

Note: It is recommended that all measurements be checked with a story pole or gauge rod from the bench mark when a job is started, and that the correct number of courses are laid to the specified height when the corner is completed.

A hog in the wall usually is associated with the laying of bricks rather than concrete block because of the size of the units. It is possible, however, to have a hog in a concrete block wall if the footings are extremely unlevel.

EXPANSION JOINTS

Structures are constantly moving. The movement may be caused by settling of the building. However, it is more often caused by expansion and contraction of the building materials resulting from temperature and moisture changes. The movement may be small, but if it is not prevented or allowances are not made, it will likely result in cracking of masonry walls. For example, a brick wall 100' long expands or contracts approximately 7/16" for every 100° of change in temperature, according to the BIA.

To allow for this movement, expansion joints are built into the masonry wall at selected points. An *expansion joint* is a continuous vertical joint built into

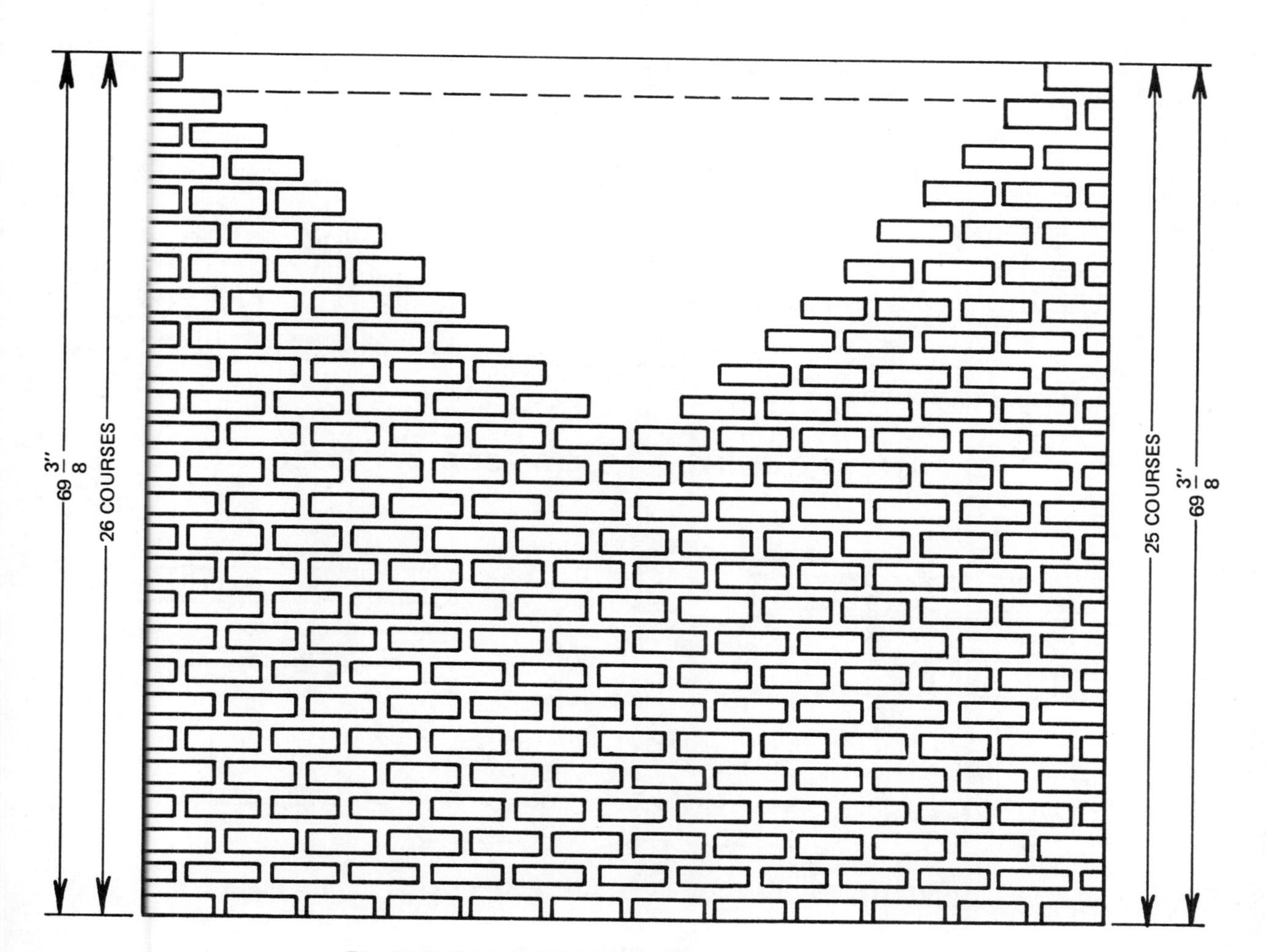

Fig. 25-4 Brick wall demonstrating a "hog in the wall"

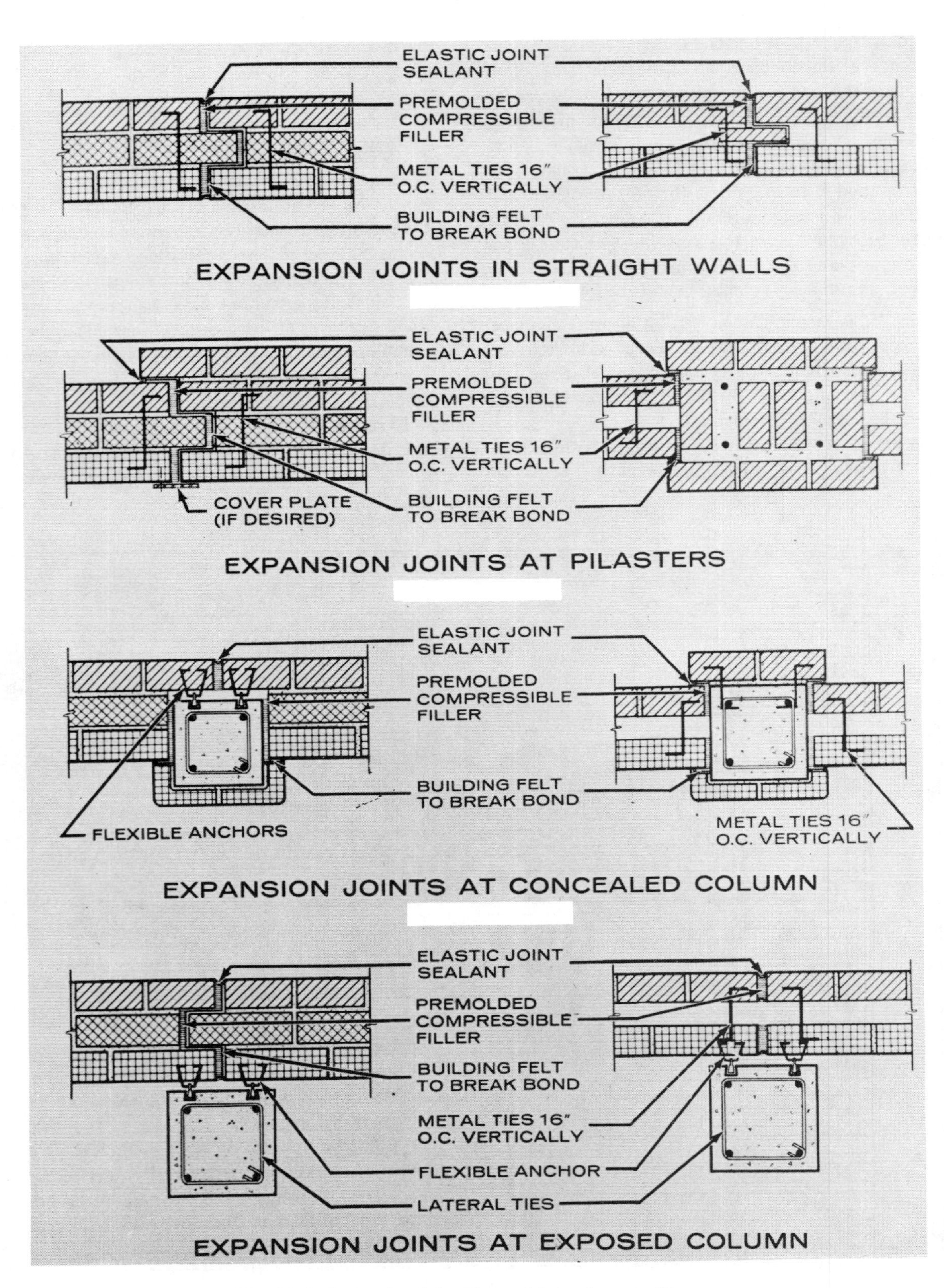

Fig. 25-5 Expansion joints located in various positions

the masonry wall in places where the forces acting on the wall cause pressure, possibly creating cracks or breaks in the wall. The expansion joint works to absorb the shifting without being noticeable. It is important that the full width of the expansion joint be kept consistent throughout its entire height.

The skill of the mason plays a very important role in the construction of an expansion joint. Expansion joints must be installed correctly to function correctly. If the joint is clogged with mortar or contains foreign matter, the joint most likely will not function properly. Expansion joints may be installed in various places, Figure 25-5.

Forming the Expansion Joint

Expansion joints may be formed in a number of ways. There is an excellent rubber control joint, Figure 25-6, which is sometimes installed in the slot on the end of window sash blocks. This makes a very neat job. These joints, recommended by many construction engineers, allow for slight movement of the structure. Neoprene compound flanges act as a seal to the outside edge of the block, making caulking unnecessary. This also enhances the appearance of the wall.

Copper water stops have been used successfully for many years. They consist of short pieces of copper sheets which overlap at the joints. Premolded compressible elastic (rubber or plastic) fillers may be inserted in back of the copper to absorb the stress caused by movement of the wall. Fiberboard and similar materials are not suitable for expansion joints because they do not return to their original size after being compressed.

So that the expansion joint is as unnoticeable as possible, care should be taken by the mason to see that it is plumb and the same width or thickness as a normal mortar joint. If the joint is exposed, it should be filled in neatly with a caulking compound unless premolded rubber is used and it is flush with the surface of the wall.

INTERSECTING WALLS

On many jobs, there is a need to build intersecting walls which branch off the main wall at a 90° angle. This can be accomplished in two different ways. The intersecting wall may be tied into the main wall by bonding the masonry unit into a hole which has been left in the main wall, Figure 25-7. Another method is to butt one wall against the other and use metal ties to bond the two walls together.

The first method mentioned, which makes use of a physical tie, was used for many years as the only method of tying walls together. In this method, 1

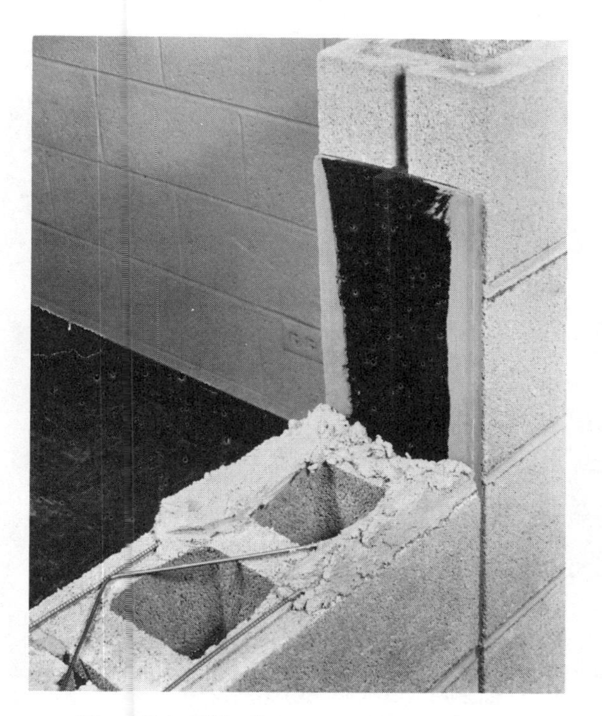

Fig. 25-6 Wide flange rapid control joint

Fig. 25-7 Hole left in concrete block wall to receive an intersecting wall. The tie used is known as a physical tie.

masonry unit is usually left out at the point at which the walls are to be tied together. At a later time, the intersecting wall is built. Every other course ties into the various holes to strengthen the wall.

If there is a settlement or movement between the walls, a crack may develop at the point of intersection or the intersecting masonry unit may break completely. Because of this, the other method mentioned was developed. In this method, the intersecting wall is butted against the main wall with a mortar joint at the point of contact. This joint is usually called a *stack joint.* The mortar is then raked approximately 3/8" and filled with a caulking compound until it is flush with the unit. When a crack occurs, all that is necessary is recaulking of the joint.

Intersecting walls which carry the load of floors and roofs of buildings must also be able to stretch or flex. The wall, however, must be tied together very securely for lateral support. Heavy metal tie bars are a very good method of tying these walls together, Figure 25-8. The tie bars are usually heavy-gauge steel which is 1/4" thick, 1 1/2" wide, and approximately 28" long. They usually have 2" right-angle bends on each end to lock into the masonry unit. The end of the tie bar should be filled completely with mortar before laying the next course over it. Tie bars should be spaced vertically every 24".

Metal reinforcement wire also can be used to tie intersecting walls. The best type is the reinforcement wire that is made especially for this purpose. It is welded in the shape of a T and is the same diameter as standard reinforcing wire, Figure 25-9.

Intersecting walls which do not bear loads, such as partitions, can be tied together with strips of metal lath or hardware cloth, Figure 25-10. This is 1/4" wire mesh which is galvanized and is placed across the joint between the two walls. When one wall is built first, the hardware cloth is allowed to extend at least 8". Metal lath or hardware cloth should be placed in the wall every other course for a secure tie. When using hardware cloth, it is recommended that a caulking joint be applied against the point at which the two walls meet.

Caution: Always fold the hardware cloth down flat against the wall to avoid the possibility of someone being injured by the wire.

Later the exposed portion of the wire is tied onto the mortar joint of the wall being built (intersecting wall) and mortared in securely. Wire mesh can be cut with ordinary snippers.

All tying methods discussed in this unit are effective when used properly and may be used with any type of masonry unit. At no time should an intersecting wall be built without some type of tie. All ties should be mortared in securely.

Fig. 25-8 Section of metal tie bar used to tie a concrete block bearing wall to the main wall

Fig. 25-9 T-shaped metal reinforcement wire used to tie intersecting concrete block walls together

Fig. 25-10 Intersecting block walls tied together by use of metal lath. Notice that the mortar joint has been raked out to receive caulking compound.

ACHIEVEMENT REVIEW

Select the best answer from the choices offered to complete each statement. List your choice by letter identification.

1. The term *hog in the wall* describes
   a. a wall which is not level.
   b. a wall which has a bulge in the brickwork.
   c. a wall which is level but contains 1 more course on one end than the other.
   d. a wall which is very poorly built and on which the mortar joints are not struck properly.

2. Scale number 8 on the modular rule represents
   a. standard brick courses.
   b. tile courses.
   c. concrete block courses.
   d. Roman brick courses.

3. All courses of the various masonry materials found on the modular rule tie together every
   a. 8″.
   b. 12″.
   c. 16″.
   d. 24″.

4. The mason's spacing rule differs from the modular rule in that it
   a. has more listings of masonry units on the rule.
   b. allows the adjustment of standard brick by the application of a larger mortar joint or a thinner mortar joint.
   c. is much easier for the mason to use and understand.
   d. is used only to mark off doors and windows.

5. The spacing rule offers the added advantage of
   a. larger numbers, which make the rule easier to read.
   b. wood construction, so that it does not rust or deteriorate.
   c. courses numbered in red from the bottom of the rule to the top.
   d. the fact that it will fold and fit into the mason's pocket.

6. The major reason that the expansion joint does not function correctly is that
   a. masonry debris sometimes remains in the joint.
   b. a copper water stop is selected instead of a rubber strip for filler material.
   c. it is not installed at the proper location in the wall.
   d. it is not built perfectly plumb with the wall.

7. Fiberboard is a poor choice as a filler for an expansion joint because
   a. it is not thick enough to fill the joint.
   b. when it compresses, it will not return to its original size.
   c. it absorbs moisture and resists movement of the wall.
   d. it is very costly and difficult to install in the wall.

8. When an intersecting bearing wall is being built, the strongest type of metal wall tie to install is
   a. the reinforced masonry T wire.
   b. hardware cloth.
   c. the veneer tie.
   d. the T bar.

9. When building intersecting walls that specify a physical tie, it is recommended that
   a. the walls be tied together by interlocking the masonry units of the intersecting wall into the main wall.
   b. all mortar joints which butt against the main wall be pointed with a high-strength mortar.
   c. the mortar joint against the main wall be raked and filled with caulking compound.
   d. no joint be formed where the intersecting wall and main wall meet.

10. When raking out a mortar joint at the intersection of a wall to receive caulking, the correct depth is
    a. 1/8".
    b. 3/8".
    c. 3/4".
    d. 1".

PROJECT 19: BUILDING AN 8" X 8" X 16" CONCRETE BLOCK WALL WITH A ROWLOCK BRICK WINDOWSILL

OBJECTIVE

- The student will be able to lay an 8" x 8" x 16" concrete block wall 5 courses high and install a brick rowlock sill. The student will use the modular rule when constructing the sill.

  Note: Windowsills such as this are typical of windowsills on wooden or metal windows in a concrete block garage.

EQUIPMENT, TOOLS, AND SUPPLIES

1 mortar pan or board
Mixing tools
Mason's trowel
Brick hammer
Ball of nylon line
2 line blocks or line pin and nail
Plumb rule
Convex striker
Flat slicker jointer
Modular rule
Chalk box
Brick set chisel
Brush
Supply of clean water
1 piece of wire mesh or hardware cloth approximately 48" long and 4 1/2" wide

The student will estimate the required number of bricks, block, lime (masonry cement), and sand according to the plans.

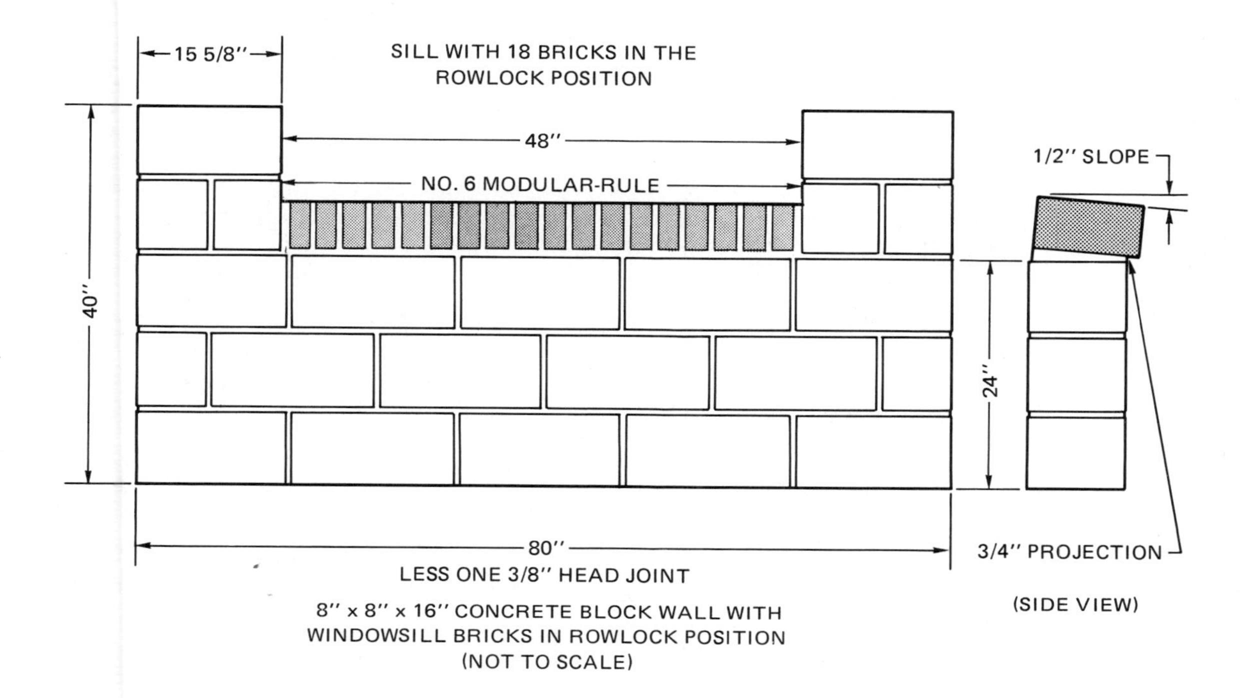

8" x 8" x 16" CONCRETE BLOCK WALL WITH WINDOWSILL BRICKS IN ROWLOCK POSITION (NOT TO SCALE)

SUGGESTIONS

- The concrete block wall should be completed according to the plans before any of the sill bricks are laid.
- Holes in the cells of the concrete block directly underneath the point at which the sill is to be laid should be filled with masonry scraps; or wire mesh should be laid across the holes to assure a solid joint under the sill.
- Select dry bricks for the sill, as they set more quickly.
- Select bricks of approximately the same height, width, and length for the sill.
- Spot check the brick sill periodically with the modular rule as it is laid to be sure that no brick pieces will have to be used.
- Level the back of the brick sill every 3 bricks to be sure that it is level.
- The last brick laid in the sill will require a buttered mortar joint on each side of the brick and the adjoining bricks. Be sure to allow 1 extra joint for the closure brick when the sill is checked with the modular rule.

PROCEDURE

1. Stock the necessary materials in the work area and mix the mortar.
2. Strike a chalk line and lay out the first course as shown on the plans.
3. Lay the concrete block wall to a height of 8" per course or to number 2 on the modular rule.
4. Make all concrete half-block with the brick set if manufactured halves are not available. Wear eye protection.
5. Lay the concrete block wall up to 24" or 3 courses high. Lay out the brick sill with the rule according to the plan.

6. Using square-ended block for the jamb, continue building the block wall to the height shown on the plan.
7. Strike all block work with the convex striker and brush before installing the sill.
8. Fill in any holes in the block underneath the sill with masonry scraps, hardware cloth, and mortar.
9. Butter a mortar joint on the side of a brick and lay it in the rowlock position against the jamb of the opening. Slope the brick to the front of the wall a minimum of 1/2" to drain water. Project the bottom of the sill bricks 3/4" from the outside face of the wall. Set a brick in the rowlock position on the opposite end of the sill in the same way.
10. Attach a line to the wall with the aid of a line block or a line pin and nail about 1" above the sill far enough back so it will not fall out. Lay a dry half-brick on each of the preset sill bricks, pinning the line to the face of the brick sill.
11. Lay the brick sill, working from both ends to the center. Check with the modular rule as often as necessary to be sure the sill will accommodate whole brick.
12. Form a mortar joint on both sides of the closure brick and the adjoining brick to prevent penetration of moisture and to form a sound joint. Check the back of the sill with the level for proper alignment and parge to ensure a watertight joint.
13. Strike vertical head joints on the front of the sill bricks with a convex jointer. Strike joints on the top of the sill with a flat slicker jointer. (The flat joint helps to prevent water from remaining on the sill.) Brush the work.
14. Recheck the sill to be sure the frame is level before having the work inspected.

PROJECT 20: TYING AN INTERSECTING 8" X 8" X 16" CONCRETE BLOCK WALL AND A 6" x 8" x 16" CONCRETE BLOCK WALL WITH WIRE MESH EVERY 2 COURSES HIGH

OBJECTIVE

- The student will be able to lay out and build an 8" x 8" x 16" block wall and lay a 6" x 8" x 16" stub wall on a 90° angle off the main wall. The student will rake out the mortar joint in the intersecting wall angle and tie the wall with wire mesh every 2 courses.

  Note: Walls such as this are found in many office buildings where partitions which are not load bearing are used.

EQUIPMENT, TOOLS, AND SUPPLIES

1 mortar pan or board
Mixing tools
Mason's trowel
Brick hammer
Brick set chisel
Convex striker and brush
Plumb rule
Chalk box
2' square
Tin snips
Modular rule
Short section of nylon line
2 line blocks
Supply of clean water
Length of wire mesh

The student will estimate the number of concrete block and mortar materials from the given plan.

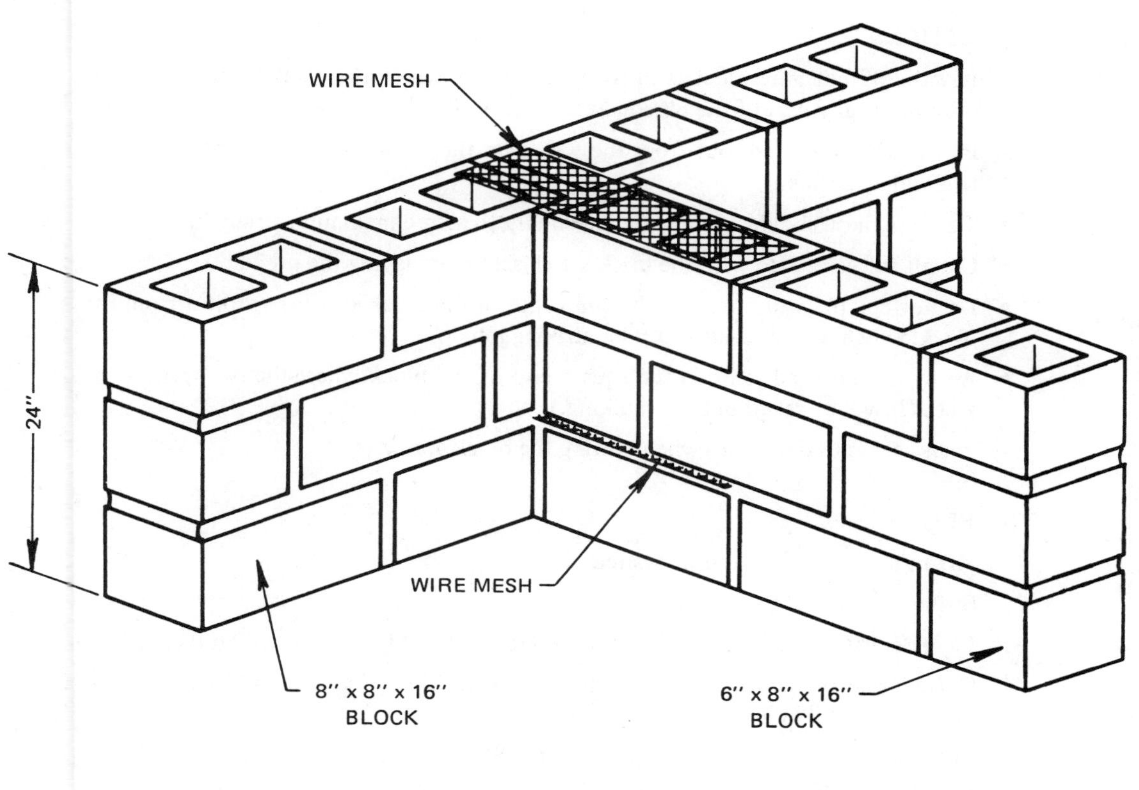
WIRE MESH
24"
WIRE MESH
8" x 8" x 16"
BLOCK
6" x 8" x 16"
BLOCK

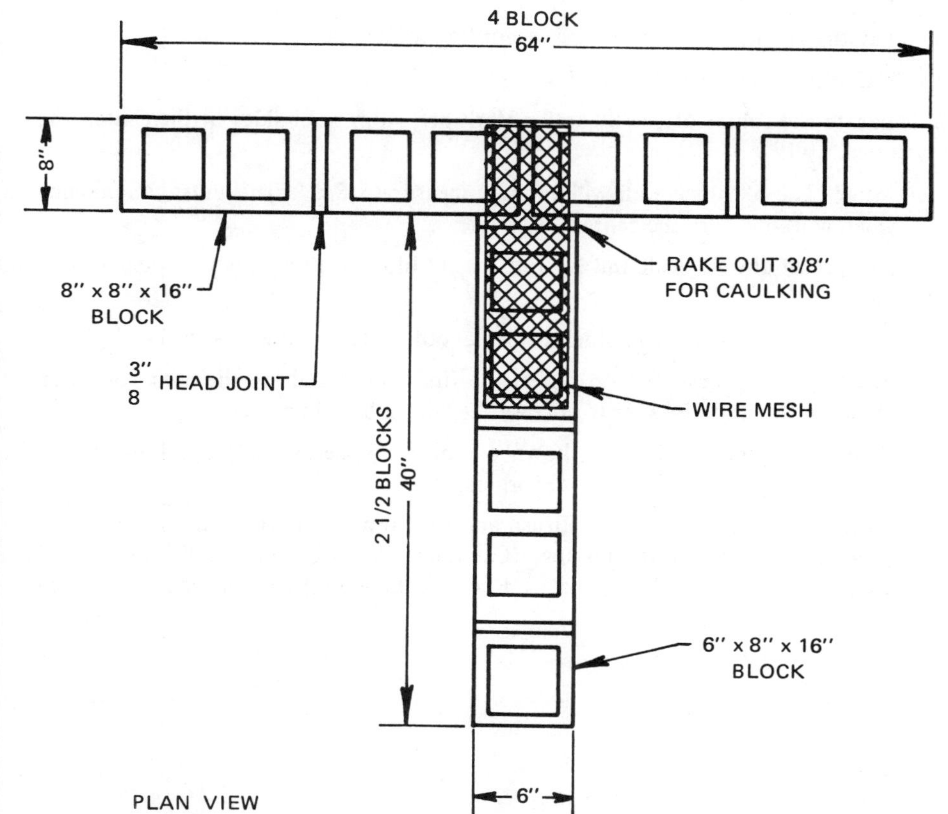
4 BLOCK
64"
8"
8" x 8" x 16"
BLOCK
3/8" HEAD JOINT
RAKE OUT 3/8"
FOR CAULKING
WIRE MESH
2 1/2 BLOCKS
40"
6" x 8" x 16"
BLOCK
6"
PLAN VIEW

SUGGESTIONS

- Lay the 8″ concrete block wall to the height shown on the plans before laying out or building any part of the 6″ concrete block wall.
- Project the wire mesh out from the wall over the second course and bend it down to avoid an accident.
- The wire mesh should be mortared in solidly to assure maximum strength.
- Cut all halves needed with the brick set chisel before laying the project.
- Do not cut the wire mesh to the full width of the block as it may project from the block, which would result in a poor striking job.
- Apply a solid head joint to the square end of the block where the two walls intersect. (It will be raked out for caulking.)
- Practice good safety rules when cutting the block and wire.

PROCEDURE

1. Mix mortar and stock the work area.
2. Strike a chalk line for the project.
3. Lay the first course of 8″ block, using a line and line block as shown on the plan.
4. Cut a piece of wire mesh, being sure that it ties into the 8″ wall at least 4″ and into the 6″ wall at least 8″.
5. Cut the halves needed for the project. (2, 8″ block halves and 3, 6″ block halves will be required.)
6. Lay the second and third courses according to the plan.
7. Strike and brush the 8″ wall. Use the convex jointer.
8. Locate the center of the 8″ wall. Mark off, from both sides of the center, the 6″ block wall.

   Note: The 6″ block wall will actually measure 5 5/8″. This must be remembered when working from the centerline.
9. Square off the 8″ wall and lay out the 6″ block stub wall with a pencil or chalk line.
10. Lay the first course of 6″ block to the layout lines. Be sure it is level and plumb.
11. Bend the wire mesh flat on top of the first course and lay the next course of 6″ block on the wall. Be sure to enclose the wire solidly in mortar.
12. Continue laying the wall as shown on the plans, periodically checking the height and maintaining plumb and level corners.
13. Strike the joints in the wall. Brush and rake out the mortar joint in the angle to a depth of 3/8″ to receive caulking. (Caulking will not be done on this project as it is usually the work of the carpenter.) Recheck the work before having it inspected.

Unit 26
Installing Anchor Bolts, Brick Corbeling, and Wall Copings

OBJECTIVES

After studying this unit, the student will be able to

- set an anchor bolt to receive a wood plate or a steel beam in a masonry wall.
- describe the process of corbeling bricks for a masonry wall.
- list the different types of copings used on masonry walls.

The art of masonry is not confined to building corners and laying units to the line. There are many fine details and arrangements of bricks and concrete block that apprentices will learn in the course of their daily work. Many of these detailed practices may be defined as basic masonry construction, such as installing anchor bolts, brick corbeling, and wall copings.

SETTING ANCHOR BOLTS

Anchor bolts are used to fasten a member of a structure, such as a wooden plate, to the structure itself. Sometimes the carpenter installs the anchor bolts, but it is usually the job of the mason. Nailing wood plates to a wall is not an acceptable method of attaching it to a structure. An anchor bolt must be

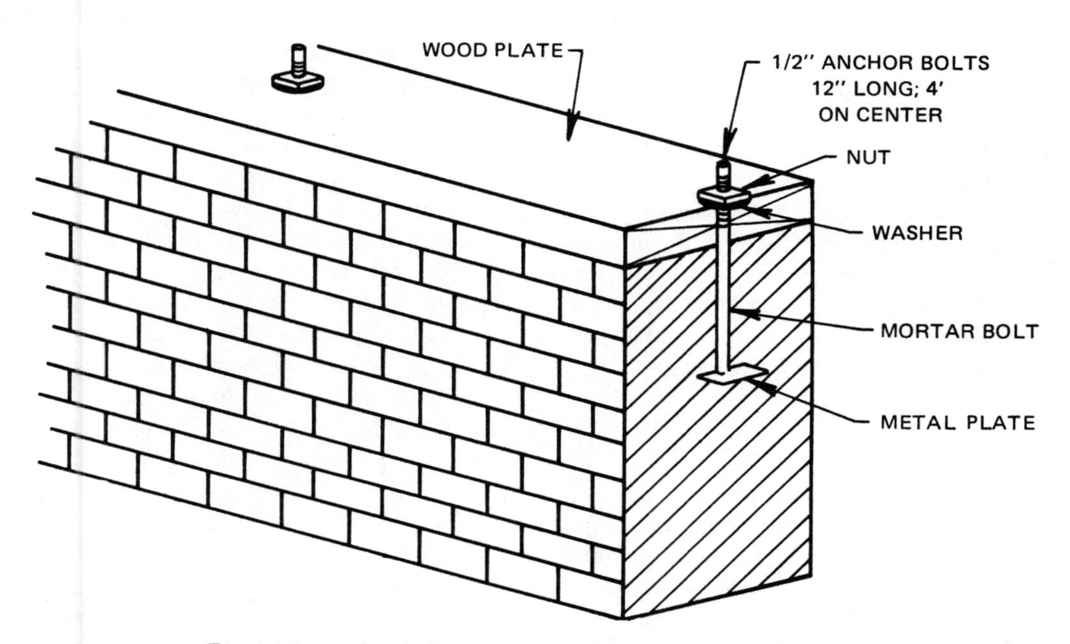

Fig. 26-1 Anchor bolts are used to secure plate to brick walls.

used. After anchor bolts are installed, they should be mortared into place so that they will not loosen when the nut is tightened against the plate.

Anchor bolts are available in different sizes to suit the requirements of specific jobs. To hold down the wood plate on an average-sized job, a bolt 1/2" in diameter and 12" long should be used. An L-shaped bend on the bottom of the bolt helps to hold the bolt in the mortar, Figure 26-1. Bolts should be spaced according to plans or drawings.

The bolt should normally be set in the center of the wall unless otherwise specified. The position of the top of the bolt should be determined before the wall is completed, since it may be necessary to cut a notch out of the bricks to allow the L-shaped section of the bolt to fit snugly in the brickwork. The carpenter on the job should be consulted for the correct height of the bolts as it may vary from job to job. If more than 1 bolt is to be set in the wall, be sure they are all spaced the same distance from the outside face of the wall. This can be done easily by using a line.

Form a base under the bolt either with mortar or scraps of bricks or block to ensure a solid base. Fill mortar around the bolt, ramming it tight with a wooden stick or trowel blade to be sure a solid joint is formed. The bolt should be installed plumb as it will be easier for the carpenter to fit the wood plate over the bolt. Try to keep as much mortar as possible off the threaded part of the bolt where the nut will be installed. Figure 26-2 shows a bolt set in a masonry wall.

The mason should never stuff empty cement bags into the hollow concrete block cell on the last course and then fill around the bolt with mortar. This is poor workmanship and may eventually cause the bolt to loosen in the wall.

Installing bolts which anchor steel beams is very critical as the holes are predrilled in the steel, so there is no room for error. In most cases, the carpenter or building engineer is on hand to help the mason install them. The responsibility for correct placement of the bolts for steel erection is usually the responsibility of the carpenter.

BRICK CORBELING

Corbeling is the projection of masonry units to form a shelf or ledge, Figure 26-3. When corbeling, the mason must especially follow good bonding practices and use well-filled mortar joints to ensure that the projection will remain stable. Corbeling can be strictly decorative or used to thicken a wall to suit a particular job condition or requirement.

Fig. 26-2 Bolt installed in masonry wall for garage door jamb

In years past, the tops of many brick buildings were identified by the masonry trim work designed by corbeling bricks in intricate patterns. This type of fancy masonry work was called *gingerbread work*. This type of work is now very costly due to high labor rates. The trend now is to simplify work to keep the costs down as much as possible. However, on some public buildings and restoration work, this type of work is still done. Inner walls inside chimneys and fireplaces still use corbeling to a great degree.

Headers are used extensively in corbeling as they lend the wall greater strength by tying further into the wall than they extend beyond it. The maximum projection should not exceed one-half the unit's height or one-third the bed height.

As a rule, the total projection of the corbeling should not extend more than one-half the thickness of the wall. If metal supports or angle irons are used, the figure may be adjusted.

Bricks should be dry and free from chips or cracks as most edges are exposed to view. A dry brick sets more quickly which eases the operation for the mason. The bricks must be level and plumb. Well-filled mortar joints give maximum strength and reduce the amount of striking and jointing required. Corbeling should not be done so quickly that the mortar and bricks cannot take a firm initial set and remain in position.

Masonry work that is projected past the face of the wall is more susceptible to moisture penetration than regular walls. Once the mortar begins to deteriorate, it proceeds at a very rapid rate. Masonry work, as a rule, should last at least 40 years before any repair needs to be considered.

Fig. 26-3 Corbeling bricks to form a decorative cornice. Notice that on the corbeling, all bricks are in a rowlock position to give it greater strength.

Excessive speed in this type of work may defeat the purpose of the job and result in shoddy work. For this reason, the contractor estimating a job which includes corbeling work usually allows extra time and money for it.

COPINGS

Coping is a material used to cap or cover a masonry wall to prevent moisture penetration. Coping is available in many shapes and forms, depending on the desired effect and the materials with which it must blend on the job. Coping can be an interesting architectural high point of a building.

Types of Coping

A course of bricks in the rowlock position can be classified as coping if it projects from both sides of the wall. Limestone coping has a very attractive appearance and sets fairly quickly. White Indiana limestone, when used to trim the top of walls, adds a variation of color and texture to a brick structure, Figure 26-4. Solid concrete block are sometimes used as coping but do not add much in the way of design or beauty to the building.

Terra-cotta tile, Figure 26-5, is used when economy is important. Each section of tile has a projecting bell-shaped edge on one end which fits over the end of the tile laid next to it. The tile is very hard and, therefore, resists the effects of weather very well. Builders of large buildings such as plants and industries many times select this type of coping since it is quick to install and is long lasting. A piece of burlap bag can be used to clean the terra-cotta.

Long sections of walls which are equipped with coping should have some room for expansion. Pliable filler materials can be inserted in the head joints of the coping and caulked much as a vertical expansion joint is installed in a masonry wall.

Regardless of the type of coping selected, it should be laid in mortar as other masonry is laid. Lay a piece of coping in mortar on each end of the wall, being sure that it is level and plumb. Attach a line and lay the remainder of the coping to the line. Fill any holes that may have formed under the projecting edge of the coping.

Installing Coping

Coping that is improperly designed or set may result in moisture damage to the building. Always use a full mortar joint under the coping. A good practice is to caulk the mortar joint which is exposed to the weather with some type of material that expands and contracts with changes in temperature. Flashing can also be used to be sure that no moisture penetrates by laying the flashing on the wall and then bedding the mortar over the flashing, Figure 26-6.

Fig. 26-4 White Indiana limestone coping on a brick building. The coping projects past the ends of the structure so that water may drop off. Otherwise a leak in the coping could allow the water to run onto the masonry work, possibly resulting in efflorescence and deterioration.

Fig. 26-5 Terra-cotta coping before installation. Each section shown is eventually cut on the seam with a sharp mason's hammer to form two pieces.

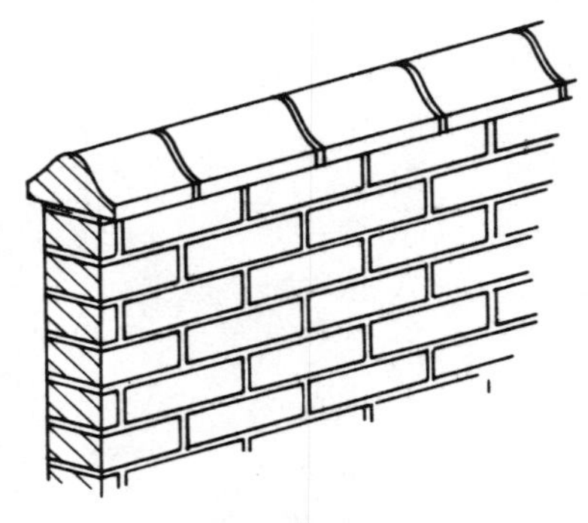

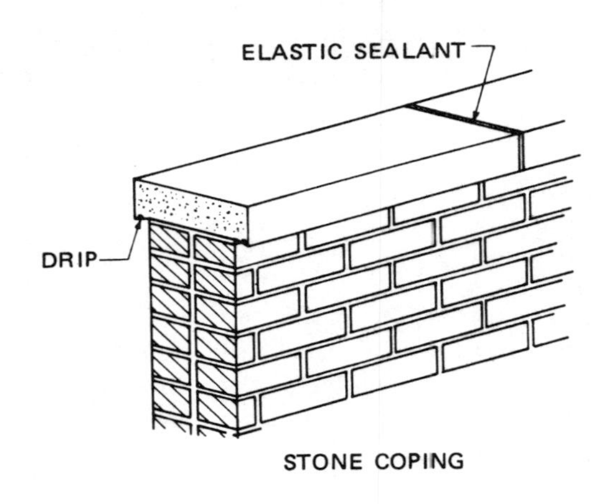

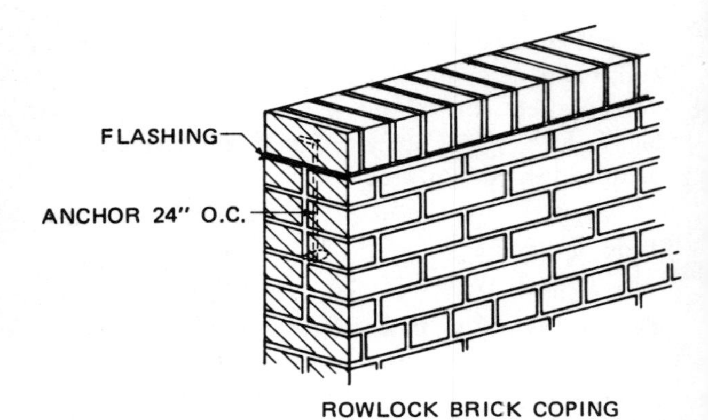

Fig. 26-6 Typical copings

ACHIEVEMENT REVIEW

Use basic math to answer the following questions. Show all of your work.

1. A foundation measuring 60′ x 20′ is to be built. Anchor bolts are to be spaced 4′ apart to anchor the wooden plates. How many bolts are needed? (Add the lengths and widths of the foundation for a total distance.)

2. A decorative corbeling ledge is planned for a building. The wall is 12″ thick before any corbeling is done. The contract for the job specifies that the corbeling total 6′. If each corbeling course projects 3/4″, how many courses of corbeling are required to reach the total projection of 6″.

3. Limestone coping is to be laid on a wall measuring 110′ in length. If each section of coping is 2′ long, including the mortar joint, how many sections of coping are required for the wall?

4. Coping projects over walls on both sides to protect them from moisture penetration. If a wall measures 12″ in width, how wide must the coping be to project 2″ on each side of the wall?

5. A large industrial plant is being built. Plans include limestone coping installed on top of the walls. The building measures 300′ x 100′. The specifications require that there be an expansion joint every 50′ in the coping. How many expansion joints are required?

PROJECT 21: CORBELING 3 COURSES ON A 12″ WALL

OBJECTIVE

- The student will be able to build a 12″ brick wall, corbel a ledge, and return to the original wall line to complete the project.

  Note: This type of corbeling is typical of decorative work at the top of some buildings.

EQUIPMENT, TOOLS, AND SUPPLIES

1 mortar pan or board
Mixing tools
Mason's trowel
Brick hammer
Plumb rule and 2′ level
Convex striker
12″ square
Chalk box and pencil
Brick set chisel
Mason's modular rule
Brush
Slicker striker
Ball of nylon line
Nail and line pin
Supply of clean water
1, 8″ x 8″ x 16″ concrete block on which to set level

The student will estimate the number of bricks needed from the given plans.

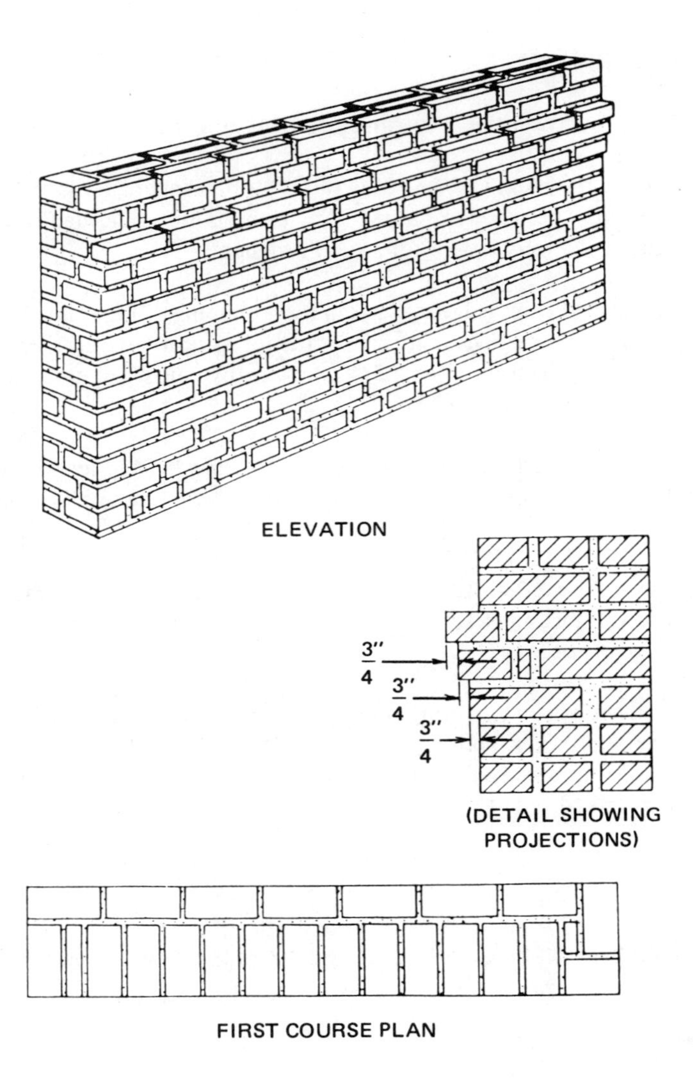

SUGGESTIONS

- Strike the chalk line a little longer than is actually needed.
- The plan shows a 2" piece to be laid against the header course as a starter piece. If this is too difficult, cut a 6" or three-quarter brick piece to use in place of the 2" piece and the header next to it.
- Use solid joints between all bricks, being especially careful on the header courses.
- Strike all the joints at the start of the corbeling since the mortar joints under the corbeling may harden too much to strike later.
- In all projecting courses, the bottom of each course should be checked to be sure it is level since it is very noticeable.
- Wear eye protection when cutting any material.
- Select straight, unchipped bricks for the corbeling.

PROCEDURE

1. Space the mortar pan or board and bricks approximately 2′ from the project.
2. Mix the mortar and fill the pan.

3. Strike a chalk line for the front line of the project. Dry bond the project.
4. Mark off the returns with the square and pencil. Lay a brick in mortar on each end, being sure it is level and plumb with the layout line.
5. Attach a line and lay the first course as shown on the drawing. All brick courses will be laid to number 6 on the modular rule.
6. Lay a short lead on each end of the project. Continue to fasten the line and lay bricks until the corbeling height is reached (9 courses from base). Strike the work with a convex striking tool.
7. The first corbeling course is a header course projecting 3/4" from the face of the wall. Lay the corner bricks, attach the line, and fill in the course. Be sure to level the bottom of the course as well as the top. See the detail of the projection on the plan.
8. Lay the second corbeling course. The second corbeling course is a stretcher which also projects 3/4" beyond the first corbeling course, or 1 1/2" beyond the original face of the wall. Back up the course with a header course.
9. Lay the third corbeling course as a stretcher projecting 3/4" beyond the second corbeling course, or a total projection of 2 1/4" beyond the original face of the wall.
10. Lay the last 2 courses by returning to the original wall line as shown on the plan.
11. Carefully fill all holes under the projecting bricks. Strike the mortar joints on the last projecting course with a flat slicker so that the wall can better resist water penetration. Brush and recheck all work before having the work inspected.

SUMMARY, SECTION 6

- Masonry units over openings are supported by lintels. The most common types of lintels are the steel angle iron and the reinforced concrete block.
- Lintels should always be set level and plumb over any opening.
- Extreme care should be taken when lifting heavy lintels or injury may occur. When in doubt as to whether the weight can be lifted safely, get help.
- A pier is a free-standing vertical mass of masonry and is used mainly to carry beams or support parts of a structure.
- Piers differ from columns in that columns are usually higher in proportion to their thickness.
- Pilasters are similar to piers but are tied to the wall of the structure. They can be used to carry beams or for decorative purposes.
- Steel-bearing plates are set in a high-strength mortar over walls and piers to help distribute the weight of beams.
- Chases should be carefully designed so that they do not reduce the strength of the wall.
- Solid masonry bearings are important at the point where joists or concrete rests on the masonry work. The purpose of a bearing is to distribute the weight of a structure on the wall underneath more evenly.
- The proper tie should be installed when installing doors and window frames.
- Chimneys carry smoke and fumes safely away from buildings.
- Chimneys should have flue lining installed in the smoke chamber to prevent the heat from burning through to the masonry work and to form a more efficient chimney.

- The mason should be careful about installing the proper-sized footing for a chimney since the weight is concentrated in a small area.
- Hard, burned brick should be selected for a brick chimney.
- The chimney should be built a minimum of 2′ above the roof to assure a good draft and discourage any downdraft from the roof.
- The flashing should be built securely into the chimney to avoid water penetration.
- Use the size recommended by the manufacturer when installing a furnace flue in a chimney.
- All wood or other burnable materials should be kept at least 2″ from the chimney for fire protection.
- Use portland cement mortar (Type M or S) for the wash coat on the chimney.
- The two rules used by the mason are the modular rule and the spacing rule.
- The spacing rule is used to evenly divide mortar joints when the modular scale or grid will not accommodate them.
- The trade term *hog in the wall* indicates that one end of a wall has 1 more course of bricks than the other, but both are built to the same height. This is usually caused by larger joints being formed on one end.
- A *hog in the wall* can usually be prevented by checking the corner carefully with a story rod from a level point or from the benchmark.
- Expansion joints are built into masonry work to permit some movement of the wall without cracking.
- The full width of the expansion joint must be kept clean of mortar and foreign matter if the joint is to function properly.
- Workmanship is very important in the installation of expansion joints. If the joint is not perfectly plumb, it presents a very poor appearance.
- Intersecting walls can be tied by bonding the masonry units into each other or by using various types of metal ties.
- If a straight (stack) joint is being used against the intersecting wall, the mortar should be raked out to a depth of 3/8″ and filled with a pliable caulking material.
- Anchor bolts should be set plumb in the masonry wall and to the correct height specified with mortar surrounding them.
- It is important that well-filled mortar joints be used in corbeling and that the bottom of the course as well as the top is level.
- The total projection of work which is corbeled should not exceed more than one-half of the thickness of the wall.
- When installing coping on exceptionally long walls, the mason should allow for expansion joints to relieve the stress.

SUMMARY ACHIEVEMENT REVIEW, SECTION 6

Complete each of the following statements referring to materials found in Section 6.

1. The deciding factor in selecting a lintel for a particular job is ________________.
2. Steel lintel bearings for residential construction should be a minimum of ________________.

3. Concrete block lintels should bear a minimum of ______________ on each side of the opening.

4. The main difference between a pier and a colunm is ______________.

5. Steel-bearing plates should be bedded in ______________.

6. Pilasters differ from piers in that ______________.

7. A vertical or horizontal slot left in a masonry wall to provide space for wires or pipes is called a ______________.

8. The last 8" of masonry work in many basement walls is required to be brick or solid block. This is done to ______________.

9. Nailing blocks are installed in masonry walls for the purpose of ______________.

10. There are two major types of drafts in a chimney. They are ______________ and ______________ drafts.

11. In order to prevent burning out of masonry work, tile or terra-cotta ______________ is installed in chimneys.

12. When the carpenter is installing wood framing around a chimney, a space of ______________ is recommended.

13. The point where the chimney goes through the roof must be made watertight by installing ______________.

14. The wash coat which is applied to the top of a chimney to prevent water from seeping in is composed of ______________.

15. The mason uses two different types of rules in his work, the ______________ and the ______________.

16. If a wall is built which is level on each end but with one end containing 1 more course, it is said to have a (an) ______________.

17. To allow for movement in a building which is caused by temperature changes, the mason can install a (an) ______________.

18. When using wall ties to tie an intersecting wall and a main wall together, it is advisible to rake out the mortar joint in the place where the two walls meet and fill the space with ______________.

19. Wood plates located over masonry units are held securely in place by the use of ______________.

20. The ledge formed by the projection of 1 unit over the unit underneath it is called a (an) ______________.

21. The type of work mentioned in the previous question must not extend more than one-half ______________.

22. The capping material laid on masonry walls to prevent moisture penetration and prolong the life of the wall is called ______________.

23. Three major types of coping in use today are ______________, ______________, and ______________.

24. Coping is always wider than the masonry wall on which it rests because ______________.

Section 7
Scaffolding

Unit 27
Types of Scaffolding

OBJECTIVES

After studying this unit, the student will be able to

- erect a section of tubular steel sectional scaffold.
- discuss the advantages offered by adjustable scaffolding.
- explain how suspended scaffolding is erected and when and how it is used.

When masonry work has been built as high as the mason can comfortably reach, scaffolding must be erected to complete the work. *Scaffolds* are platforms which set the worker a certain distance off the ground.

For many years, wooden scaffolding was the only type of scaffolding available. Wooden scaffolding is not in common use today for several reasons. The high cost of lumber, the time required to erect it, and the superior strength of steel scaffolding have limited its use.

Today, almost all scaffolding is constructed of some form of steel. The most commonly used is the steel sectional scaffold. This type of scaffold is used on most residential work and on moderately sized buildings. The main reason that steel scaffolding is preferred over scaffolding of other materials, such as wood, is the safety factor. Steel does not decay or split, and the load capacity of steel scaffolding is extremely high. It is possible to overload a steel scaffold to the breaking point. However, the section which actually breaks is the wooden planking rather than the steel framework.

New designs in scaffolding have been developed to satisfy modern-day construction needs, speed the work process, and reduce strain on the mason.

This unit deals with the three major types of scaffolding used in masonry work. All types discussed are of steel construction, although some types are also available in wood.

TUBULAR STEEL SECTIONAL SCAFFOLDING

Sectional steel scaffolding is assembled by the mason from steel frames. The most popular size of tubular steel scaffolding is 4′ to 5′ wide x 5′ high. Sections which are 4′ high are available but require more sections for use. Steel supporting braces, welded horizontally between the top and bottom of a section, can be used as ladder rungs if all are kept in a straight vertical line when constructing the scaffold, Figure 27-1. The frames are tied together by braces. These braces fit on threaded bolts, pins, or slip locks and are welded into the tubular frame of the scaffold. Each section of scaffolding is set on the next by slipping the hollow post on top of a steel pin, or sprocket, which fits into the hollow tubing of the section underneath, Figure 27-2. In the trade, these pins are known as *nipples.* It is important to keep the tubular sections into which the pin fits free from mortar and dirt to speed up the process of erecting the scaffolding.

By law, scaffolding must have certain safety features to protect masons and other people underneath the scaffolding from injury, Figure 27-3. These guidelines were established by the Occupational Safety and Health Act passed by Congress in 1970. The purpose of this congressional act (commonly referred to as OSHA) is to establish safe working practices in all places of employment in the United States.

Two of the most important safety items on scaffolding are guardrails and toe boards. Guardrails are placed around the outside edge of the scaffolding to prevent the worker from falling off the scaffolding.

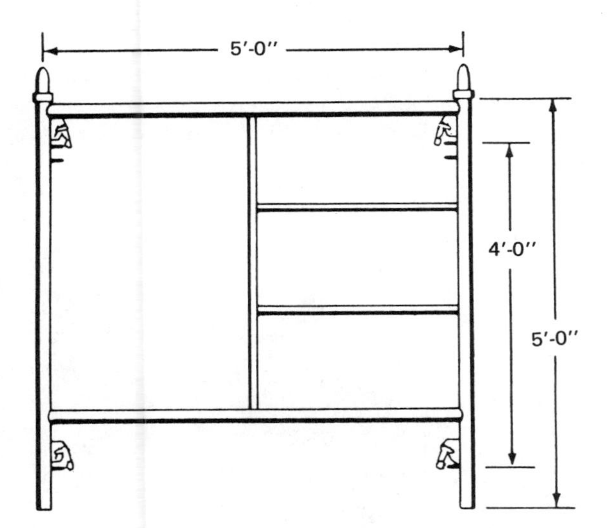

Fig. 27-1 Steel sectional scaffolding

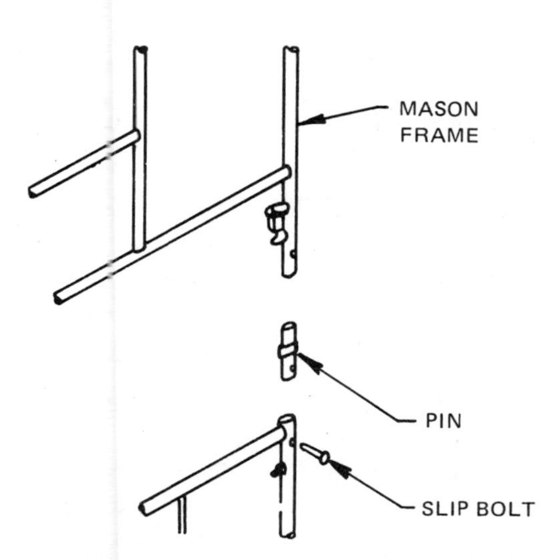

Fig. 27-2 Sections of steel scaffolding held together by pins

Toe boards are installed at the bottom outside edge to prevent masonry materials or scraps from dropping off the scaffolding and injuring persons below.

A tubular steel scaffold with more headroom has been developed to allow a walking area for the mason and laborer. It has become increasingly popular since it permits the laborer to walk or push wheelbarrows along the scaffold without having to bend over. This type of scaffold is usually 5′ wide and approximately 6 1/2′ high. Since the structure is taller than other scaffolds, the mason does not have to stop to erect new scaffolding as often.

Platform extenders (more commonly known as *scaffold hangers* or *brackets*) are fastened on the front of the scaffolding frame, allowing the mason to work at the best possible height, Figure 27-4.

Caution: The mason should always check the scaffold before beginning work to be sure that the hangers are fastened securely and that the planks are lapped well over one another. Scaffolding which holds materials should always be stocked with all necessary equipment and supplies before the mason starts work, since the weight of the material counterbalances the weight of the mason.

A masonry building may be completely encased in scaffolding as the work progresses. Power equipment, such as hydraulic lifts on a portable powered vehicle,

Fig. 27-3 Scaffolding section showing features required by the Occupational Safety and Health Act. Features shown include the guardrail, toe boards, toe board clips, and climbing ladder with 12-inch spacing of steps.

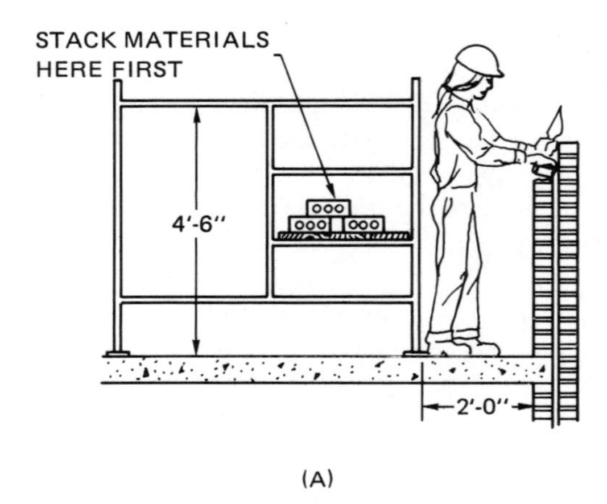

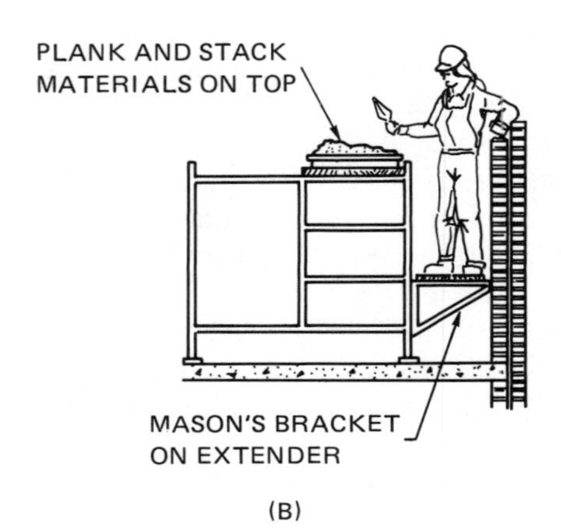

Fig. 27-4 The bracket and extender allow the mason to work at the most convenient height. Note: The mason should never place materials on the bracket. Until a wall reaches a height of about 4 feet above floor level, masons can lay brick overhand standing on the floor (A). Above the 4-foot height, brackets can be attached to tubular scaffolding, creating a platform on which masons can stand (B).

can be used to lift materials as high as 4 stories to place the materials on the scaffolding, Figure 27-5.

Many accessories for steel sectional scaffolding make it adaptable to many situations. Wheels may be inserted so that the scaffolding becomes a *rolling* scaffold, Figure 27-6. This type of scaffold is acceptable for striking and pointing joints but is not recommended to be used as a base from which to lay units. When the scaffold is loaded with materials, there is too great a possibility of the scaffold moving, thereby possibly causing the mason to fall and be seriously injured. The rolling scaffold is popular with workers who are working on a job which does not require the permanent, conventional scaffold. Screw jacks with baseplates are also available to level the sectional steel scaffold after it has been built.

Fig. 27-5 Mechanical tractor lifting material onto steel sectional scaffolding. Note: The guardrails are put into place after the scaffold is loaded with materials.

The Advantages of Steel Scaffolding

Ease of Erecting. There are several advantages to using steel scaffolding. The option of interchanging parts and frames is a great time saver. Fatigue is minimized by the placement of the masonry materials at the best working level. No special tools are needed to erect the scaffold. Built-in coupling pins and wing nuts for the vertical frame assembly and an improved slip lock fastener for attaching cross braces speed the process of erection. Standard parts are available from suppliers to fit the major types of scaffolds with a minimum of delay. If more scaffolding is needed, it is usually rented.

Durability. Steel scaffold lasts indefinitely with reasonable care. Scaffolding should not be tossed to the ground from high heights. The mason should not use unnecessary rough treatment when disassembling the sections after the job is completed. Pins or scaffold nipples should be tied together in bundles with a piece of wire and stored for use on the next job. If wing nuts are being used, the threaded bolt and the nut should be coated with grease. This is done so that the person erecting the scaffold can assemble and disassemble it without a wrench. Wing nuts should not be tightened or loosened with a blow of a heavy hammer as the nut or bolt may be permanently damaged.

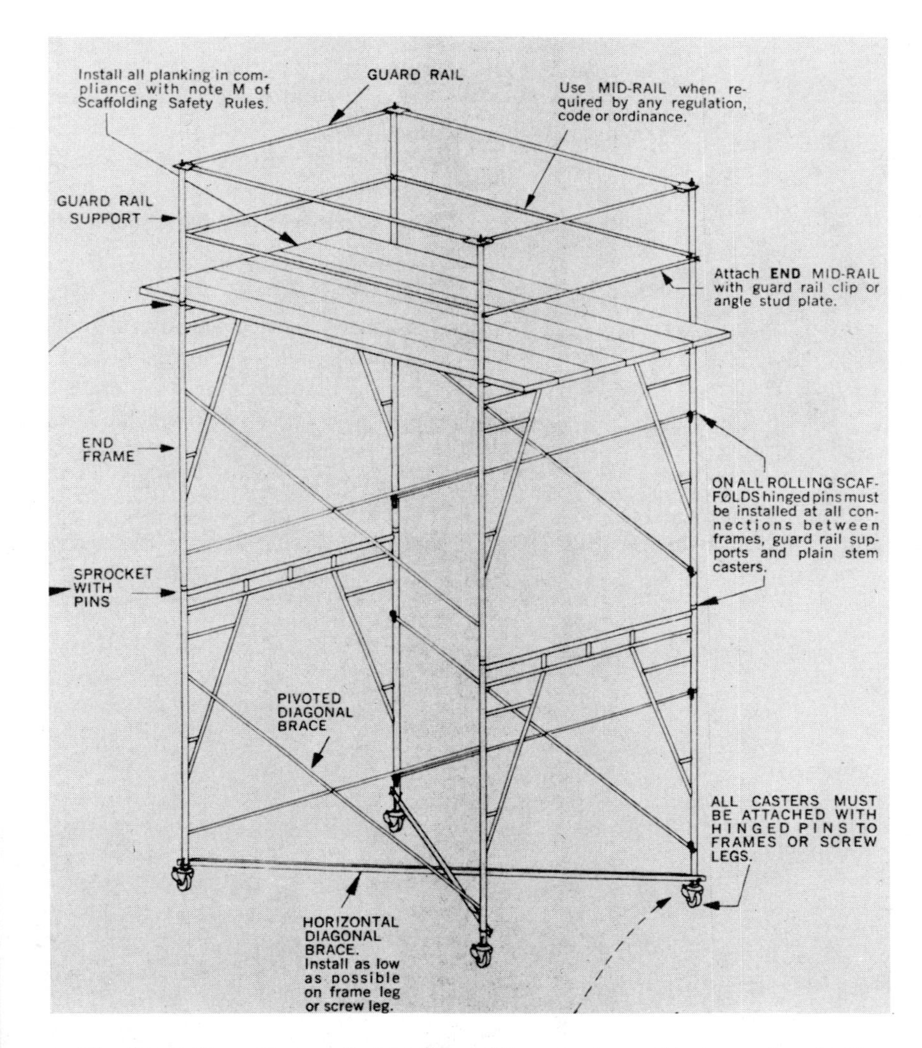

Fig. 27-6 The rolling scaffold. Note: The wheels on this type of scaffold should always be locked into position while the scaffold is in use.

Steel scaffolding is the property of the contractor and is costly to replace. It should be treated as if it were personal equipment.

Strength. The superior strength of sectional steel scaffolding is a major advantage of its use. The strength and complete protection from fire give the mason a feeling of security while working on the scaffold.

ADJUSTABLE TOWER SCAFFOLDING

The *adjustable tower scaffold,* Figure 27-7, consists of vertical metal towers braced to the structure with alternating bays of metal scaffolding between each tower section. The main feature of this type of scaffold is the winch mechanism which raises the working platform as the work progresses. In this way, the mason can always be at his or her best working position.

Scaffolding of this type is a relatively new concept in the masonry business. One of the best adjustable scaffolds, designed by a mason, is the Hoist-O-Matic trademark sold by Patent Scaffolding Company. In this scaffold, 6′ wide base panels are erected perpendicular to the proposed wall line. A sliding, locking cross arm is then installed as the load-bearing member of the platform to support materials, equipment, and workers. Additional scaffold height can be obtained from the base unit by adding posts and braces. This is done as the mason continues working, thus eliminating the loss of time and labor usually involved in rescaffolding.

Fig. 27-7 Masons walking on an adjustable scaffold. The masons attempt to stock materials to a level higher than themselves for convenience.

As the wall is being built, workers raise the platform using a portable mechanical winch which is attached to the posts. An operator moves the winch wich elevates the scaffold as necessary to keep the mason at the best working height, Figure 27-8.

The Advantages of Adjustable Scaffolding

Scaffolding of this type offers many advantages to the mason and the contractor. There is no need for the mason to relocate several times a day while the scaffold is being rebuilt or restocked with materials. Material platforms are wide enough to enable movement of materials without hindering the mason. An overhead canopy can be easily built for wintertime enclosure, thereby allowing the job to continue. The mason is always working at the most convenient height, reducing fatigue and increasing production.

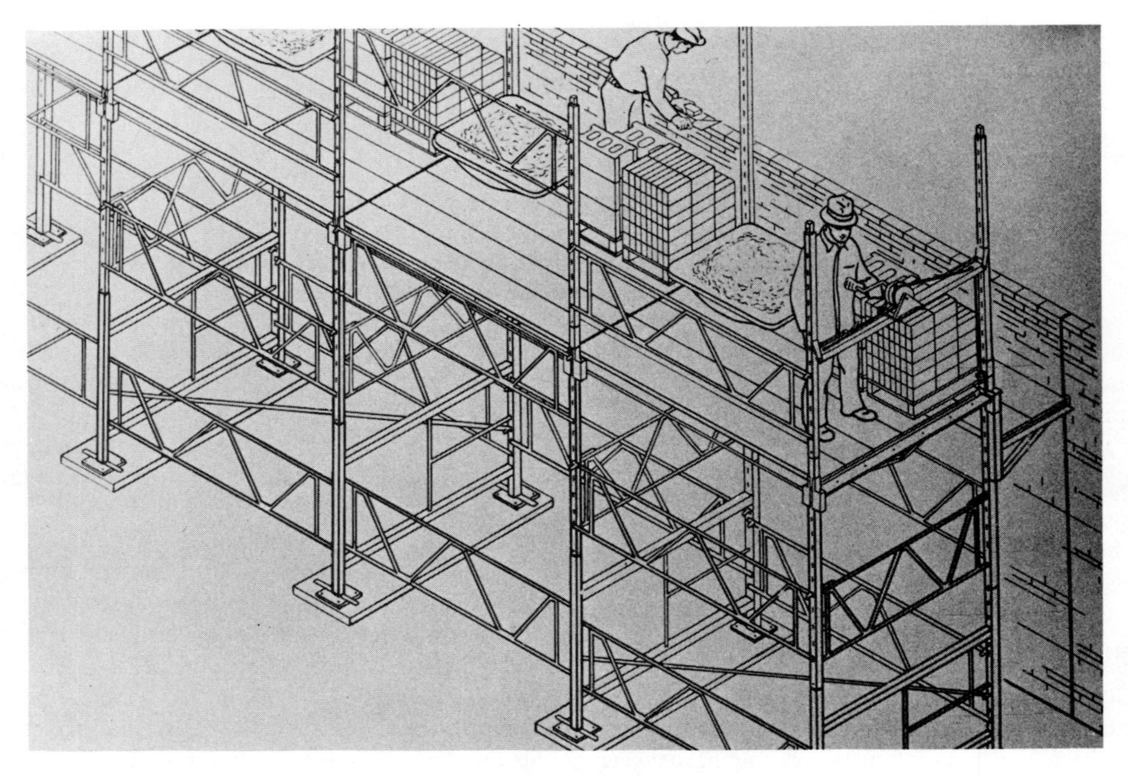

Fig. 27-8 Mason raising the adjustable scaffold with the winch

Fig. 27-9 Suspended scaffolding extends around the entire structure. Notice the planking above the scaffolding to protect the workers.

SUSPENDED OR SWINGING SCAFFOLDING

Suspended scaffolding is usually the best choice for very tall, multistory buildings, Figure 27-9. Steel beams are usually embedded and bolted into the concrete framework of the roof with steel cables dropped to the ground. Affixed to these cables are heavy-duty drum mechanisms with crank handles. Stop bolts at the end of the steel outrigger I-beams prevent the shackles from slipping off.

The planking of the scaffold is inserted in the metal frame below the drum mechanism and overlapped with lengths of scaffolding planks reaching to the next drum. A heavy steel frame is installed underneath the wood planking on each set of suspended hangers. This frame makes the scaffolding extremely strong and safe. Guardrails and toe boards are standard equipment. An overhead canopy attachment provides protection from the elements or objects falling from overhead.

Scaffolding machines provide a platform approximately 5′ wide which are adaptable for use on buildings of normal height.

As the work progresses the scaffold is cranked up, keeping the mason at the best working level. Each

Fig. 27-10 Masons working on suspended scaffolding. Safety rails, protective wire screen, and overhead planking all add to the mason's safety. Notice the winch with which the scaffolding is raised.

drum mechanism has 2 steel locking devices which engage in a large steel gear or sprocket to prevent the wire cable from unwinding or slipping.

Suspended scaffolding is the most economical way to build masonry work on a very tall building. It is one of the safest types of scaffolds that can be used since it is highly unlikely that the steel outrigger beams would become disattached, causing the scaffolding to collapse. Also, masons are constantly facing their work, which means easier working conditions for the masons and improved quality of the work being done, Figure 27-10.

Accidents are less likely to occur with suspended scaffolding because guardrails, toe boards, and overhead planking are used for the workers protection. Suspended scaffolding is usually rented and installed by scaffolding specialists.

Regardless of the type of scaffold used on a particular job, it should conform to good safety standards and meet federal OSHA regulations and specifications. Always check scaffolding carefully before beginning work.

ACHIEVEMENT REVIEW

Select the best answer from the choices offered to complete each statement. List your choice by letter identification.

1. The most important factor to consider when erecting any steel scaffolding is the
   a. required amount of time to build the scaffolding .
   b. safety of the masons involved.
   c. weight of the scaffolding.
   d. requirements for the job.

2. Adjustable scaffolding was developed to
   a. speed the process of constructing the scaffold.
   b. reduce the costs of constructing scaffolding.
   c. give the mason the opportunity to work at the best possible height.

3. Steel sectional tubular scaffolding is fastened together vertically by using
   a. steel braces.
   b. bolts.
   c. steel pins or nipples.
   d. wire cables.

4. A major safety law was passed in 1970 by the United States Congress. The law is referred to as
   a. FHA.
   b. VA.
   c. HEW.
   d. OSHA.

5. One of the following scaffolds is *not* recommended for the mason to use when laying masonry units. This scaffold is the
   a. steel sectional scaffold.
   b. rolling scaffold.
   c. suspended scaffold.
   d. adjustable tower scaffold.

6. Scaffolds which are built higher than a few feet should be equipped with 2 major safety devices. They are
   a. wire screening and safety ties.
   b. toe boards and guardrails.
   c. reflector pads and warning buzzers.

7. Suspended scaffolding is recommended for very tall buildings because
   a. it is more economical to construct.
   b. the height can be continually adjusted.
   c. more materials can be stocked on the scaffold.
   d. it is quicker to erect.

8. Scaffolding can be of various heights but should be a standard width of
   a. 4′.
   b. 5′.
   c. 6′.
   d. 8′.

Unit 28
Safety Rules for Erecting and Using Scaffolding

OBJECTIVES

After studying this unit, the student will be able to

- list safety rules for erecting and working on scaffolding.
- explain how to safely construct a foot board.
- describe in detail how a steel sectional scaffold is built.

A considerable amount of the mason's time is spent working on scaffolding. The danger of an accident resulting from poor scaffolding erection or faulty safety practices while working on the scaffold is higher than that of other practices in masonry work. Masons must be constantly aware of other workers while working on scaffolding. Materials should never be thrown to the ground from a scaffold, where they might cause injury.

Regardless of the type being used, scaffolds must be built properly to support the load placed on them. The National Safety Council recommends that scaffolds be built to support at least four times the anticipated weight of the materials and workers placed on them. This measure gives the mason protection in case additional loads are placed on the scaffold.

Scaffolding should be periodically inspected after it has been erected. Heavy rains and other severe weather conditions may cause sections of the scaffold to become disattached from the wall. The mason should always inspect every scaffold before beginning work. If the mason is not satisfied, work should not begin on the scaffolding until the problem is corrected. Contractors are bound by federal law to provide a safe working environment for their employees. Immediately report any unsafe conditions to the employer or safety engineer.

SAFETY RULES

Figure 28-1 shows a listing of safety rules established by the Scaffolding and Shoring Institute for persons erecting and working on scaffolds.

SAFETY IN STRUCTURAL COMPONENTS

Footing

The mason should make sure there is good footing under the scaffold. The board on which the scaffolding rests is known as the *scaffold sill.* The mason should be sure that it is level and on solid ground before placing any scaffolding on it.

Caution: A hollow block on which to rest a scaffold leg should never be used. The weight of the scaffold after it has been loaded could cause the block to break through its web, possibly causing the scaffold to collapse.

Metal baseplates equipped with screws are available for the legs of the scaffold, Figure 28-2. The screws are tightened or loosened to adjust the scaffold until it is level. The metal plates assure that the scaffolding rests firmly on the scaffold sill.

Foot Boards

Foot boards (also known as *blocking boards*), Figure 28-3, are erected to give the mason more height. Technically, however, they are not classified as scaffolding. When building a wall to the first level of the scaffold, it is sometimes necessary to lay a wooden plank on several block or bricks so that the mason is able to reach high enough to build. This is especially true when 5′ scaffolding is being used.

There are safety rules which must be followed when erecting a foot board. Select a wood plank that

SCAFFOLDING SAFETY RULES

as Recommended by

SCAFFOLDING AND SHORING INSTITUTE

(SEE SEPARATE SHORING SAFETY RULES)

Following are some common sense rules designed to promote safety in the use of steel scaffolding. These rules are illustrative and suggestive only, and are intended to deal only with some of the many practices and conditions encountered in the use of scaffolding. The rules do not purport to be all-inclusive or to supplant or replace other additional safety and precautionary measures to cover usual or unusual conditions. They are not intended to conflict with, or supersede, any state, local, or federal statute or regulation; reference to such specific provisions should be made by the user. (See Rule II.)

I. **POST THESE SCAFFOLDING SAFETY RULES** in a conspicuous place and be sure that all persons who erect, dismantle or use scaffolding are aware of them.

II. **FOLLOW ALL STATE, LOCAL AND FEDERAL CODES, ORDINANCES AND REGULATIONS** pertaining to scaffolding.

III. **INSPECT ALL EQUIPMENT BEFORE USING**—Never use any equipment that is damaged or deteriorated in any way.

IV. **KEEP ALL EQUIPMENT IN GOOD REPAIR.** Avoid using rusted equipment—the strength of rusted equipment is not known.

V. **INSPECT ERECTED SCAFFOLDS REGULARLY** to be sure that they are maintained in safe condition.

VI. **CONSULT YOUR SCAFFOLDING SUPPLIER WHEN IN DOUBT**—scaffolding is his business, **NEVER TAKE CHANCES.**

A. **PROVIDE ADEQUATE SILLS** for scaffold posts and use base plates.

B. **USE ADJUSTING SCREWS** instead of blocking to adjust to uneven grade conditions.

C. **PLUMB AND LEVEL ALL SCAFFOLDS** as the erection proceeds. Do not force braces to fit—level the scaffold until proper fit can be made easily.

D. **FASTEN ALL BRACES SECURELY.**

E. **DO NOT CLIMB CROSS BRACES.** An access (climbing) ladder, access steps, frame designed to be climbed or equivalent safe access to the scaffold shall be used.

F. **ON WALL SCAFFOLDS PLACE AND MAINTAIN ANCHORS** securely between structure and scaffold at least every 30′ of length and 25′ of height.

G. **WHEN SCAFFOLDS ARE TO BE PARTIALLY OR FULLY ENCLOSED,** specific precautions must be taken to assure frequency and adequacy of ties attaching the scaffolding to the building due to increased load conditions resulting from effects of wind and weather. The scaffolding components to which the ties are attached must also be checked for additional loads.

H. **FREE STANDING SCAFFOLD TOWERS MUST BE RESTRAINED FROM TIPPING** by guying or other means.

I. **EQUIP ALL PLANKED OR STAGED AREAS** with proper guardrails, midrails and toeboards along all open sides and ends of scaffold platforms.

J. **POWER LINES NEAR SCAFFOLDS** are dangerous—use caution and consult the power service company for advice.

K. **DO NOT USE** ladders or makeshift devices on top of scaffolds to increase the height.

L. **DO NOT OVERLOAD SCAFFOLDS.**

M. **PLANKING:**
1. Use only lumber that is properly inspected and graded as scaffold plank.
2. Planking shall have at least 12″ of overlap and extend 6″ beyond center of support, or be cleated at both ends to prevent sliding off supports.
3. Fabricated scaffold planks and platforms unless cleated or restrained by hooks shall extend over their end supports not less than 6 inches nor more than 12 inches.
4. Secure plank to scaffold when necessary.

N. **FOR ROLLING SCAFFOLD THE FOLLOWING ADDITIONAL RULES APPLY:**
1. **DO NOT RIDE ROLLING SCAFFOLDS.**
2. **SECURE OR REMOVE ALL MATERIAL AND EQUIPMENT** from platform before moving scaffold.
3. **CASTER BRAKES MUST BE APPLIED** at all times when scaffolds are not being moved.
4. **CASTERS WITH PLAIN STEMS** shall be attached to the panel or adjustment screw by pins or other suitable means.
5. **DO NOT ATTEMPT TO MOVE A ROLLING SCAFFOLD WITHOUT SUFFICIENT HELP**—watch out for holes in floor and overhead obstructions.
6. **DO NOT EXTEND ADJUSTING SCREWS ON ROLLING SCAFFOLDS MORE THAN 12″.**
7. **USE HORIZONTAL DIAGONAL BRACING** near the bottom and at 20′ intervals measured from the rolling surface.
8. **DO NOT USE BRACKETS ON ROLLING SCAFFOLDS** without consideration of overturning effect.
9. **THE WORKING PLATFORM HEIGHT OF A ROLLING SCAFFOLD** must not exceed four times the smallest base dimension unless guyed or otherwise stabilized.

O. For **"PUTLOGS"** and **"TRUSSES"** the following additional rules apply.
1. **DO NOT CANTILEVER OR EXTEND PUTLOGS/TRUSSES** as side brackets without thorough consideration for loads to be applied.
2. **PUTLOGS/TRUSSES SHOULD EXTEND AT LEAST 6″** beyond point of support.
3. **PLACE PROPER BRACING BETWEEN PUTLOGS/TRUSSES** when the span of putlog/truss is more than 12′.

P. **ALL BRACKETS** shall be seated correctly with side brackets parallel to the frames and end brackets at 90 degrees to the frames. Brackets shall not be bent or twisted from normal position. Brackets (except mobile brackets designed to carry materials) are to be used as work platforms only and shall not be used for storage of material or equipment.

Q. **ALL SCAFFOLDING ACCESSORIES** shall be used and installed in accordance with the manufacturers recommended procedure. Accessories shall not be altered in the field. Scaffolds, frames and their components, manufactured by different companies shall not be intermixed.

9-73—10M

Fig. 28-1

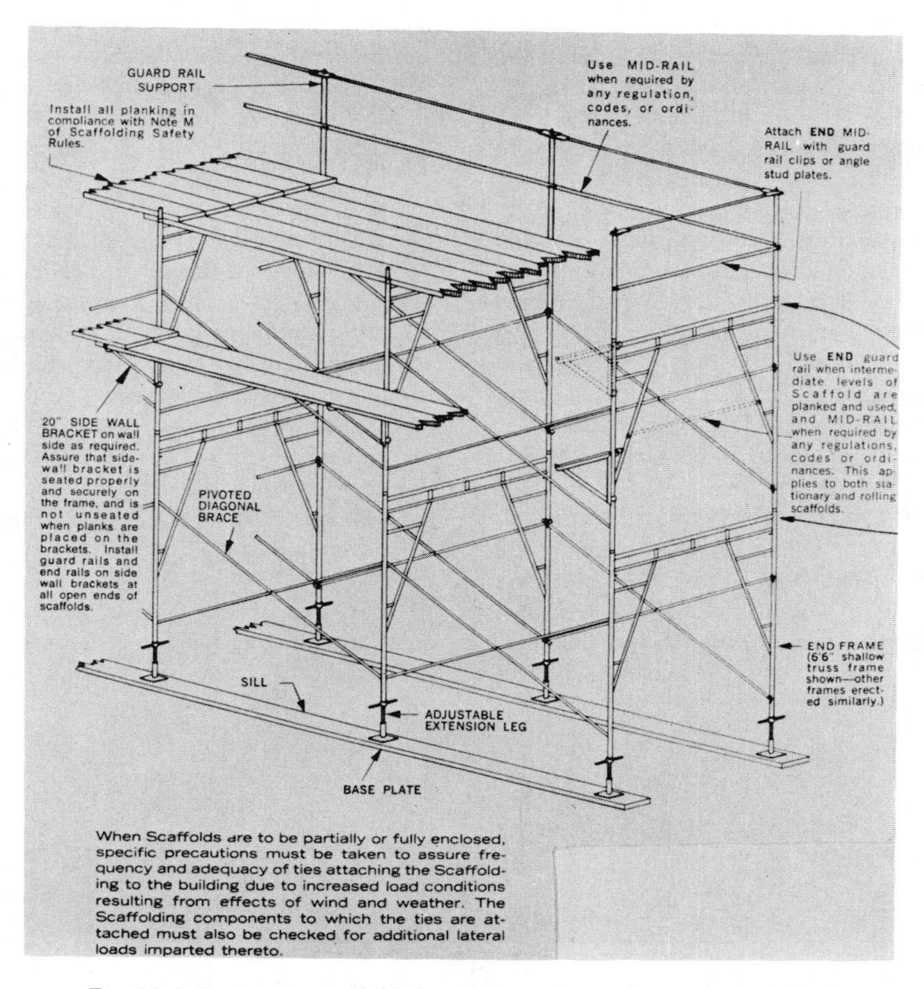

Fig. 28-2 Stationary scaffold showing metal baseplate and wooden sill

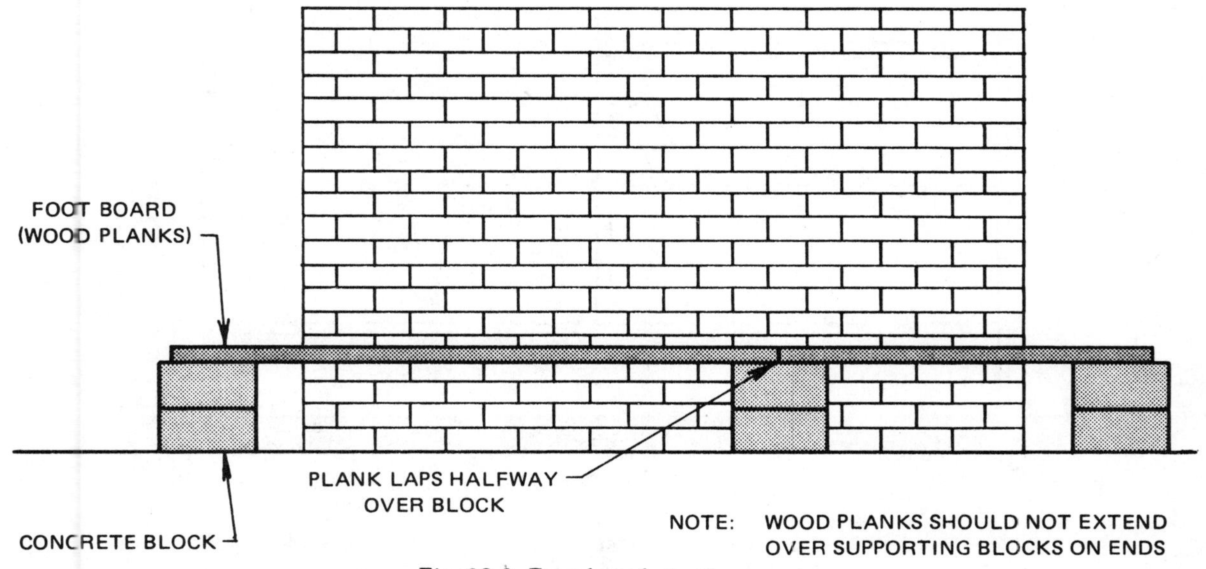

Fig. 28-3 Foot board structure

is free from splints and breaks, and one that is not excessively warped or twisted, to ensure a stable footing. An insecure footing could slip out from under the masons' feet while they are lifting a heavy load and cause severe injury. The idea that masons must work high on the scaffold to be injured is false. Many injuries occur at ground level.

Divide the supporting materials under the foot board evenly along the length of the plank to avoid a sag in the middle. Be sure that the blocking material is placed level on the ground. In places where two boards meet, lay each board halfway over the block.

> **Caution:** A foot board should never be allowed to extend over end block. This constitutes a safety hazard.

Do not build foot boards very high, as they will not be stable support. As a rule, foot boards are usually built 8" from ground level and are never higher than 16" from the ground. Space the foot board approximately 2" from the wall to allow sufficient working room and room for mortar droppings. Do not attempt to make a scaffold out of foot boards. Their purpose is to allow only a small amount of additional height for the mason to prevent undue strain.

Putlogs

On occasion, there may be a ditch near the wall being constructed; therefore, there is no base on which to set the scaffold frame. In these cases, putlogs are used, Figure 28-4. *Putlogs* are crosspieces of heavy timber or steel beam, one end of which rests in a hole left in the wall, with the other end resting on solid ground or on a suitable base. This provides a strong footing on which to set the scaffold frame. After the masonry work has been completed and the scaffold disassembled, the putlog is withdrawn from the hole which is filled by the mason.

Braces

All braces must be fastened securely. Extra braces may be added if necessary. If the nuts are missing from the bolt where the braces fit, the brace should

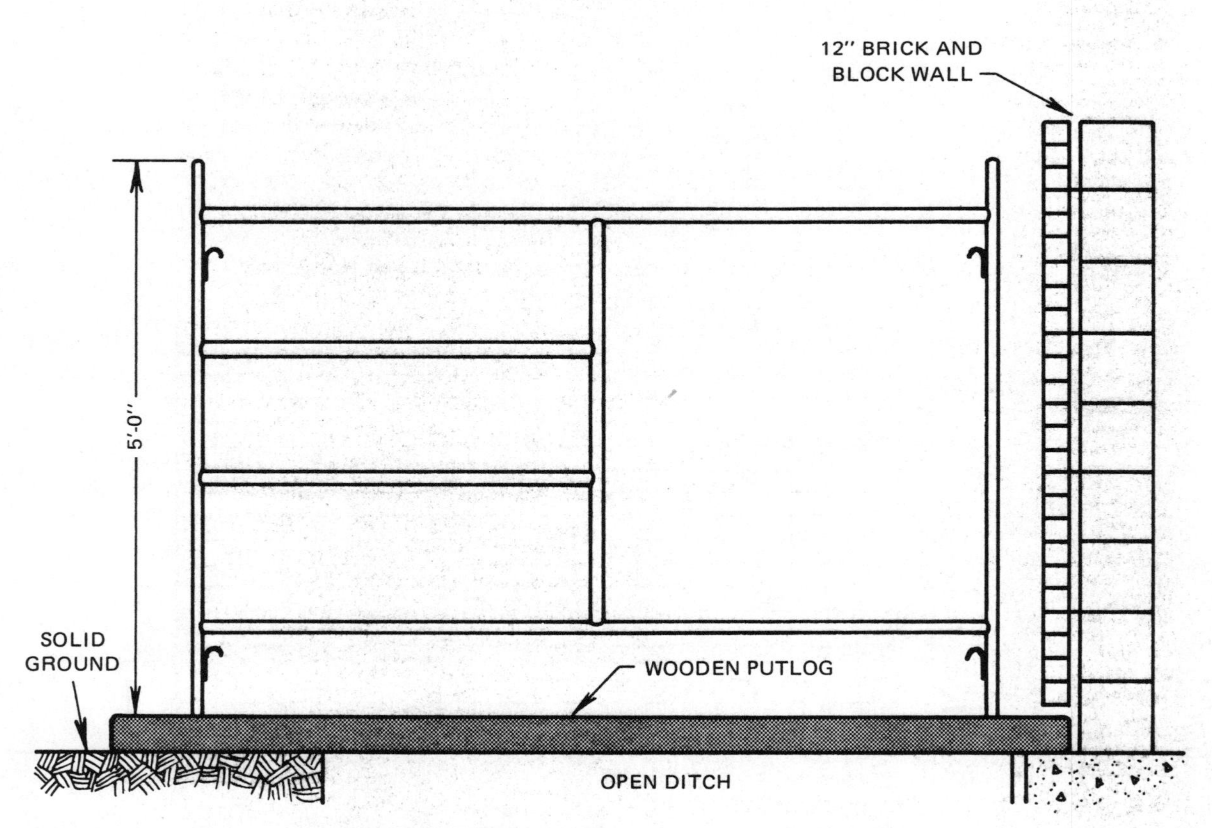

Fig. 28-4 Scaffold frame resting on putlog. After the wall is completed, the putlog is removed and the hole is filled by the mason.

not be wired; it will not be secure. New nuts should be obtained and installed to hold the braces in position.

A mason should not climb on cross braces; the rungs built into the scaffold or a ladder should be used. Steel ladders which attach to scaffolding are available. Wooden ladders are also acceptable for scaffolds, providing they are wired to the top of the scaffold.

Ties

When scaffolding is higher than 2 sections, it should be tied to the structure every 26′ of height with a heavy gauge wire. The wire is usually fastened to the frame of the structure and then walled into place. This is accomplished by wrapping the wire around one of the interior columns and laying it in the mortar joint through the masonry wall. The wire then cannot be pulled out. Ties should also be installed every 30′ in length. After the masonry work has been completed and the scaffold is ready to be torn down, the wire can be cut as the scaffold is disassembled, and the hole can be filled with fresh mortar.

Planking

The scaffolding is no stronger than the planking used on it. Reasonably straight, warp-free boards should be selected. Rough, crooked sawmill lumber should be avoided as it does not provide stable footing. Poplar is the most commonly used wood due to its strength and light weight.

Scaffold planks are usually at least 2″ thick. All planks should be overlapped at least 12″ and extended at least 6″ beyond the center of the support; or, cleats may be installed at both ends to prevent them from sliding off the scaffold. The mason should secure the planks to the scaffold with nails when necessary. All holes in the floor of scaffolding should be repaired before work is begun. Pull all loose nails out of the planking; do not merely bend them over with a hammer.

The planks should be laid as close together as possible to prevent the possibility of materials or workers falling through. The total planking space should be a minimum of 4′ wide to allow sufficient working room. For best results, it is recommended that the scaffold be built 5′ wide. At the completion of the job, all planking is removed from the scaffolding with care. Do not throw it to the ground. Stacking it in a neat pile will prevent excessive warping until it is ready to be used on the next job. Wooden scaffold planking should last for many years with reasonable care.

Planking laid on top of the metal frames provides protection from falling objects. Planking should always be used for overhead protection on swinging scaffolding. Unsafe areas underneath the scaffold should be marked with rope or trestles to protect other workers or people passing by.

Scaffold Brackets

When scaffold brackets are attached to the scaffolding, they must be installed correctly. Brackets should support only the weight of the mason; materials should never be placed on them. Materials are stocked on the main section of the scaffold.

Scaffolding brackets should never touch the wall being built, since the scaffold will probably be moving somewhat as the work is being done. The movement could cause the brackets to knock the masonry work out of alignment. Provisions to install the brackets the correct distance from the wall must be made when the scaffold is begun by proper spacing of the scaffold frames from the wall with the bracket attached. Scaffolding which is not plumb when it is erected will cause the brackets to be knocked either in or out and may necessitate rebuilding or moving the frames before the masonry can be laid.

When building a single section of scaffolding and laying the wooden planks, be sure that they extend at least 8″ further than the brackets. A cleat should be nailed to the bottom of the planks to prevent them from becoming disattached from the brackets.

Guardrails, toe boards, and midrails should be on all steel scaffolds at an elevation of 6′ or more. Protect all open sides and ends of a scaffold with guardrails. Protective screen can be fastened on the side which faces a street if the work is being done in a populated area.

SPECIAL PROBLEMS OF ROLLING AND SUSPENDED SCAFFOLDING

Rolling and suspended scaffolding present some unique problems and, therefore, require safety regulations in their use.

As a rule, the mason does not use rolling scaffolds, but they are sometimes used on small repair jobs or when pointing joints. Rolling scaffolds are limited in height to four times their narrowest base dimensions, unless the base is widened by outriggers or more end frames.

Rolling scaffold should have the caster brakes locked in place when it is being worked on. Brackets are never used on rolling scaffolding. Rolling scaffolds should never be parked on a grade or slope without placing blocks against the wheels. A mason should never ride on rolling scaffolding.

Because of the nature of its construction, suspended scaffolding also has particular safety rules. Before using the scaffolding, inspect all wire cables to be sure that the cables are not frayed or worn. Before installation, be sure that the roof of the building will sustain and support the weight of the scaffold and the load to be placed on the scaffold. The working parts of the drum mechanism must be checked frequently to be sure that they operate freely. The working parts must also be kept well greased. Overhang of the outriggers must not exceed the distance recommended by the manufacturer. Be sure that toe boards and guardrails are installed and that all boards are lapped sufficiently over one another.

Scaffolding should be raised in small intervals to keep the entire scaffold reasonably level so that planking does not slip out of the metal frame. Overhead protection should be provided at all times by laying planking on top of the metal frames. Hard hats must be worn at all times if there is any overhead work being done.

GENERAL PRECAUTIONS IN SCAFFOLDING

Electric power lines near scaffolds are potentially very dangerous, since masons working on the scaffolding could receive an electrical shock. Use extreme caution when working around power lines. Call the local power company if there is any question of safety.

Do not prop piles of building materials above the scaffold level as a base from which to complete the work. Makeshift devices can easily tip over and cause a serious accident.

Materials stocked on the scaffold should always be placed directly over the scaffolding frame for greater strength. Stack and interlock the masonry materials to a reasonable height so they do not topple over. When a mechanical tractor is placing materials, a worker should be on the scaffold to direct their placement. Scaffolding should not be overloaded. Even steel scaffolding has limits as to what it will support. Each scaffold manufacturer gives a safe load limit for their scaffolding.

Caution: Any accidents should be reported immediately to the supervisor so that prompt medical attention can be given.

Practical joking can be hazardous in any phase of construction work but is absolutely forbidden when working on or around scaffolding. Scaffolding work requires strict adherence to safety regulations, common sense, and a serious attitude about the work.

ACHIEVEMENT REVIEW

A. Select the best answer from the choices offered to complete each statement. List your choice by letter identification.

1. The most popular type of scaffolding for general masonry work is
   a. suspended scaffolding.
   b. adjustable tower scaffolding.
   c. steel tubular sectional scaffolding.
   d. wooden scaffolding.
2. A building is being constructed with a ditch around the wall, making it impossible to set the scaffold close enough to the wall. This situation can be overcome by placing
   a. a brace against the wall.
   b. a putlog in the wall.
   c. brackets on the scaffold.
   d. filling in the ditch.
3. The purpose of brackets or scaffold hangers is to
   a. strengthen the scaffold.
   b. tie the scaffold to the wall for stability.
   c. prevent the worker from falling from the scaffold.
   d. act as a platform on which the mason works.

4. When erecting scaffolding with one vertical section on top of another, the sections should be tied to the wall every
   a. 10′.
   b. 15′.
   c. 20′.
   d. 26′.
5. In thickness, scaffolding planks should be at least
   a. 4″.
   b. 2″.
   c. 1″.
   d. 3/4″.
6. The mason can be protected from overhead hazards on scaffolding by using
   a. wire screen fastened to the overhead section of the scaffold.
   b. metal braces tied to the scaffold frame.
   c. wooden planking laid over the metal frame.
   d. canvas on the top of the scaffolding frame.
7. If a mason is working on a scaffold and a crane begins to swing accidentally knocking objects overhead, the proper procedure for the mason is to
   a. call the supervisor and complain.
   b. call the safety inspector.
   c. leave the scaffold immediately.
   d. cover the scaffolding with planking.
8. If a mason is injured on scaffolding, the proper procedure is to
   a. fill out an accident report.
   b. treat the victim.
   c. call the supervisor.
   d. move the victim from the scaffolding.
9. Swing scaffolding should be raised very gradually because
   a. the cables may break under the strain.
   b. stacks of masonry materials on the scaffold may be upset.
   c. it involves too many laborers to raise the scaffolding.
   d. boards on the bottom of the scaffold may slip out of the hangers.

B. Identify each lettered part of the scaffolding illustration.

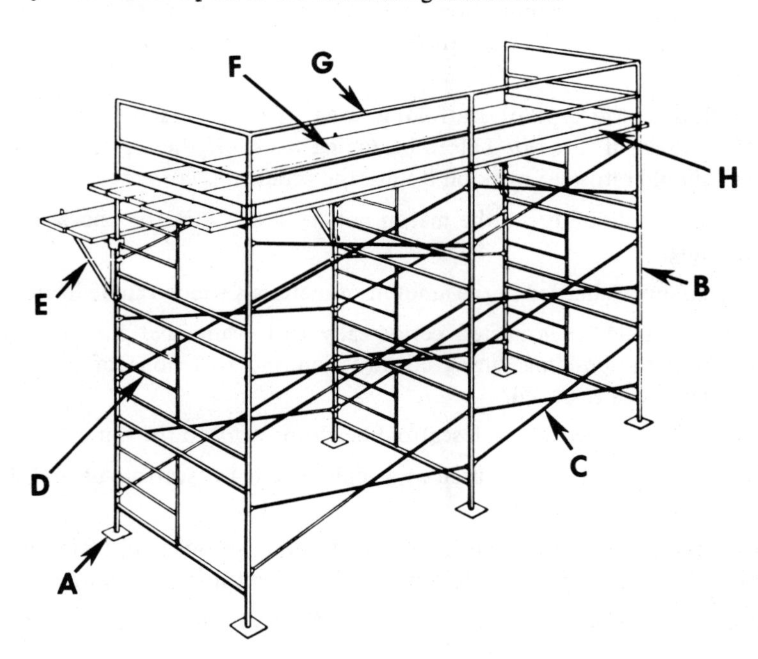

SUMMARY, SECTION 7

- Steel scaffolding has replaced wooden scaffolding for the most part because of speed of erection, cost of lumber, and requirements of state and federal safety regulations.
- The most commonly used scaffolding is the tubular steel sectional scaffolding.
- The OSHA federal safety laws were established to protect all employees from hazards.
- The employer is responsible for seeing that safe working conditions are maintained at all times on the job.
- Scaffolding must be built to conform to OSHA regulations and requirements.
- Two major types of tubular steel scaffolding are used, the standard frame measuring 5′ x 5′ and the type with more height which allows the worker to walk freely through the center.
- Platform extenders (more commonly called brackets) and scaffold hangers should carry only the weight of the mason, not the materials.
- Rolling scaffolding is not recommended for general masonry work, only for quick, small jobs such as pointing joints.
- Wheels must be locked at all times when the mason is working on rolling scaffolding.
- The mason has a responsibility to care properly for steel scaffolding since it is expected to last for a period of years.
- The adjustable tower scaffolding offers the advantage that the mason may remain at the best working height. It also has an advantage in that the mason does not have to move while the scaffold is built higher.
- Suspended or swinging scaffold is used on high-rise structures where the erecting of standard tubular steel sectional scaffolding is not profitable or safe.
- Suspended scaffold is hung from steel cables which are fastened to steel outriggers anchored in the roof of the structure. It is considered a very safe scaffold if built correctly.
- Regardless of the type of scaffolding used, the mason should always inspect the scaffold to see if it is safe before beginning work. If the scaffolding is unsafe, the mason should refuse to work on it until the problem is corrected.
- It is the responsibility of the mason not to throw, kick, or knock anything from scaffolding.
- All scaffolding is limited in the amount of materials which can be loaded on it.
- Scaffolds should be inspected periodically to be sure that they have not moved away from the building during construction and that none of the safety attachments have been removed.
- Foot boards are used to reach scaffold height or work to the top of a wall.
- Foot boards should not be used as a replacement for scaffolding and should never be built over 16″ in height.
- All scaffolding set on ground level should have a wooden sill plate under it to distribute the weight of the scaffolding and to prevent it from sinking into the ground.
- Braces should be attached with nuts or slip fasteners, never wired to the frame.

- Cross braces are not to be climbed upon, but are there only to hold the scaffolding rigid and secure.
- Scaffolding more than 2 sections high should be tied to the structure with heavy wire.
- Wooden planking should be free from splits and twists and lapped 12" over each other.
- Scaffolds should be spaced far enough from the wall so that no part bumps against the wall when the masons are working on it.
- Federal and state regulations require that guardrails and toe boards be on all scaffolds that are a height of 10' or more.
- Extreme caution should be taken that no contact is made with electrical power lines when they are located near the scaffold or work area.
- Overhead protection in the form of planking laid on steel frames should be used whenever there is a danger of objects falling from overhead.
- Never ride on rolling scaffolding.
- Suspended scaffolding should be inspected frequently for frayed cables, loose boards, or other unsafe conditions.
- Practical joking is potentially dangerous and will not be tolerated on the construction site.

SUMMARY ACHIEVEMENT REVIEW, SECTION 7

Complete each of the following statements referring to material found in Section 7.

1. Steel scaffolding has replaced wooden scaffolding because of three main reasons, which are ______________, ______________, and ______________.
2. The most frequently used type of scaffold for masonry work is the ____________.
3. In 1970, Congress passed the Occupational Safety and Health Act, which is commonly referred to as ______________.
4. Scaffolding built in excess of 10' vertically must have 2 protective devices, which are ______________ and ______________.
5. Brackets are installed on scaffolding to provide ______________.
6. Scaffolding which is equipped with wheels is known as ______________.
7. The outstanding feature of a tower scaffold is ______________.
8. Adjustable tower scaffolding is raised by ______________.
9. Suspended or swinging scaffold is selected for use on high structures because it is ______________.
10. Suspended scaffold is supported by ______________.
11. Overhead protection for suspended scaffold is provided by ______________.
12. Regardless of the type of scaffolding, the first thing the mason should do before beginning work is to ______________.
13. If the mason requires a platform on which to step so that he may build to the point at which the scaffolding begins, a (an) ______________ is built.

14. Scaffolding built on the ground should always have a (an) ______________ under it to prevent the scaffold from sinking and to permit better distribution of weight.

15. The reason that a scaffold frame is never set on top of a hollow concrete block is because ______________.

16. Scaffolding should be tied to the structure whenever it is higher than __________.

17. A section of wood or steel which runs the length of a ditch and on which scaffolding rests is known as a (an) ______________.

18. Wood planking selected for scaffolding boards should be free of ______________.

19. If electrical power lines are located so near a scaffold that they interfere with the building of the scaffold, the proper procedure is to ______________.

20. Masonry materials stocked on scaffolding should be placed directly over the ______________.

Section 8
Cleaning Masonry Work

Unit 29
Cleaning Brick and Concrete Block

OBJECTIVES

After studying this unit, the student will be able to

- describe methods the mason uses to keep masonry work clean.
- list poor cleaning techniques which damage masonry work.
- describe the step-by-step procedure in cleaning brickwork with muriatic acid.
- explain how concrete block are cleaned.

The finishing touch on all masonry work is the removal of dirt and stains. All brickwork should be thoroughly cleaned to enhance the color of the masonry units. In the construction of masonry walls, a skilled mason generally keeps the surface reasonably clean of mortar and other agents which cause stains. However, even the highly skilled mason may find it very difficult to keep the work perfectly clean. For this reason, the cleaning of masonry work is considered an essential part of every masonry job.

The National Apprenticeship and Training Standards for Bricklaying, established by the United States Department of Labor, requires that all apprentices have work experience totaling 150 hours in pointing, cleaning, and caulking during the course of their apprenticeship. This experience should be done under the supervision of a competent journeyman mason.

GENERAL INFORMATION

There are different methods involved in the cleaning of bricks and block. The mason should be acquainted with these methods, including the proper tools, supplies, and safety measures. The mason should also practice techniques for keeping the work clean while it is being constructed so that cleaning of the work does not have to be extensive.

The correct clothes are important to protect the mason from injury and effects of the cleaning agent. A hard hat should be worn on the job at all times; this includes cleaning work. When cleaning, the mason is facing the masonry wall and is subjected to falling objects. Since masons are constantly moving when cleaning, it is impossible that they always be protected from overhead danger. A rubber apron or rubber pants are best to keep the clothing dry and to protect the worker from the acid or cleaning agent. Rubber boots or galoshes keep the feet dry. Rubber gloves which are long enough to cover a good portion of the arm should be worn. Standard, shorter rubber work gloves do not give enough protection from the cleaning agents.

Most masons do not wear safety goggles when cleaning because of poor vision from the spray of the solution. However, it is recommended that the apprentice wear safety goggles during all cleaning work.

BRICKWORK

The most common mistakes when cleaning brickwork include using a cleaning agent which is too strong or ignoring manufacturer's directions when mixing the agent, and applying the compound to too great an area without rinsing.

When using acid, neglecting to rinse the wall often enough with water may result in an acid burn. When determining the proper intervals for washing down the wall, the mason must consider the time of year, dryness of masonry units, and heat and humidity in the air.

When cleaning and washing masonry work, be sure to do the following:

- Refill the bucket before it is completely empty.
- Replace the cleaning agent when it becomes dirty.
- Follow safety rules at all times.

Keeping Brickwork Clean during Construction

To keep brickwork clean during construction, the mason stocks the mortar pans and boards far enough away from the wall (usually a minimum of 2′) to avoid splashing mortar on the work. Poor tempering practices also contribute to mortar smears on walls.

The mason should also be careful when laying the units in the wall. Cutting off the mortar from the bricks without smearing it with the trowel greatly reduces the amount of washing required. Striking the work too soon causes the masonry units to be smeared unnecessarily. Mortar joints should be tooled when thumbprint hard. After tooling with the striking iron, all particles of mortar remaining are scraped off with the blade of the trowel before brushing. Never brush the masonry work while the mortar joints are still wet, or they will smear. Always avoid any motion that will result in rubbing or pressing mortar particles into the brick face. A good, medium-bristle brush such as a soft stove brush works well. If the bricks are too wet when they are laid, the mortar does not absorb properly and the joints bleed and smear on the outside surface of the wall.

Mixing the mortar to the right consistency prevents excessive smearing on the face of the wall. Clean materials should be used. When bricks arrive on the job, they should be stored out of the way of concrete, tar, and other materials that may stain them prior to laying them in the wall. Never pile masonry units under the scaffolding while work is in progress.

Additional staining may be caused by rain beating against the scaffold boards and splashing dirt and mortar on the wall. Many times this happens at night after the jobsite has been abandoned. This problem can be prevented to a certain extent by turning the scaffolding boards on their edges with the clean side to the wall at the end of the day's work. Always cover walls at the end of the day to prevent the mortar joints from washing out and rainwater entering the wall.

Tools and Equipment

The tools used in cleaning masonry work are few but essential. A wooden paddle and nonmetallic scraper tool or chisel removes the large particles and does not leave rust or stains on the wall. A rubber bucket or heavy-duty plastic pail is needed to hold the cleaning solution. A water hose with a nozzle attached to adjust the spray of the water and a good stiff-bristled brush are also necessary. The handle should be long enough to protect the hands from the cleaning solution. Also useful are a couple of brick pieces with which the mason can rub stubborn particles of mortar loose from the wall.

Testing the Cleaning Agent

The best method of cleaning new brickwork is the least severe method. Acids are necessary on some occasions. However, if they are used when they are not really needed, they can cause severe problems. Similarly, detailed instructions on how to use a given product should always be superceded by the specific manufacturer's recommendation.

Before the actual cleaning of a brick wall begins, all cleaning procedures and solutions should be tested and evaluated on a test wall area of approximately 20 sq ft in size. The size of the test area could be larger but never smaller than this. The testing is important because small quantities of certain minerals found in some fired clay units, and materials added to color brick, such as manganese, may react with some solutions and cause staining. Chemical cleaning solutions are generally more effective when the outdoor temperature is 50°F or above.

Cleaning New Brickwork

Figure 29-1, developed by the BIA, lists cleaning methods for new masonry. It can be used as a general cleaning guide for all new masonry.

The three cleaning methods for new masonry are bucket and brush hand cleaning, high-pressure water cleaning, and sandblasting. When using any of these methods, be sure to wear proper clothing and eye protection, Figure 29-2.

Bucket and Brush Hand Cleaning. This is by far the most popular of all the methods used for cleaning new brick masonry. This is because of the simple procedures required and the easy availability of muriatic acid and proprietary cleaning compounds. The following is a recommended step-by-step procedure for cleaning brickwork with muriatic acid solution.

1. The cleaning operation should be one of the last phases of the job. Do not start before the mortar is thoroughly set and cured. If a prolonged time period passes, there may be a problem, since after a period of 6 months to 1 year, mortar particles are very difficult to remove. As a rule, 1 week of good weather should pass before the wall is cleaned. This allows the mortar time to cure well.

2. Dry clean the wall first with a wooden paddle. Large particles may require removal with the chisel, wood scraper, or a piece of brick, Figure 29-3.

3. Protect any metal, glass, wood, limestone, and cast stone surfaces. Mask or otherwise protect windows, doors, and fancy trim work from the acid solution. Do not allow metal tools to contact acid solutions.

Brick Category	Cleaning Method	Remarks
Red and Red Flashed	Bucket and Brush Hand Cleaning High Pressure Water Sandblasting	Hydrochloric acid solutions, proprietary compounds, and emulsifying agents may be used. *Smooth Texture:* Mortar stains and smears are genereараly easier to remove; less surface area exposed; easier to presoak and rinse; unbroken surface, thus more likely to display poor rinsing, acid staining, poor removal of mortar smears. *Rough Texture:* Mortar and dirt tend to penetrate deep into textures; additional area for water and acid absorption; essential to use pressurized water during rinsing.
Red, Heavy Sand Finish	Bucket and Brush Hand Cleaning High Pressure Water	Clean with plain water and scrub brush, or *lightly* applied high pressure and plain water. Excessive mortar stains may require use of cleaning solutions. *Sandblasting is not recommended.*
Light Colored Units, White, Tan, Buff, Gray, Specks, Pink, Brown and Black	Bucket and Brush Hand Cleaning High Pressure Water Sandblasting	*Do not use muriatic acid!!* Clean with plain water, detergents, emulsifying agents, or suitable proprietary compounds. Manganese colored brick units tend to react to muriatic acid solutions and stain. Light colored brick are more susceptible to "acid burn" and stains, compared to darker units.
Same as Light Colored Units, etc., plus Sand Finish	Bucket and Brush Hand Cleaning High Pressure Water	Lightly apply either method. (See notes for light colored units, etc.) *Sandblasting is not recommended.*
Glazed Brick	Bucket and Brush Hand Cleaning	Wipe glazed surface with soft cloth within a few minutes of laying units. Use soft sponge or brush plus ample water supply for final washing. Use detergents where necessary and acid solutions only for *very difficult mortar* stain. For dilution rate, see Step 1d, *Select the Proper Solution,* under Bucket and Brush Hand Washing. Do not use acid on salt glazed or metallic glazed brick. Do not use abrasive powders.
Colored Mortars	Method is generally controlled by the brick unit	Many manufacturers of colored mortars do not recommend chemical cleaning solutions. Most acids tend to bleach colored mortars. Mild detergent solutions are generally recommended.

Fig. 29-1 Cleaning guide for new masonry

Fig. 29-2 Proper protective clothing and eye protection for cleaning masonry

4. Prepare the acid. Use a clean, stain-free commercial grade of muriatic (hydrochloric) acid. Mix not more than 1 part acid to 9 parts of clean water in a nonmetallic container. Pour the acid into the water; do not pour the water into the acid.

> **Caution**: If any of the acid solution gets in the eye or on bare skin, immediately flush out with plenty of clean water.

5. Presoak the wall to remove loose particles or dirt. Flush the wall with plenty of water with the hose, Figure 29-4. Start from the top of the wall and work down. A saturated wall does not absorb dissolved mortar particles. The area immediately below also is soaked with water to prevent the acid solution from drying into the wall.

6. Scrub with the brush starting at the top. Begin with a small area, not more than approximately 20 sq ft, Figure 29-5. Heat, direct sunlight, warm masonry, and drying winds affect the drying time and reaction rate of the acid solutions.

Fig. 29-3 Removing mortar particles with wood scraper

Fig. 29-4 Prewetting the wall with water from a hose. The wall is thoroughly saturated before applying acid solution.

7. To remove stubborn spots, try using a piece of brick to rub the wall, Figure 29-6. After rubbing the spot with the brick, rescrub with more acid solution to work the particle or stain loose.

8. Rinse the wall thoroughly with water from the hose starting at the top and letting it run to the bottom of the wall. Acid solutions generally lose their strength after 5 to 10 minutes of contact with mortar particles. Failure to completely

flush the wall with water and remove the acid solution may result in the formation of *white scum*, or the acid may burn the wall. For stubborn places, repeat the washing action to assure a clean wall.

When an acid solution is not desired as a cleaning agent, detergents or proprietary compounds may be the answer.

Detergent or soap solutions are usually used to remove dirt, mud, and soil that has gotten on the wall during construction. A suggested solution is 1/2 cup (c) of trisodium phosphate (such as Calgon®) and 1/2 cup laundry detergent dissolved in 1 gal of clean water.

When using *proprietary compounds* (cleaning compounds made by manufacturers and sold by brand names), be sure that the one selected is suitable for the brick and follow the manufacturer's specific recommended directions for dilution. Many proprietary cleaning compounds perform in a satisfactory manner for their intended cleaning job. However, their formulas are not disclosed and may be subject to change. It is suggested, therefore, that the mason conduct a test on a sample section of wall that does not show too much before cleaning the entire wall.

High-Pressure Water Cleaning. To cut down on labor costs, many masonry contractors are using high-pressure water as a method of cleaning masonry work. Some of the high-pressure water systems feature a high-pressure gun and nozzle equipped with a control switch. This permits the operator to apply solutions to a wall over 100′ from the base unit. There are other systems that utilize two hoses, one with plain water and the other with a cleaning solution.

Some manufacturers recommend that only cleaning agents that will not damage the guns be used in them. Special guns and hoses permit the use of muriatic acid solutions.

The equipment should be as portable as possible with the units on wheels for easy movement. More elaborate equipment consists of water storage tanks mounted on trucks. High-pressure water cleaning generally is considered to do a good job.

Sandblasting. Dry sandblasting has been used for many years on restoration work. It is one method that eliminates the danger of mortar smears, acid burn, and efflorescence. There is the danger, however, of the brick face being damaged or scarred by the wrong use of the equipment.

In this process, air pressure forces the sandblasting materials against the work to remove old paint or finish. The abrasive materials used are mined silica sand, crushed quartz, granite, white urn sand (round particles), crushed nut shells, and other softer abrasives. There are various degrees of cutting or cleaning, and, therefore, many types of abrasive materials. The worker should always direct the abrasives at the brick and not the mortar joints.

Sandblasting is still a very popular method of cleaning and restoring old brickwork. Care must be taken by the cleaning contractor that areas other than the

Fig. 29-5 Scrubbing the wall with a long-handled brush. Eye protection is worn at all times.

Fig. 29-6 Rubbing a stubborn spot with a brick piece. The wall is washed afterwards with an acid solution.

masonry work be protected during the cleaning process, particularly old molding in the building.

Sandblasting should be done by a qualified person, usually one who specializes in the field. Basically, it is done with a portable air compressor, blasting tank, blasting hose, and nozzle. It is very important that the operator wear proper clothing and eye protection.

Chemicals and Steam Cleaning. Chemicals and high-pressure steam are used in special cases to remove applied coatings, such as paint, from masonry. Like sandblasting, it is a highly specialized field. Frequently, the proper cleaning agent can be determined only after an analysis of the various factors involved in a particular project.

CONCRETE BLOCK

Keeping Block Clean during Construction

Ideally, mortar is not cleaned from concrete masonry units with an acid wash. Therefore, care should be taken to keep the work as clean as possible during construction. The task of cleaning the walls in preparation for painting or finishing is the responsibility of the masonry contractor. The cleaner the work is kept during construction, the less work there is to be done at the completion of the job.

Treat joints carefully during construction to eliminate unnecessary cleaning. When a mortar joint is

Fig. 29-7 Rubbing a concrete block wall with a piece of block to remove mortar particles

rubbed flat (when striking a flush joint), there is sometimes a tendency to smear the mortar over too large an area, causing the joint to appear larger than it should. Immediately after joints are rubbed and pointed, they are brushed with a stiff-bristled brush. This should remove most of the excess mortar which clings to the edges. Once mortar smears, it tends to

PROCEDURE FOR STAIN REMOVAL			
A	Scrub with brush and chemicals or detergents	H	Apply liquid to surface by brush
A_1	Scrub with brush and water	H_1	Apply liquid to surface by brush at 5 to 10 minute intervals until thoroughly soaked
B	Wash thoroughly with clear water		
C	Cool until brittle. Chip away with chisel	J	Allow stain to disappear by aging
		K	Stir liquids together
D	Mix solids	L	Put paste on trowel. Sprinkle crystals on top of paste. Apply to surface so crystals are in contact with block and paste is on outside.
D_1	Stir solids and liquid to thick paste		
D_2	Apply paste to stain to thickness of 1/8 to 1/4 inch		
D_3	Place heated concrete brick over top to drive liquid in	M	Soak up fast and soon with poultice material until no free oil remains
D_4	Cover to minimize rate of evaporation		
D_5	Let dry as needed for periods up to 24 hours. Scrape off. Use wood scraper if block has tile-like finish	N	Scrap any solidified matter off surface
		O	Let harden. Remove large particles with trowel, putty knife or chisel
		P	Let stand. Remove with scraper and wire brush.
E	Repeat as needed	R	Allow to age three days
E_1	If brown stain remains, treat as for iron stain	S	Absorb with soft cloth or paper towels, then scrub vigorously with paper towels
F	Apply with saturated cloth or cotton batting	T	Provide thorough ventilation
G	Dissolve solid chemical in hot water	V	Follow manufacturer's directions
G_1	Dissolve solid chemicals in water	X	Pour into paste and mix

Fig. 29-8 Materials for stain removal and procedural sequences

Stain	Appearance	Materials Needed		Procedural Sequence (Letters refer to steps listed in Figure 29-8)
		Chemicals and Detergents*	Poultice Materials	
Aluminum	White deposit	10%Hydrochloric acid	–	A-B
Asphalt	Black	1.* Ice. (Dry ice is not very effective)	–	C
		2.* Scouring powder	–	A
Emulsified Asphalt	Black	Water	–	A_1
Cutback Asphalt	Black	1.* Benzene	Talc or whiting	D_1-D_2-D_3
		2.* Scouring powder	–	A
		3.* Same	Same	E
Coffee	Tan	Sodium hypochlorite OR Glycerine 1 part Water 4 parts	–	F
Copper, bronze	Green, sometimes brown	Ammonium chloride 1 part Ammonium hydroxide as needed for paste	Talc 4 parts	D-D_1-D_2-D_5-E
Creosote	Brown	1.* Benzene	Talc or whiting	D_1-D_2-D_5
		2.* Scouring powder		A
Ink, ordinary blue	Blue	Sodium perborate Water	Whiting	G-D_1-D_2-D_5-E-E_1
Containing Prussian Blue	Blue	Ammonium hydroxide OR Strong soap solution	–	F
Red, green, violet, other bright colors and indelible synthetic dyes	Varies	Sodium perborate Water OR Sodium hypochlorite OR Calcium hypochlorite	Whiting	G-D_1-D_2-D_5-E
		OR Ammonium hydroxide OR Sodium hypochlorite OR Calcium hypochlorite	–	F
Indelible, containing silver salt	Black	Ammonium hydroxide	–	F-E
Iodine	Brown	Ethyl alcohol	Whiting or talc	H_1-D_1-D_2-D_5
		OR None	–	J
Iron	Brown yellow	Sodium or ammonium citrate 1 part Water (lukewarm) 6 parts Glycerine (lime-free) 7 parts	Whiting or Diatomaceous earth	G_1-K-D_1-D_2-D_5-E-B
		OR Same plus sodium hydrosulfite for step L	Same	G_1-K-D_1-D_2-L-D_5-B
Linseed oil	Dark gray	Trisodium phosphate 1 part Sodium perborate 1 part Liquid green soap or strong soap solution in hot water	Lime, whiting, talc or portland cement	M-G_1-D_1-D_2-D_5-B-E
Lubricating oil or grease that has penetrated	Dark	1.* Trisodium phosphate 1 lb Water 1 gal.	–	N-G_1-A
		2.* Benzene	Talc, lime or whiting	D_1-D_2-D_5-B
		OR Amyl acetate (for small areas only)	Asbestos fiber	D_1-D_2-D_3-D_5
Mortar smears	Gray	–	–	O
Paint, at least 3 days old	Varies	Trisodium phosphate 2 lb. Water 1 gal OR Commercial paint remover	–	H-P
freshly spilled		Same		S-R-H-P
Perspiration stains from hands, and hair oil stains	Brown or yellow	Trichloroethylene	Talc	T-D_1-D_2-D_4-D_5
Plant growth, mold and moss	Green, brown or black	Ammonium sulfamate (from garden supply stores)	–	V, and B if needed
Smoke and fire	Brown to black	Trichloroethylene	Talc	T-D_1-D_2-D_4-D_5
Soot and coal smoke	Black	Soap Water Pumice	–	A
Wood tar and smoke	Dense black	1.* Scouring powder Water	–	A
		2.* Sodium hypochlorite	–	F
Tobacco	Brown	1.* Calcium chloride Water	–	D_1
		2.* Trisodium phosphate 2 lb Water 5 qt.	–	G_1-X
		3.* –	Talc	D_1-D_2-D_5-E-B
Wood rot	Chocolate	1.* Glycerine 1 part Water 4 parts	–	A
		2.* Trichloroethylene	Talc	T-D_1-D_2-D_4-D_5

***Numbers indicate that materials are to be used in sequence.**

Fig. 29-8 (Continued)

embed itself into the surface of the block, causing noticeable stains. Paint only magnifies the mortar smears, chipped block, and other evidence of poor work.

Unless covered, concrete block should not be stocked near an area where it will be subjected to such things as spillage from tar buckets, paint, and dirt.

Procedures in Cleaning Block

The tools needed to clean mortar from concrete block are fairly simple. A piece of brick, broken block, or rubbing stone can be used to rub the particles of mortar from the wall, Figure 29-7. Larger particles of mortar may be removed with metal scrapers, putty knives, chisels, or the mason's trowel. The work is then brushed. The use of acid solutions is not usually recommended for concrete block. However, if there is a difficult stain on the block, such as efflorescence, an acid wash may be used.

If a stubborn stain must be removed from concrete block, check with the foreman or superintendent before attempting to remove it chemically. The type of block and material to be removed determine the procedure and cleaning methods to be used. Colored concrete block products could be damaged by using the wrong cleaning agent. Once the wall has been damaged, correcting the problem is usually not a simple procedure. Figure 29-8 shows the various stain-removing materials and the procedure to follow when removing stains from concrete block.

ACHIEVEMENT REVIEW

Select the best answer from the choices offered to complete each statement. List your choice by letter identification.

1. The first procedure in washing a brick wall with muriatic acid is
   a. scrubbing the wall with an acid solution.
   b. soaking the wall with water.
   c. dry cleaning the wall with a wooden paddle or scraper.
2. A metal scraper should not be used on masonry work when an acid solution is being used because
   a. it will scratch and scrape the bricks.
   b. it will catch in the mortar joint and cause permanent damage.
   c. it may cause rust spots or stains.
   d. the mason could be injured by the sharp metal edge on the scraper.
3. A common trade name for muriatic acid is
   a. sulphuric acid.
   b. hydrochloric acid.
   c. sure kleen.
   d. acetic acid.
4. When mixing muriatic acid for cleaning brick, the correct proportions of mix are
   a. 1 part acid to 3 parts clean water.
   b. 1 part acid to 5 parts clean water.
   c. 1 part acid to 7 parts clean water.
   d. 1 part acid to 9 parts clean water.
5. To protect the mason from the acid solution when cleaning masonry work, the clothing should be
   a. canvas or heavy twill.
   b. rubber.
   c. plastic.
   d. light cloth.
6. Temperature and weather conditions have a definite effect on how much of the wall is washed before it is rinsed with water. However, the average area measures
   a. 5 to 10 sq ft.
   b. 10 to 20 sq ft.
   c. 20 to 30 sq ft.
   d. 30 to 40 sq ft.
7. Acid solution loses its ability to remove mortar smears after
   a. 5 to 10 minutes.
   b. 10 to 20 minutes.
   c. 20 to 25 minutes.
   d. 30 minutes.

8. When cleaning rough-textured bricks, it is absolutely essential that
   a. a solution of acid or cleaning agent which is stronger than that for smooth bricks be used.
   b. pressurized water be used.
   c. the bricks not be scraped or rubbed as the texture may be damaged.
   d. muriatic acid not be used.

9. The major drawback to proprietary compounds is
   a. that the formula is not printed on the container.
   b. their chemical composition.
   c. that they are too slow in removing dirt and mortar from walls.
   d. that there is danger to the mason from the high acid content.

10. Particles of mortar are usually cleaned from concrete block by
   a. washing with an acid solution.
   b. washing only with clean water.
   c. rubbing them with a piece of block.
   d. scrubbing with a wire brush.

11. The last step in cleaning concrete block is to
   a. fill any holes.
   b. brush the wall.
   c. chisel off any large particles.
   d. rinse the work with water.

Unit 30
Removing Various Stains

OBJECTIVES

After studying this unit, the student will be able to

- identify stains on masonry work.
- select the proper treatment for removing stains.
- explain the importance of the proper use of cleaning agents.
- list protective clothing items to be worn when working with cleaning solutions.

On occasion, there may be dirt and stains other than mortar deposited on masonry work. These must be removed by the mason when the building is cleaned. There are different types of chemicals which are used for this purpose.

> **Caution**: Strict attention should be given to the directions for the use of any chemical, as they have a harmful effect if not used properly.

In many cases, small buildings or buildings which contain hard, smooth surfaces are cleaned successfully with regular soaps or detergents. This method works particularly well on glazed bricks or glazed tiles since mortar or other stains do not cling to this type of surface as easily as they do on a sand or rough-finished brick. The soap or detergent does not burn or damage the face of the masonry unit, as might other cleaning agents discussed in the previous unit. This method is more costly, however, since it is considerably slower. Because of this, regular soaps or detergents cannot be used on very large structures.

Although the mason is responsible for cleaning and washing masonry work, stains such as tar, paint, or welding splatters are the responsibility of the subcontractor and general contractor working on that particular job. The apprentice should be aware of materials that may stain and damage the masonry work, and report them to his or her employer at once so that the proper precautions may be taken. This unit discusses various stains, originating both internally (within the brick) and externally, and methods which may be used to correct the problem. Figure 30-1 lists various cleaning agents and their sources.

INTERNALLY CAUSED STAINS

Efflorescence

Generally, *efflorescence* refers to a white, powdery substance sometimes seen on masonry wall surfaces. It is composed of 1 or more water-soluble salts originally present in the masonry materials which have been brought to the surface by water and deposited on the surface by evaporation of the water. In many cases, it can be removed by applying clean water to the wall and then scrubbing with a brush. If this procedure does not remove all of the efflorescence, the surface is scrubbed with a solution of muriatic acid mixed no stronger than 1 part commercial acid to 12 parts water (by volume). It is very important that the wall be presoaked with water before any washing is done, and that the wall be thoroughly rinsed with water from a hose equipped with a nozzle after it has been washed with the acid solution.

Vanadium Stains

It is generally agreed that greenish stains are caused by salts present in the metallic element known as *vanadium.* While the stain is usually green, it is at times a brownish green and, more rarely, brown. The

Agent	Supply Source
1. Aluminum Chloride	Pharmacist.
2. Ammonia Water	Supermarket. Household ammonia water.
3. Ammonium Chloride	Pharmacist. Salt-like substance.
4. Ammonium Sulfamate	Nursery and garden stores. Past use was as a base for weed killers. Not now readily available. Substitute any brand weed killer solution.
5. Acetic Acid (80%)	Commercial and scientific chemical supply firms.
6. Hydrochloric Acid	Hardware stores. Muriatic acid is generally available in 18° and 20° Baumé solutions.
7. Hydrogen Peroxide (30 - 35%)	Some commercial and scientific chemical supply firms.
8. Kieselguhr	Commercial, scientific chemical and swimming pool supply firms. Diatomaceous earth.
9. Lime-free glycerine	Drug stores. Used as a hand lotion base.
10. Linseed Oil	Hardware and paint stores.
11. Paraffin Oil	Hardware stores.
12. Powdered Pumice	Hardware stores. A sanding or polishing material.
13. Sodium Citrate	Pharmacist. Appears like enlarged salt granules.
14. Sodium Hydroxide (Caustic Soda)	Supermarket. Available in brand name substances such as Drano.
15. Sodium Hydrosulphite	Pharmacist or photographic stores. A white salt or "hypo" of photographic fixing agent.
16. Talc	Drug stores. Inert powder available as "purified talc." Bathroom talcum powder may be substituted.
17. Trichloroethylene	Commercial/scientific chemical supply firms and possibly some service stations or supermarkets. A highly refined solvent for dry cleaning purposes.
18. Trisodium Phosphate	Paint stores, some hardware stores, supermarkets. Strong base type powdered cleaning material sold under brand names. Also available in brand name substance such as Calgon.
19. Varsol	Service Stations. A refined solvent by the brand name Varsol.
20. Whiting	Paint manufacturers, possibly some large paint stores. A powdered chalk. Substitute kitchen flour, if purchase is difficult.

Fig. 30-1 Sources of cleaning and masking agents

amount of vanadium in a brick is very small, about 0.01%. It is not known in what form the vanadium is present in the raw materials, in the fired bricks, or on the surface of the stained bricks. If this could be determined, the problem of removal would be greatly simplified. Research is currently being conducted to identify the sources of the compounds involved.

The following are three facts about the chemistry of vanadium which the mason should know.

- Vanadium salts may be divided into two classes, which include *colorless salts* which crystalize in alkaline or neutral solutions, and *colored salts* which are obtained from an acidic solution. The colorless salts are quickly soluble in water, while the colored salts are slightly soluble.
- The reaction of the colorless salts is practically instantaneous; the colored salts change very slowly.
- Vanadium salts react much more rapidly with acid than they do with alkaline substances.

Green-stained bricks often show no sign of a stain until they are washed in an acid solution, at which time the salts mix with the acid solution and then become evident on the face of the wall as dry, colored salts. It is impossible to determine if vanadium stains are going to appear on masonry. Therefore, it is good practice to test the effect of an acid wash on masonry units by applying it to a sample wall before washing the entire structure.

If green stains appear on the surface of the wall following the acid wash, the following procedure, which provides for neutralization of the acid, should be followed.

1. Flush the wall thoroughly with water.
2. Wash or spray the wall with a solution of potassium or sodium hydroxide, consisting of 1/2 lb hydroxide to 1 qt water (2 lb per gal). A paint brush may also be used to apply the solution. Allow this to remain on the wall for 2 or 3 days in order to neutralize the acid which causes green staining. An easy way to use sodium hydroxide is in the form of Drano®. The mixture that has been used successfully in testing by the BIA is 1, 12 oz can per quart of water. The sodium hydroxide, or Drano®, leaves the white salt which can be washed off with a hose. Various proprietary compounds, such as compounds 5, 6, and 7 in Figure 30-1, have proved successful in some cases.
3. The white salt left on the wall by the hydroxide may be hosed off the wall after 2 or 3 days or allowed to set until the first heavy rain removes it.

To date, research has not developed any single method for removal of green stains which can be recommended as best for all conditions.

Manganese Stains (Brown Stains)

Under certain conditions, manganese stains occur on mortar joints of brickwork made up of units colored with manganese dioxide. It appears as a tan, brown, nearly black, or, sometimes, gray stain. The *brown stain* has an oily look and may streak down the face of the brick. The salts are deposited when the solution reaches the mortar joints and becomes neutralized by the cement or lime.

During the burning process in the manufacture of some brick, the manganese coloring agents experience several chemical changes. This results in compounds that are not water soluble, but are soluble in weak acid solutions. Since brick can absorb acid, such weak acid solutions can prevail in brick washed with muriaticacid. Rainwater is also acidic in some highly industrialized areas.

To solve this problem, do not use muriatic acid solutions on tan, brown, black, or gray brick. There are proprietary cleaning compounds available for cleaning manganese brick. Advice of the brick manufacturer should be followed if there is a problem.

Permanent removal of manganese stains may be difficult. After the first removal, it sometimes returns. The following method has been very effective in removing brown stain and preventing its return.

1. Carefully mix a solution of acetic acid (80% or stronger), hydrogen peroxide (30-35%), and water in the following proportions by volume: 1 part acetic acid, 1 part hydrogen peroxide, and 6 parts water.

Caution: Although this solution is very effective, it is a dangerous solution to mix and use. Consult with your supervisor or instructor before attempting to mix or use this solution. Otherwise, serious injury could result.

2. After wetting the wall, brush and spray the solution on the wall. Do not scrub. The reaction is very rapid and the stain disappears quickly. After the reaction is complete, thoroughly rinse the wall with water.
3. A proprietary compound, Brick Klenz®, is sometimes effective in keeping the stain from reappearing. Brush or spray a solution of 1 part Brick Klenz® to 3 parts water by volume. Do not scrub it; allow it to remain on the brick surface.

An alternate solution suggested for new and light colored *brown stain* is oxalic acid crystals and water. Mix 1 lb crystals to 1 gal water.

Using the Poultice

A *poultice* is a paste consisting of a solvent and an inert material. The inert material may be talc, whiting, fuller's earth, or bentonite. The solution or solvent used depends upon the stain to be removed. Enough of the solution or solvent is added to a small quantity of the inert material to make a smooth paste. The paste is smeared on the stained area with a trowel or spatula and allowed to dry. It is then scraped off.

The solvent in the poultice dissolves the stain on the bricks. The resulting solution moves to the surface of the poultice where the solvent evaporates. The stain which is left on the loose, powdery residue is then removed. If all of the stain does not come off the first time, the procedure is repeated. The chief advantage of poultices is the way in which they work.

Poultices tend to prevent the stain from spreading during treatment and tend to draw the stain out of the pores of the brick.

If the solvent being used to prepare a poultice is an acid, do not use whiting as the inert material. As a carbonate, whiting reacts with acids to produce carbon dioxide. This is not dangerous but is messy and destroys the power of the acid.

Caution The student should wear eye protection and rubberized clothing when using any of these cleaning agents. If any of the cleaning agents are splashed on the eyes or bare skin, flush with clean water. If any discomfort persists, see a doctor immediately.

Paint Stains

For fresh paint, apply a commercial paint remover or a solution of trisodium phosphate and water (2 lb trisodium phosphate to 1 gal water). Allow the mixture to stand and remove it with a scraper and wire brush. Wash with clean water. For very old, dried paint, organic solvents similar to the above may not be effective. In these cases, scrubbing with steel wool or sandblasting may be required.

Iron Stains

Iron stains, quite common, sometimes cover entire walls. These stains are easily removed by spraying or brushing with a strong solution of 1 lb oxalic acid to 1 gal water Ammonium bifluoride added to the solution (1/2 lb/gal) speeds the reaction. The ammonium bifluoride generates hydrofluoric acid which etches the brick. The etching will be evident on very smooth bricks, and, therefore, the solution should be used with caution.

Another method is to mix 7 parts lime-free glycerine with a solution of 1 part sodium citrate and 6 parts lukewarm water, and mix with whiting or kieselguhr to form a thick paste. Apply the paste to the stain with a trowel and scrape it off when it dries. Repeat the process until the stain is removed. Wash well with clean water. A poultice made from a solution of sodium hydrosulphite and an inert powder (such as talc) also has been used for removal of iron rust stains.

Copper and Bronze Stains

Mix together dry 1 part ammonium chloride or sal ammoniac and 4 parts powdered talc. Add ammonia water and stir until a thick paste is formed. Place the mixture over the stain and leave until dry. An old stain may require several applications.

Welding Splatters

When metal is welded too close to a wall or pile of bricks, some of the molten metal may splash onto the bricks and melt the surface. The oxalic acid-ammonium bifluoride mixture which is recommended for iron stains is particularly effective in removing welding splatters. Scrape as much of the metal from the bricks as possible. The solution is then applied in a poultice. The poultice is removed when it has dried. If the stain has not disappeared, sandpaper is used to remove as much of it as possible and a fresh poultice is applied. For stubborn stains, several applications may be necessary.

Smoke Stains

Smoke stains are usually difficult to remove. A thorough scrubbing with scouring powder (preferably one containing bleach) and a stiff-bristled brush works well. Some alkaline detergents and commercial emulsifying agents may be brushed or sprayed on. These do a good job when they are given time to work. These have the added advantage that they can be used in steam cleaners. For more stubborn stains, a poultice using trichloroethylene usually draws the stains from the pores.

Caution: When using trichoroethylene, make sure the work area is well ventilated since the fumes are harmful.

Oil and Tar Stains

Oil and tar stains are effectively removed by commercial emulsifying agents. For heavy tar stains, the compounds can be mixed with kerosene to remove the tar and then with water to remove the kerosene. Sometimes, a steam-cleaning apparatus is used to remove tar without the use of kerosene. In a small area or in a situation where the job must be very clean, a poultice using benzene, naphtha, or trichloroethylene is more effective in removing oil stains.

Dirt Stains

Dirt can be very difficult to remove from a textured brick. Scouring powder and a stiff-bristled brush are effective if the texture of the unit is not too rough.

Scrubbing with the oxalic acid-ammonium bifluoride solution recommended for iron stains has proved effective on moderately rough textures. High-pressure steam cleaning appears to be the most effective method for cleaning dirt stains.

Straw and Paper Stains

Stains from straw and paper sometimes result from wet materials used to pack bricks for shipment. This stain can be removed by applying household bleach and allowing it to dry. Several applications may be necessary to remove the stain. A solution of oxalic acid-ammonium bifluoride cleans the stain more rapidly.

Plant Growth

Sometimes exterior masonry which is exposed to sunlight and remains constantly damp develops plant growth, such as moss. Applications of ammonium sulfamate (marketed under the manufacturer's brand name and available in gardening supply stores) made according to directions which come with the compound have been used successfully to remove such growths.

Stained masonry which is being recleaned after a long period of time may require a more severe method. High-pressure steam and sandblasting are two of the most successful methods employed. The mason seldom uses these methods, as they are usually left to professional cleaning companies.

ACHIEVEMENT REVIEW

Select the best answer from the choices offered to complete each statement. List your answer by letter identification.

1. The best method to remove efflorescence is to apply a solution of
   a. sulphuric acid.
   b. potassium.
   c. household cleanser.
   d. muriatic acid.

2. Manganese stains can be removed with
   a. muriatic acid.
   b. a combination of acetic acid, hydrogen peroxide, and water.
   c. whiting.
   d. trichloroethylene.

3. Green stains on masonry walls are caused by
   a. efflorescence.
   b. manganese.
   c. vanadium salts in the bricks.
   d. iron in the bricks reacting to the mortar.

4. The chief advantage of a poultice is that
   a. it is easier to apply than a solution.
   b. it draws the stain to the surface and stops it from spreading.
   c. it presents no danger to the mason who is applying the poultice.
   d. it permanently prevents the stain from reappearing.

5. Oil and tar stains are removed from masonry using certain compounds. Heavy tar stains are removed by adding to these compounds
   a. muriatic acid.
   b. household bleach.
   c. amonium sulfamate.
   d. kerosene.

6. Plant growth on masonry work is removed by using a solution of
   a. ammonium sulfamate.
   b. muriatic acid.
   c. whiting.
   d. talc.

7. Paint stains are removed with a commercial paint remover or a solution of
   a. muriatic acid and water.
   b. household bleach and water.
   c. trisodium phosphate and water.
   d. oxalic acid and water.

8. When in doubt about the effects of a cleaning compound on a masonry wall, the best practice is
   a. to clean only a small area and observe the results.
   b. to use judgment as to the type of cleaner to be used.
   c. not to attempt to remove the stain.
   d. to apply household bleach.

SUMMARY, SECTION 8

- The purpose of washing brickwork with a cleaning agent is twofold; to remove the mortar stains and to bring out the full color of the bricks.
- Cleaning time and labor can be drastically reduced if care is taken during construction to keep the masonry work clean.
- Damage to the masonry wall can be caused by poor cleaning practices and can result in rebuilding of the work.
- Whenever there is any question of the effect of a cleaning agent on the wall, it is good practice to first clean a small area and observe the results.
- Metal tools should not be used when cleaning with acid solutions since they may leave rust stains on the bricks.
- A rubber, wood, or good-quality plastic container should be used for storage of cleaning solutions.
- The mason should be dressed in waterproof clothing and safety eye glasses or goggles whenever working with cleaning compounds or solutions.
- When cleaning masonry work on a scaffold, be aware of any person nearby or underneath the scaffold.
- Presoak the wall in water before applying any acid or cleaning solution.
- Immediately after washing a certain section of wall (the size of the section depends on weather conditions), rinse with plenty of water until the water running down the wall is clear.
- Rough-textured brick walls should be rinsed with a pressurized source of water, such as a garden hose with a nozzle, to remove dirt from the deep pores.
- When cleaning brickwork, the mason must be careful not to mix a solution which is too strong, thereby burning the wall, and covering too great an area before rinsing, thereby allowing dirt to be absorbed in the wall.
- Wear a long pair of rubber gloves to cover part of the arm to prevent infections in open cuts. Rinse any area exposed to a cleaning agent immediately with plenty of clean water.
- Concrete block usually is cleaned without the use of water by rubbing the area with a scrap of block or brick unless there is a special problem. A good brushing usually removes all dirt.
- Filling holes in the wall with fresh mortar usually is done at the same time the concrete block wall is rubbed for stain removal.
- Always read directions before using any cleaning agent.

- Soap and detergents can be used to wash smooth bricks or tile, but this can be a very time-consuming task since they do not destroy mortar the way acid does.
- The best way to remove efflorescence from a masonry wall is to apply a solution of water and muriatic acid.
- Green staining is caused by the salts in vanadium and can be removed with sodium hydroxide applied with a paint brush.
- Manganese stains are very difficult to remove. The best method is to brush on an acetic acid-hydrogen peroxide solution and wash off with plenty of water.
- The chief advantage of a poultice is that it tends to draw the stain out to the surface pores of the brick, where it is easier to treat.
- A fresh paint stain can be removed with a paint remover or a solution of trisodium phosphate and water.
- Iron stains are removed with a combination of ammonium chloride and 4 parts powdered talc.
- A welding splatter can be removed with an oxalic acid-ammonium bifluoride mixture.
- A smoke stain can be removed by using a scouring powder containing bleach.
- Common dirt can be removed by using water and scouring powder.
- Straw and paper stains can be removed with a household bleach, or a solution of oxalic acid and ammonium bifluoride.
- Plant growth can be removed with ammonium sulfamate (sold in garden stores).
- Stubborn stains which cannot be removed with a cleaning agent may require the use of high-pressure steam or sand-blasting. These methods are done only as a last resort and are left to professional cleaners rather than to the mason.

SUMMARY ACHIEVEMENT REVIEW, SECTION 8

Complete each of the following statements which refer to material found in Section 8.

1. Brickwork is cleaned not only to remove mortar and dirt but to _______________.
2. Before applying any cleaning agents on brick walls, the walls should be _______________.
3. Using too strong an acid solution may cause the wall to _______________.
4. Washing too large a section with a cleaning solution without rinsing the wall may result in _______________.
5. When there is a question of how the masonry wall will react to a cleaning agent, it is good practice to _______________.
6. Metal tools or containers are not recommended for use around acid or cleaning agents because _______________.
7. For best results, acid solutions should be removed from the brick wall by _______________.
8. Proper clothing and protective devices worn when cleaning walls consist of _______________.
9. The chief disadvantage of proprietary compounds is _______________.

10. Concrete masonry products are not usually cleaned with ________________ or ________________ unless there is a special problem.
11. If an acid solution gets into a cut or into the eyes, the victim should immediately ________________.
12. Regardless of the type of cleaning agent being used, the mason should follow ________________ as a guide to its use.
13. Efflorescence is removed most easily with a solution of ________________.
14. Green staining on masonry walls is usually caused by ________________.
15. Manganese stain is usually ________________ in color.
16. The main advantage of a poultice is its ability to ________________.
17. Paint stains can be removed with a chemical called ________________ and water.
18. Iron stains are removed by spraying or brushing the area with a strong solution of ________________.
19. Many times, smoke stains are removed from brick walls by use of a common ________________.
20. Plant growth is removed from masonry by using a chemical sold at many garden stores called ________________.

Section 9
Understanding and Reading Construction Drawings

Unit 31
Specifications

OBJECTIVES

After studying this unit, the student will be able to

- explain the relationship between specifications and working drawings as plans.
- discuss what information is contained in average masonry specifications.
- list divisions of a typical set of general specifications.
- explain the purpose and importance of specifications.

A skilled mason should be able to perform work without constant supervision. The ability to read and understand building specifications and working drawings provides this chance for independence on the job. Once the basics of plan reading have been learned, the process becomes a simple one. To perfect the mason's skills in reading plans, a course in plan reading should be taken to supplement this material.

Plan reading is a subject which covers all phases of the building trades. To apply the specifications to the actual construction, it is important to be able to picture mentally the features of the finished building before it has been constructed.

Specifications are defined as the written or printed instructions and information that are needed to complete a structure. They form part of the contract, describing qualities and methods of construction and stating any information not shown on the plans. Since specifications are part of a legal document, they can be used in a court of law as evidence. It is important to remember that specifications are the written interpretations and instructions from the architect and the owner of the proposed structure. They are an essential part of the contract and supersede the drawings when there are differing statements. When conflicts in the specifications and drawings arise, however, the job supervisor and architect should be consulted. It is helpful to learn early in a masonry career the contractual relationships among the masonry contractor, the general contractor, and the architect.

Specifications contain information concerning the entire structure, generally arranged by trade areas. A table of contents or index is usually supplied with the specifications so that information may be easily found.

SPECIFICATIONS AS THEY RELATE TO WORKING DRAWINGS

Working drawings (more commonly called *blueprints* or simply *plans*) are drawings made by an architect which contain detailed layouts of the structure and measurements. Working drawings consist of foundation drawings, floor plans, elevations, sectional views, and special details of construction. Numerous other items may be attached to a set of drawings. Some of these items include a plot plan showing the entire property, locations of the septic system, well, and driveway, and landscaping details. As a rule, the mason is concerned primarily with those parts of the plans which relate to the masonry work. The mason must have both the specifications and the plans to construct the building.

It is important for the mason to understand the difference between specifications and working drawings. Working drawings may show the location, height, length, and thickness of a brick wall along with sizes of any openings which must be left in the wall. The specifications, on the other hand, indicate the kinds of bricks, methods of bonding, and the type of mortar to be used.

Reading Specifications and Plans

Reading specifications and plans can be learned in several different ways. Observe at every opportunity on the job how the building is being constructed. Feel free to ask as many questions as necessary of the supervisor and experienced masons. Examine the specifications and plans on various projects in your spare time and relate them to the work as it is actually being done. There are many excellent books dealing with reading of specifications and plans that can be obtained for home study. A systematic study of these will help the mason acquire basic knowledge of reading and understanding plans.

Masons should realize early in their careers the importance of learning details of specifications and plans relating to masonry and other building trades. In this way, masons can make provisions for workers in other trades as they work.

SCOPE OF THE WORK

Scope of the work is a term found in practically all specifications. This entails exactly what the worker is responsible for completing. On some jobs, the mason is responsible for caulking around windows and doors. On other jobs, it is classified as the work of the carpenter. Important items, such as cleaning after the work has been completed, should be spelled out clearly in the statement on the scope of the work.

Many specifications require that the job be kept in a clean, orderly condition during construction to prevent accidents. The architect, or the architect's representative, conducts periodic inspections to be sure that the conditions set forth in the scope of the work are followed. Estimators take the scope of the work into account when they estimate the cost of the structure, as they do for all aspects of the specifications. Carefully written specifications make it possible for the estimator to price jobs accurately.

Generally, a set of specifications for masonry and stonework should include the following information.

Scope of the Work

- Type of work and materials to be used

Type of Materials

- Type of face and common brick, concrete block, and stone, and type of mortar mix specified for each
- Type of waterproofing for foundation wall and where it is to be applied to the wall
- Quality standards which the materials should meet, such as compressive strength and color range

Masonry Workmanship

- Quality of workmanship expected
- Type of mortar joint finish and, often, the type of bond desired
- Size of pieces of masonry units to be installed in the wall; smallest allowable piece (for example, no piece smaller than 2" in face brickwork). Many masonry specifications stipulate that no bats are to be used in place of headers.
- Cutting instrument to be used on masonry units, such as a chisel and hammer or masonry saw
- Care of the wall at the conclusion of the day's work, such as the stipulation that all masonry work be covered with tarpaulins or heavy plastic to protect it from the weather
- Minimum temperature that work may commence to prevent the freezing of mortar
- Specific instructions on the parging of the basement walls

Stonework

- Type of stone and bond pattern desired
- Size and type of mortar joint to be used; joint finish required, such as the raked or head joint
- Method in which the stone is to be cut, chiseled, or sawed with the masonry saw
- Type of protection for the work

Concrete Masonry Unit

- Specific shape, such as bull nose or corner block
- Type of mortar and mix

- Type of joint finish and allowable thickness
- Type of wall ties or joint reinforcement
- Type of lintel to be placed over openings, such as concrete masonry lintel, steel, or concrete poured in form

Building in and around Mechanical Work

- Methods of installation of pipe chases, heating units, and electrical work

Caulking

- Type of caulking and filler materials to be installed in back of the caulking if necessary
- Workmanship to be neat and excess caulking removed from frames

Cleaning and Pointing Masonry Work

- Type of cleaning agent to be used; instructions on how concrete block and stone are to be cleaned
- Directions indicating that all holes in mortar joints are to be pointed with fresh mortar as the work is cleaned

General specifications usually consist of sixteen divisions as developed by the Construction Specifications Institute. The following are these sixteen divisions.

General requirements and conditions
Site work
Concrete work
Masonry work
Metals
Carpentry
Moisture control
Doors, windows, glass
Finishes
Specialties
Equipment
Furnishings
Special construction
Conveying systems
Mechanical
Electrical

THE PURPOSE OF SPECIFICATIONS

Plans and specifications are essential in estimating and constructing buildings. They help avoid disputes among the owner, general contractor, subcontractor, and architect. They also supply much information that is not shown on the plans.

By consulting the specifications, conflicting opinions regarding the grade of materials to be used or workmanship may be greatly decreased or eliminated. The specifications for a project must be studied very carefully, as the misinterpretation of a single word or phrase could lead to legal difficulties.

Although masons should carefully study specifications and building plans, they should also use good judgment in their use. If there is an obvious mistake in the specifications or plans, the problem should be solved before the work is continued. The mason should consult his or her supervisor with any questions about specifications. As a rule, the masonry foreman goes to the superintendent for a ruling but if the problem is not minor, the masonry contractor should be informed before work continues.

ACHIEVEMENT REVIEW

A. Select the best answer from the choices offered to complete each statement. List your choice by letter identification.

1. Specifications differ from plans in that they
   a. act as the contract between the builder and owner of the building under construction.
   b. contain all measurements needed to build the structure.
   c. contain all critical information not included in the plans.
   d. specify where materials are obtained.
2. Whenever there is a conflict between plans and specifications, the mason foreman should
   a. use his or her own judgment and proceed to do what seems best.
   b. call the architect for a decision.
   c. call the prospective owner and discuss the problem.
   d. consult the job supervisor and the architect so that they can make a decision.

3. The term *scope of the work* indicates
   a. a contract stipulating the length of time allotted to construct the building.
   b. what work the mason is specifically responsible for completing.
   c. specific instructions concerning the type of materials to be used.
   d. information contained in drawings.

4. General requirements and conditions of a project are found in the specifications at the
   a. beginning.
   b. middle.
   c. end.
   d. preface to the work.

5. The place in which the brickwork begins in relation to the finished grade is listed in the masonry specifications under the
   a. type of materials.
   b. scope of the work.
   c. masonry workmanship.
   d. protection of the work.

B. Using the specifications format found in this unit, write the specifications for a small masonry job of your choice. Be brief but to the point in describing requirements for the proposed job. After completing the assignment, students may exchange the specifications to help determine if they are suitable.

Unit 32
Line and Symbol Identification

OBJECTIVES

After studying this unit, the student will be able to

- explain the use of lines in a set of plans.
- recognize the most commonly used symbols and explain their meaning.
- define certain abbreviations found in plans.
- explain what comprises a schedule and its importance to the mason.

Working drawings or plans consist of a group of drawings whose sole purpose is to inform various craftspersons what their specific jobs are and how they are to be completed. All phases of construction require specific information from plans. Since plans can be lengthy and complicated, architects developed a language of their own involving various types of symbols, conventions, and abbreviations. (A *convention* is a pictorial representation of something which cannot be identified on a plan by a symbol or abbreviation.) By studying and learning these, the mason can read and interpret the architect's drawings and specifications and proceed with the work as indicated.

It is extremely important that the workers understand exactly what the architect wants accomplished on the structure. It is equally important that the architect not omit details of construction. Constant communication between the architect's office and the jobsite is very time consuming and can be avoided if the plans are written and interpreted properly. Symbols and abbreviations help make plans concise and readable.

THE USE OF LINES IN PLANS

Lines as used in building plans comprise a major trade language. The thickness of the line and the way it is drawn have specific meanings, Figure 32-1.

Main Object Line

The *main object line* is a heavy continuous line which shows visible outlines or edges of what a person sees when looking at the structure.

Dimension and Extension Line

A *dimension line* shows the actual distance between two points. The line is lighter in weight than the main object line and has an arrowhead, slash, or dot at each end. The arrowhead, slash, or dot, which touches the extension lines, represents the edges of the main object or structure. The numbers in the break of the dimension line give the exact distance between the two points.

Dimension lines are used in conjunction with extension lines. An *extension line* is of the same weight (thickness) as the dimension line and extends from the edge of the main object.

Cutting Plane Line for Section Indicator

Cutting plane lines are solid lines indicating that a wall, building, or object has been cut at some point along the line. The cross sections of areas indicated by cutting plane lines are shown in greater detail elsewhere in the plans. The arrows indicate the direction in which you are looking as you view the cross section. The particular cross sections may be indicated by letters such as A, B, or C.

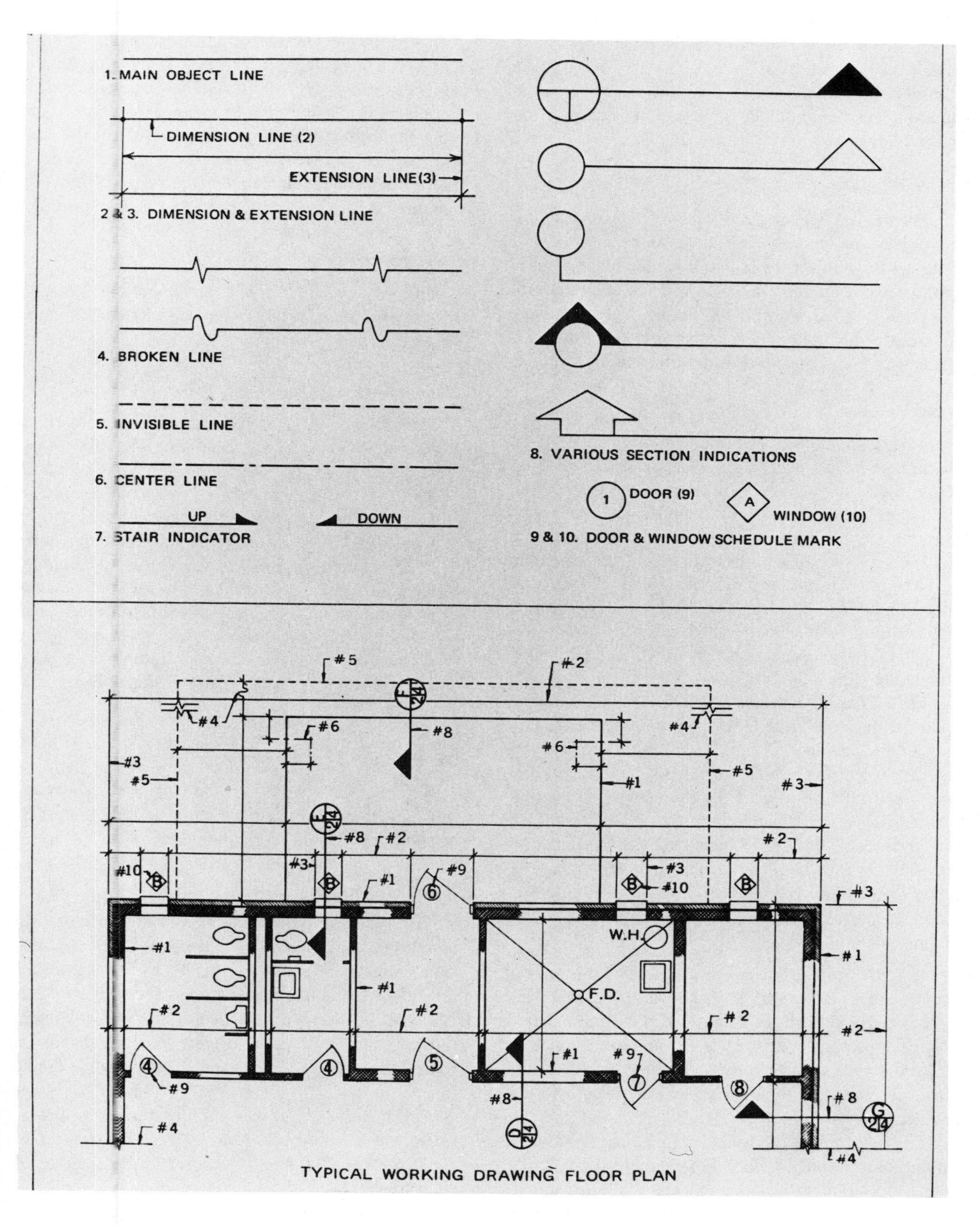

Fig. 32-1 Typical working drawing floor plan (bottom) with various lines used (top)

When the cross section to which the cutting plane lines refer is not shown on the same page of the drawing, symbols such as F/24 or G/24 are used. These symbols inform the reader that the particular cross section will be found in the plans in Section F on page 24 or in Section G on page 24.

Broken Line

The *broken line* indicates that parts of the structure or object are omitted from the plan or that the full length of the object is not drawn on the plan. This is done because an item on a plan which is shown in full by a long, constant line would require a great amount of space. One broken line allows the item to be shown in shorter form, and vital information is not omitted.

Invisible Line

The *invisible line,* consisting of a series of short dashes, indicates edges of objects which are hidden from view or which are under another part of the building. When the line continues through the actual structure, the worker can be sure that there are parts of the structure below the surface. In this case, the worker must locate another view in the drawings to determine at what level the hidden item occurs. Often, these hidden parts are explained in an elevation view or a cross section view of that area to which the hidden line refers.

For example, consider a first floor plan. Assume that the mason is viewing the plan to examine the layout of the fireplace. The foundation of the fireplace is not shown on the first floor plan, but on the basement plan. If the base of the foundation of the fireplace is larger than the fireplace itself, the size is shown by an invisible line. Steel beams are also indicated by invisible lines to help the carpenter locate the beams after the subfloor is laid.

Centerline

A *centerline* (often called a *dot and dash line*) is composed of alternating long and short dashes. This line is relatively light in weight and is used to indicate or locate center points. It is a common practice in the building trades to locate an object by measuring from the center of the object to the center of another object.

The mason uses this system of measurement when laying out masonry walls. For example, 4" concrete block does not really measure a full 4", but is actually 3 5/8" in width. If, when laying out a long space, the mason assumes that all walls are 4" in width and measures the space between walls from the outside of one wall to the outside of the next wall, the total measurement will be inaccurate. Depending on the length of the space, the difference in measurement can be great. If, however, all measurements are taken from the center of one wall to the center of the next wall, any difference in measurement is undetectable. Therefore, it is recommended that masonry units be measured from center to center whenever possible.

Stair Indicator

The *stair indicator* is a short, unbroken line with an arrowhead at one end indicating whether the stairs are ascending or descending.

Door and Window Indicators

Many times, doors and windows of various types and sizes are used on the same job. Therefore, they must be identified correctly before they are installed.

There are different ways to designate doors and windows. One of the most common methods is shown in Figure 32-1. The door type is indicated by a number inside a circle; the window is shown by a number inside a diamond. To determine the sizes and descriptions of the windows and doors as shown on the plans, the window or door schedule must be studied. As a rule, the schedule appears on a separate sheet of plans and can be found by consulting the index or table of contents.

SYMBOLS

Without the use of symbols, architects cannot show all necessary information regarding materials, methods, and location of components.

Types of Symbols

The types of symbols used include those used in elevation views and those in sectional views. Elevation symbols are easily recognized, as they look very much like the actual material or object, Figure 32-2. An *elevation view* is a vertical picture of an object showing the front, side, or rear view of an object, room, or structure as one would view it while facing it.

The materials shown in an elevation view appear differently in a sectional view. A *sectional view* shows the object as if it were sliced vertically, showing of what the object would be composed, Figure 32-3. For example, a sectional view of a masonry wall would show the thickness of the joints and the units,

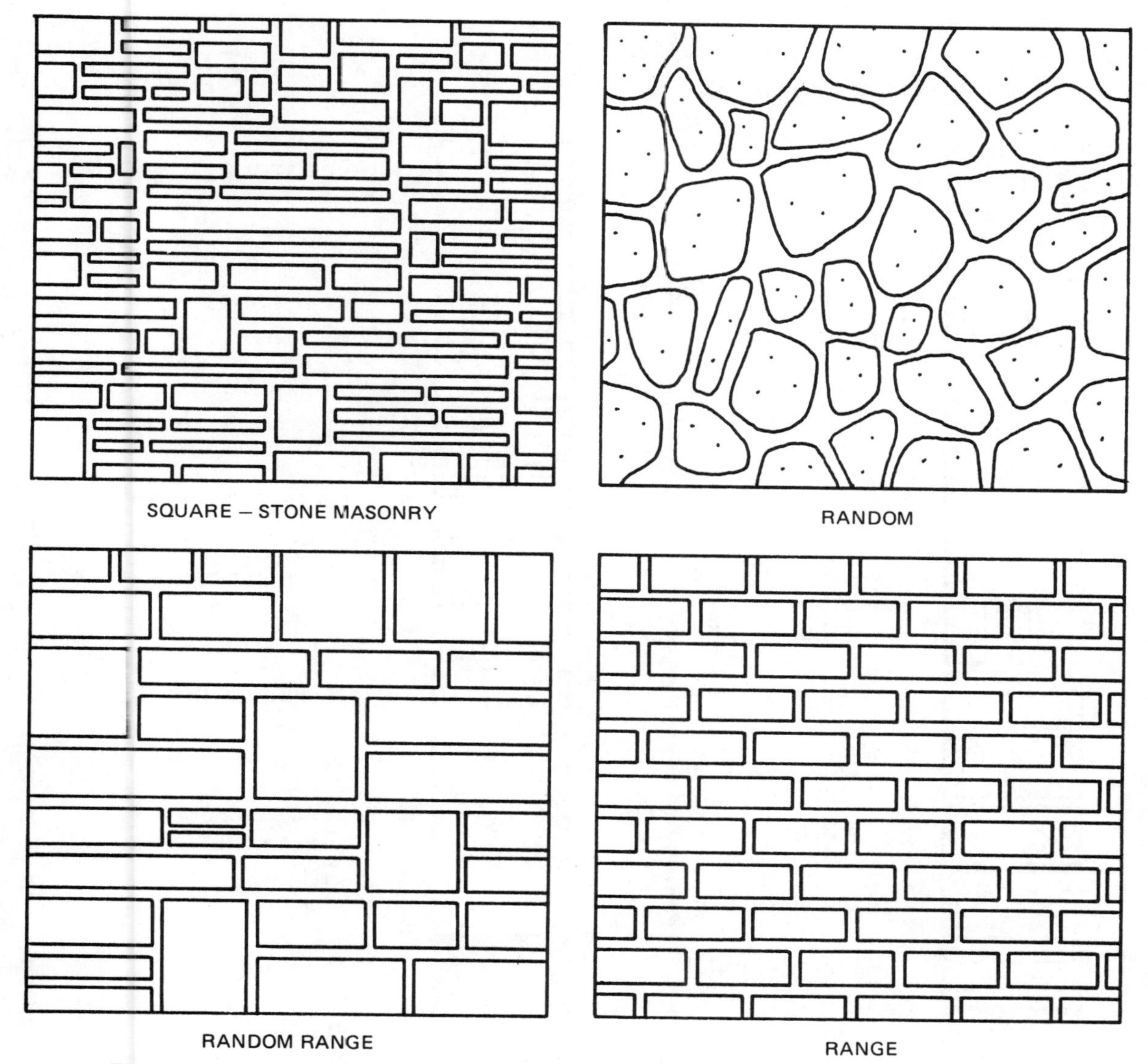

Fig. 32-2 Elevation masonry symbols. Notice both materials used and construction details.

how the wall ties are installed, and, many times, the exact height of the wall.

The mason should be familiar with some of the more common symbols for the mechanical trades, as they may affect the work when building in or around certain equipment. For example, it may be necessary for the mason to provide for electrical switch boxes, heating units, or built-in plumbing fixtures. Figure 32-4 shows typical electrical and plumbing symbols.

ABBREVIATIONS

As masons view a set of plans, they should notice the use of abbreviations. Since they are used so often, the more important ones affecting the mason should be memorized. Figure 32-5 shows abbreviations commonly used in construction work.

SCHEDULES

A *schedule,* in architecture, is a list added separately to the plans which describes such items as windows, doors, floors, and wall finishes. The mason must be able to recognize and interpret schedules correctly, or great expense could result from necessary changes made at the conclusion of the job so that the job conforms to the finished schedule.

As mentioned before, windows and doors are designated by a number or letter on plans. The same letter or number is duplicated in the schedule, with a

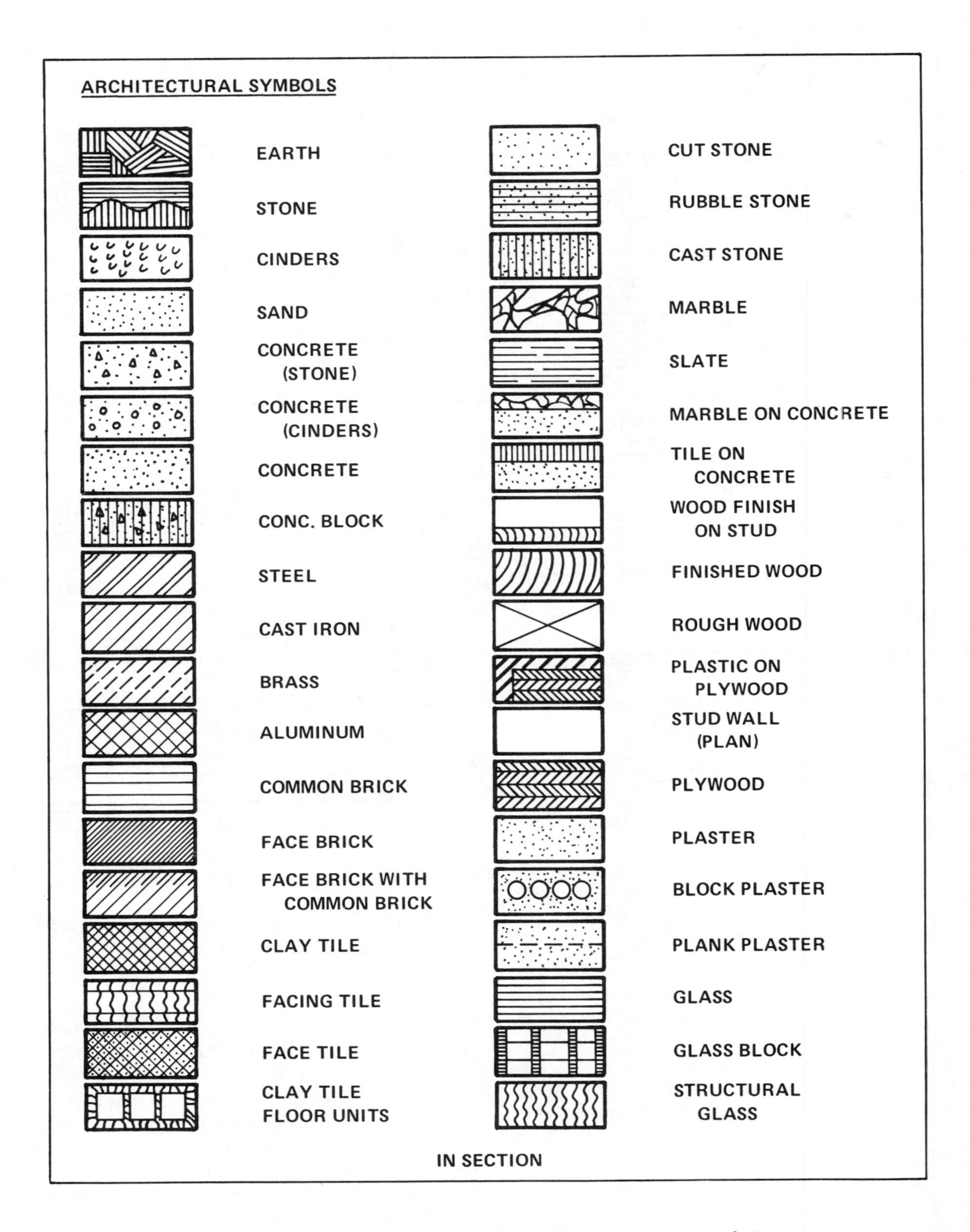

Fig. 32-3 Architectural symbols showing materials in the sectional view

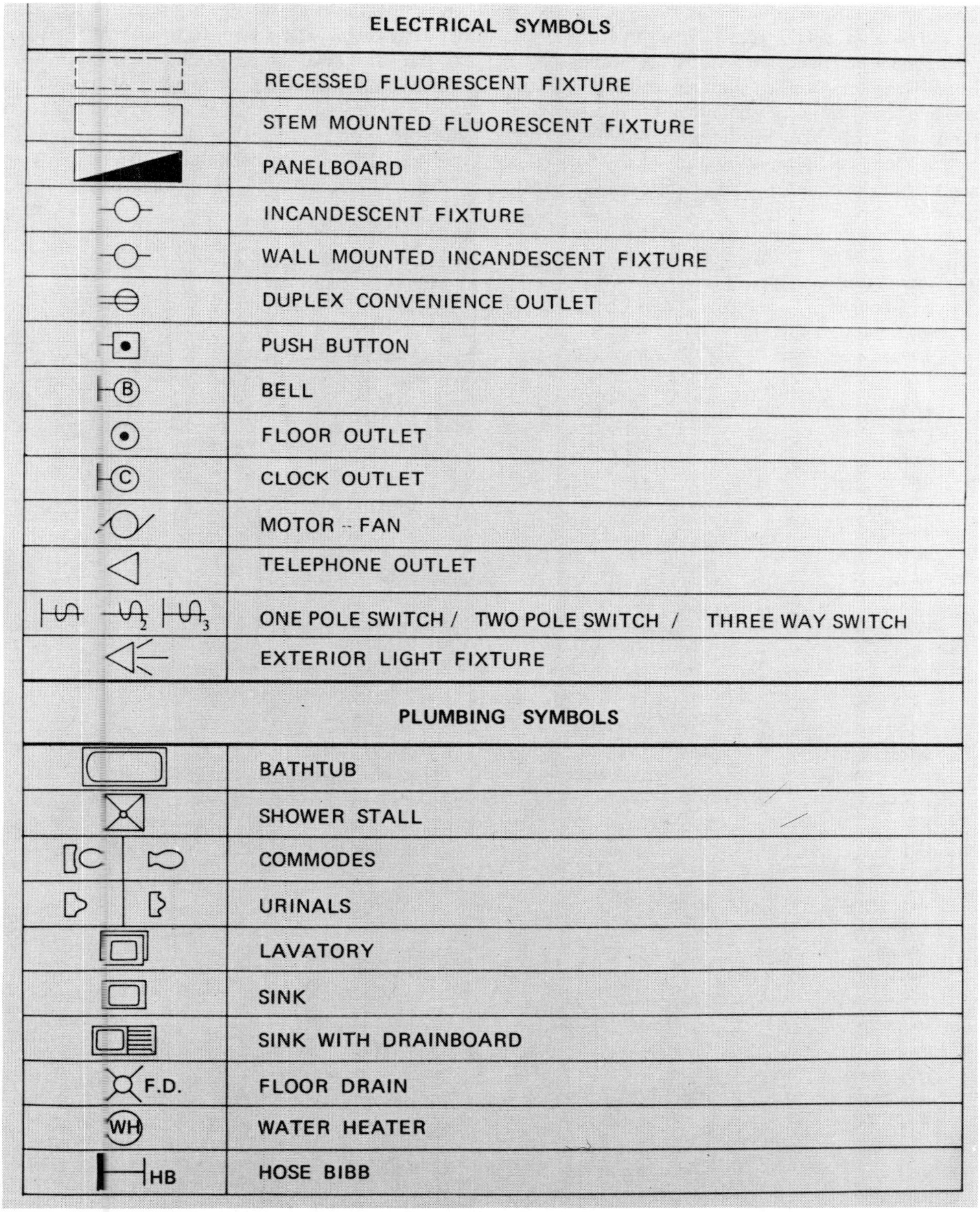

ELECTRICAL SYMBOLS
RECESSED FLUORESCENT FIXTURE
STEM MOUNTED FLUORESCENT FIXTURE
PANELBOARD
INCANDESCENT FIXTURE
WALL MOUNTED INCANDESCENT FIXTURE
DUPLEX CONVENIENCE OUTLET
PUSH BUTTON
BELL
FLOOR OUTLET
CLOCK OUTLET
MOTOR -- FAN
TELEPHONE OUTLET
ONE POLE SWITCH / TWO POLE SWITCH / THREE WAY SWITCH
EXTERIOR LIGHT FIXTURE

PLUMBING SYMBOLS
BATHTUB
SHOWER STALL
COMMODES
URINALS
LAVATORY
SINK
SINK WITH DRAINBOARD
F.D. FLOOR DRAIN
WH WATER HEATER
HB HOSE BIBB

Fig. 32-4 Typical electrical and plumbing symbols

brief description of the item, Figure 32-6. Typical door and window schedules must show their relationship to the floor plan, Figure 32-7.

It should be remembered that schedules consist of brief descriptions; workers must also consult the drawings and specifications to obtain all the necessary information. Schedules help greatly in estimating the cost of a job. For example, by studying the floor plan and schedule in Figure 32-7, it can readily be seen that 3 Type 2 windows are required. It can also be determined that the living room has oak floors, plaster walls, a plaster ceiling, and a wooden base at the floor line.

The schedule is very important to the mason when working on commercial buildings. For example, it would be very easy to build a masonry wall and strike the joints, when plaster may have been specified on the plans. This type of misunderstanding may be avoided by carefully studying the schedule.

access door	AD	exterior	EXT
access panel	AP	feet	FT
acoustical tile	AT	finished floor	FIN. FL
aggregate	AGGR	firebrick	FBRK
anchor bolt	AB	fireplace	FP
angle	L	flashing	FL
approximate	APPROX	floor	FL
barrel	bbl or BBL	footing	FTG
basement	BSMT	foundation	FND
beam or bench mark	BM	glass block	GL BL
blocking	BLKG	grade	GR
blueprint	BP	height	HB
board	BD	hot water	HW
brick	BRK	I beam	I
building	BLDG	inch	in or "
building line	BL	insulation	INS
cast iron	CI	level	LEV
cast stone	CS	lineal feet	lin ft
catch basin	CB	louver	LV
caulking	CLKG	manhole	MH
ceiling height	CLG HT	marble	MR
cement mortar	CEM MORT	masonry opening	MO
centerline	C or CL	modular	MOD
center to center	C to C	not to scale	NT
channel	CHAN	on center	OC
cinder block	CB	opening	OPG
cleanout door	COD	partition	PTN
concrete	CONC	plaster	PLAS
concrete masonry unit	CMU	plate glass	PL GL
column	COL	precast	PRCST
common	COM	reinforced	REINF
contractor	CONTR	rough opening	RGH OPNG
courses	C	schedule	SCH
cross section	X-SECT	specifications	SPEC
dampproofing	DP	wall vent	WV
detail	DET	weatherproofing	WP
elevation	el or EL	weephole	WH
excavate	EXC	window	WDW
expansion joint	EXP JT		

Fig. 32-5 Architectural abbreviations

WINDOW SCHEDULE		
NO.	SIZE	TYPE
①	3'-9" x 2'-9"	STEEL SECURITY SASH
②	1'-9" x 2'-9"	STEEL SECURITY SASH
③	2-2'-8" x 4'-6"	WOOD DOUBLE HUNG

MATERIAL LEGEND
CONCRETE
BRICK
CINDER BLOCK
WOOD STUD PART

DOOR SCHEDULE		
NO.	SIZE	TYPE
Ⓐ	3'-0" x 6'-8"	KALAMEIN - SELF CLOSING WITH FIRE CODE LABEL
Ⓑ	3'-0" x 7'-0" x 1 3/4"	WOOD - PL. GLASS PANEL
Ⓒ	3'-0" x 7'-0" x 1 3/4"	WOOD - OBSCURE WIRE GL. PNL
Ⓓ	3'-0" x 6'-8" x 1 3/4"	WOOD - OBSCURE GL. PNL.
Ⓔ	2'-8" x 6'-8" x 1 3/8"	WOOD
Ⓕ	2'-4" x 6'-8" x 1 3/8"	WOOD
Ⓖ	2'-4" x 6'-8" x 1 3/8"	WOOD - LOUVERED LOWER PNL

LINTEL SCHEDULE	
NO.	TYPE
L-1	4 - 3/4" ϕ RODS
L-2	3 Ls - 4" x 3 1/2" x 5/16"
L-3	1 L - 4" x 4" x 5/16" 2 Ls - 4" x 3' x 5/16"
L-4	3 Ls - 3" x 3" x 1/4"

Fig. 32-6 Schedule for a small job, including information on windows, doors, and lintels

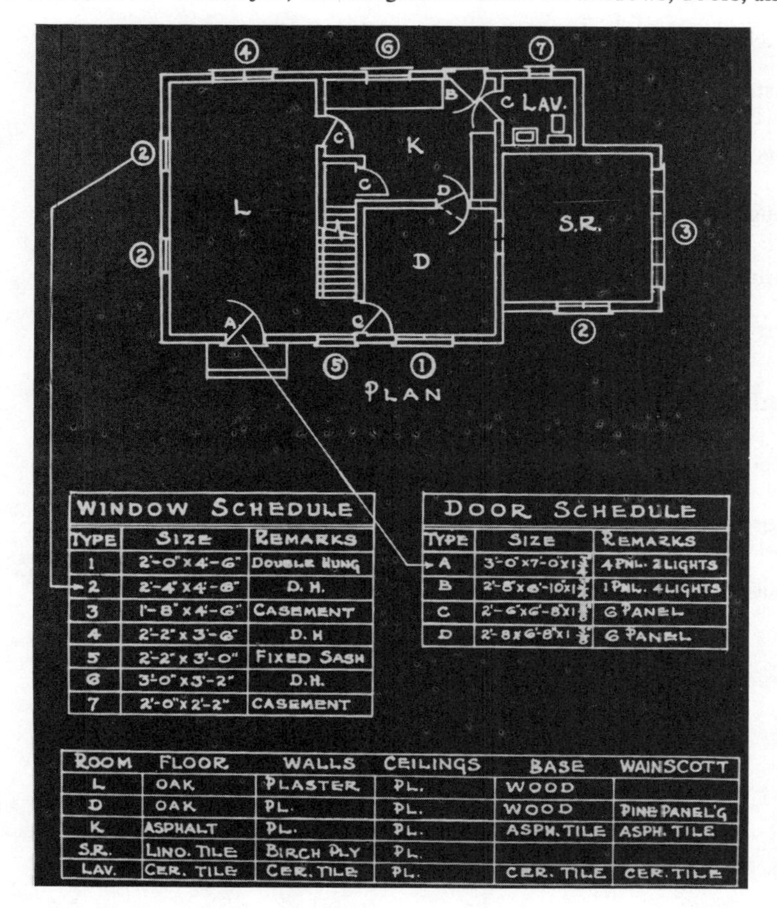

WINDOW SCHEDULE		
TYPE	SIZE	REMARKS
1	2'-0" x 4'-6"	DOUBLE HUNG
2	2'-4" x 4'-8"	D. H.
3	1'-8" x 4'-6"	CASEMENT
4	2'-2" x 3'-6"	D. H
5	2'-2" x 3'-0"	FIXED SASH
6	3'-0" x 3'-2"	D. H.
7	2'-0" x 2'-2"	CASEMENT

DOOR SCHEDULE		
TYPE	SIZE	REMARKS
A	3'-0" x 7'-0" x 1 3/4	4 PNL. 2 LIGHTS
B	2'-8" x 6'-10" x 1 3/4	1 PNL. 4 LIGHTS
C	2'-6" x 6'-8" x 1 3/8	6 PANEL
D	2'-8 x 6'-8" x 1 3/8	6 PANEL

ROOM	FLOOR	WALLS	CEILINGS	BASE	WAINSCOTT
L	OAK	PLASTER	PL.	WOOD	
D	OAK	PL.	PL.	WOOD	PINE PANEL'G
K	ASPHALT	PL.	PL.	ASPH. TILE	ASPH. TILE
S.R.	LINO. TILE	BIRCH PLY	PL.		
LAV.	CER. TILE	CER. TILE	PL.	CER. TILE	CER. TILE

Fig. 32-7 Floor plan and related schedule. Note: Arrows show where the schedule items should be applied to the floor plan.

ACHIEVEMENT REVIEW

The column on the left contains lines, symbols, and abbreviations which appear in building plans and schedules. The column on the right contains the actual item. Select the correct item from the right-hand list and match it with the proper term on the left.

Part A. Line Identification

1.	Invisible edge line	a.	
2.	Stair indicator	b.	
3.	Door and window (schedule mark)	c.	
4.	Dimension line	d.	
5.	Centerline	e.	I A
6.	Section indicators	f.	
7.	Broken line	g.	
8.	Main object line	h.	UP DOWN

Part B. Symbols (Sectional)

1.	Plaster	a.	
2.	Concrete	b.	
3.	Rough wood	c.	
4.	Marble	d.	
5.	Face brick	e.	
6.	Earth	f.	
7.	Steel	g.	
8.	Cast stone	h.	
9.	Finished wood	i.	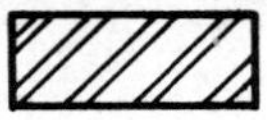

Part C. Abbreviations

1.	Basement	a. BBL
2.	Concrete masonry unit	b. MO
3.	Center to center	c. SPEC
4.	Cinder block	d. CHAN
5.	Plaster	e. GL BL
6.	Specifications	f. AGGR
7.	Modular	g. COL
8.	Brick	h. CMU
9.	Contractor	i. CB
10.	Grade	j. BM
11.	Weep hole	k. PLAS
12.	Glass block	l. BSMT
13.	Aggregate	m. C TO C
14.	Barrel	n. BRK
15.	Column	o. MOD
16.	Bench mark	p. GR
17.	Masonry opening	q. WH
18.	Building	r. BLDG
19.	Louver	s. CONTR
20.	Channel	t. LV

Unit 33
The Working Drawing

OBJECTIVES

After studying this unit, the student will be able to

- describe the major sections of a set of working drawings.
- explain the relationships between various parts of a drawing.

Every construction job, regardless of the size and design, must have a set of working drawings. *Working drawings* are completely dimensioned views with all necessary notes accompanying the drawings. Working drawings are supplied to all of the trades involved in that particular job.

A complete set of drawings for a structure usually includes three major sections: architectural, structural, and mechanical and electrical. If the project is large or complicated, there may be a set for each major section to reduce the time necessary for the various tradespersons to locate information.

Within the three major sections are subdivisions, usually assembled in the following order: plot plan, foundation plan and floor plan, elevation drawing, sectional drawing, and detailed drawing.

When studying plans, the mason must be especially careful to read all notes included in the drawings, since they supply a great amount of information supplementing the drawing. It is important to consult the specifications for any information not shown on the plans.

PLOT PLAN

The architect usually draws a plot plan for a building of any size or importance. The *plot plan* (also known as a site plan), Figure 33-1, supplies information on the following:

- Property line
- Overall building lines for the proposed structure
- Contour lines showing the rise and fall of the ground
- Indications of the location of trees, shrubs, sidewalks, driveways, septic systems, and well

Also included on the plot plan is a reference point for determining elevations, commonly called a *bench mark,* and the *grade line* or *grade level,* which shows the rough and finished levels of the ground surrounding the structure. The elevation is calculated from coastal sea level points established by the National Oceanographic and Atmospheric Administration. Information on the grade line is very important to the masons on the job because they must be able to determine at which level to stop laying concrete block for the foundation and where to begin laying the exposed masonry such as brick or stone. The location of the land on which the job is to be built is determined by the surveyor from city or county land records, deeds, or known reference points.

A complete plot plan is necessary before local planning and zoning authorities will issue a building permit. Many banks and mortgage companies require that a survey and plot plan be submitted before loans on the property can be made.

The plot plan usually indicates the location of the building on the plans in reference to the points of a compass. This greatly simplifies locating a particular view on a plan. It is a good practice to spread the plot plan out on the worktable in the same position as the building is to actually be built on the lot to eliminate possible errors.

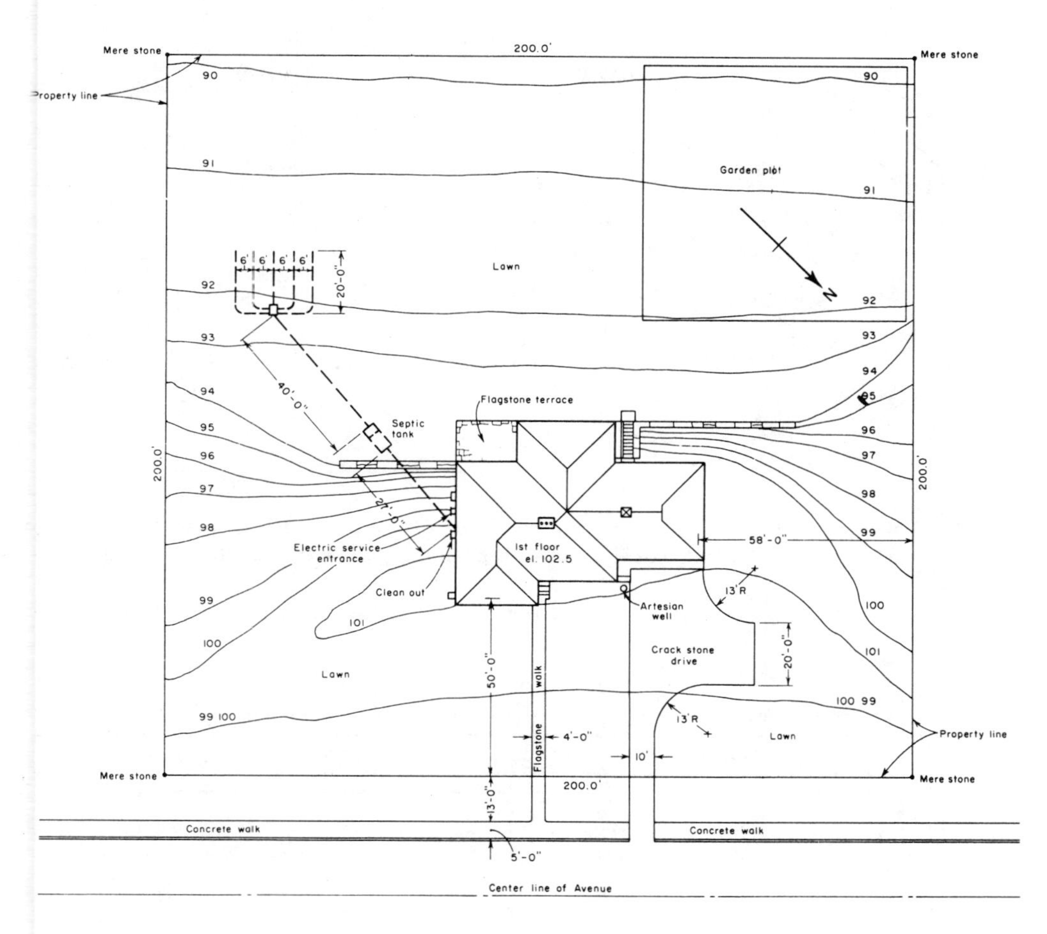

Fig. 33-1 Typical plot plan

FLOOR AND FOUNDATION PLAN

Floor plans show the measurement and location of such items as walls, windows, doors, chimneys, and electrical devices, Figure 33-2.

Most sets of plans also contain plan views for the foundation and second floor if they exist, Figure 33-3. The information found on the foundation plan is of extreme importance since all of the upper parts of the structure depend upon accurate layout and construction of the footing and foundation. All measurements and dimensions should be checked at least twice with those on the floor plans. In cases where an overall measurement is shown on the drawing of a wall, with many smaller measurements of windows and doors shown on the same wall, the mason should add all of the smaller measurements together to be sure that the total is the same as the overall measurement. Simple mathematical errors should be caught before they become problems later. This process is known as *cross-checking.*

ELEVATION DRAWING

An *elevation* is the view of the structure that the worker has from a normal standing position, Figure 33-4. Elevation views indicate certain exterior features with which the worker must be familiar, such as the floor heights, finish grades, foundation and footings, and exterior and interior walls.

A typical house elevation shows a right elevation, left elevation, front elevation, and rear elevation. On many drawings, they are indicated as north, east, south, and west elevations. All of the elevation views of the building must be studied before it is possible to visualize the structure as it will look when completed.

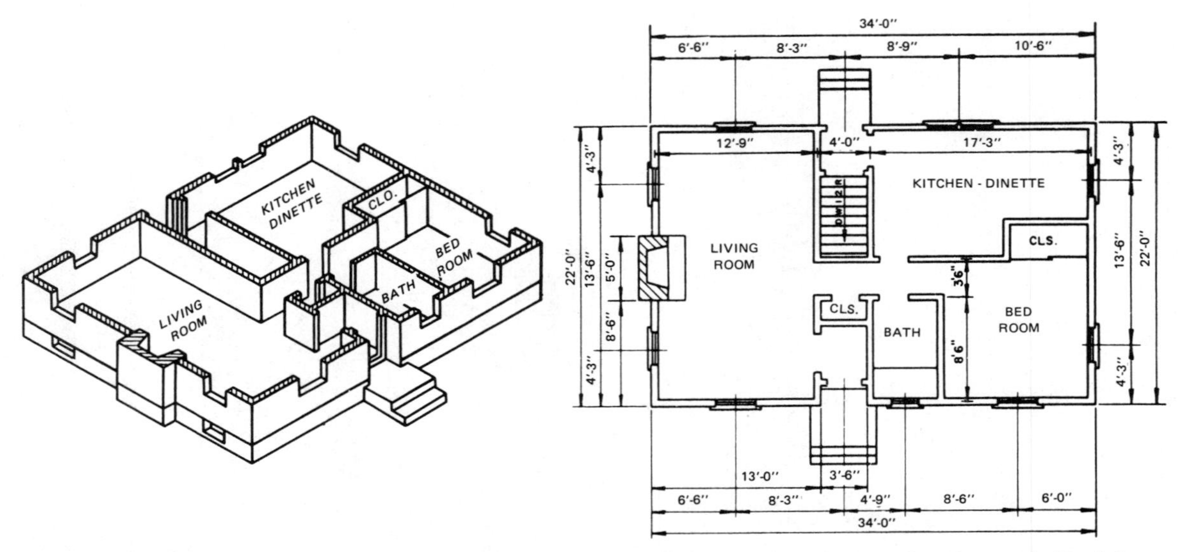

Fig. 33-2 Typical floor plan. To obtain the proper view of a floor plan, assume that the top half of the structure has been removed and that the viewer is examining the exposed section from above.

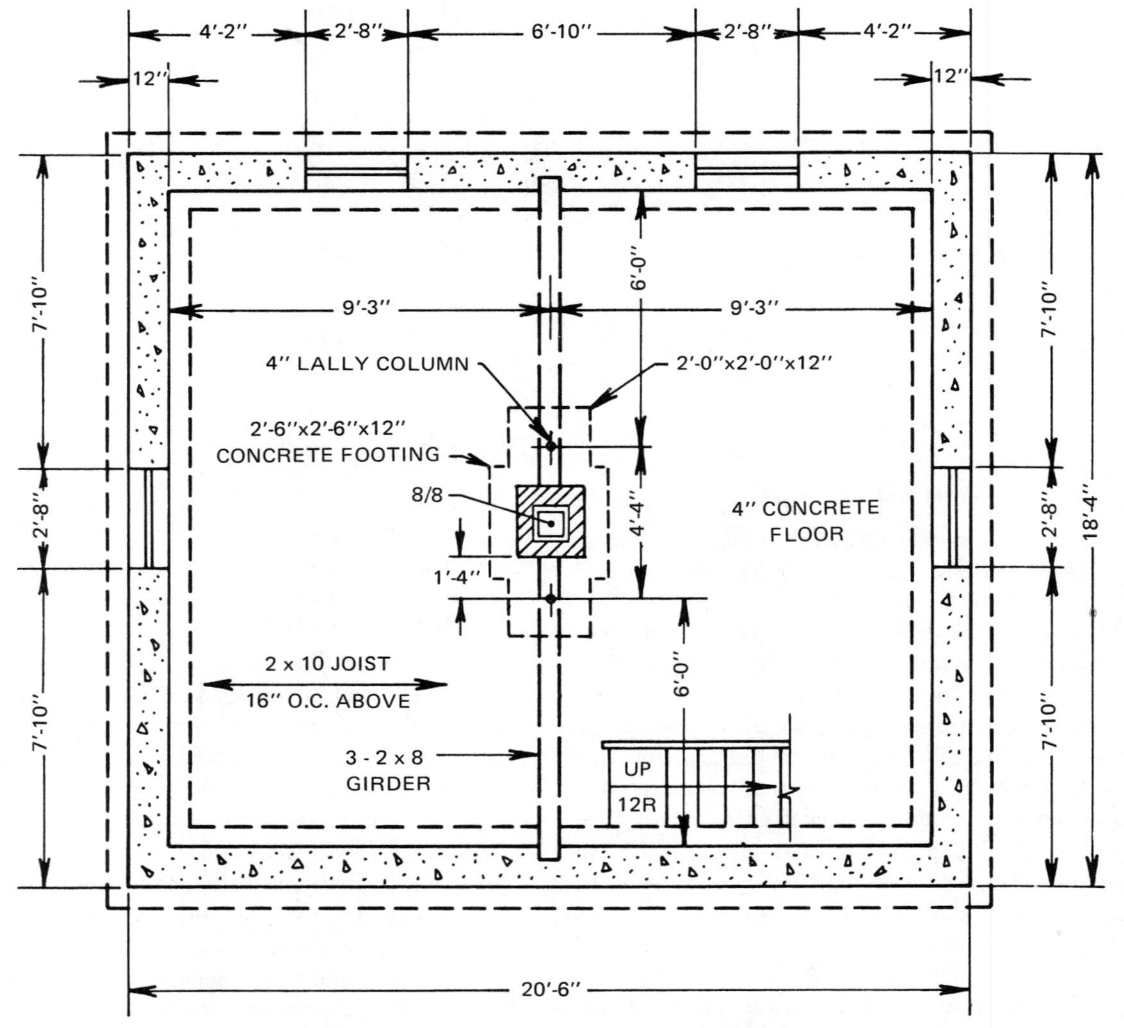

Fig. 33-3 Concrete foundation plan for a typical home

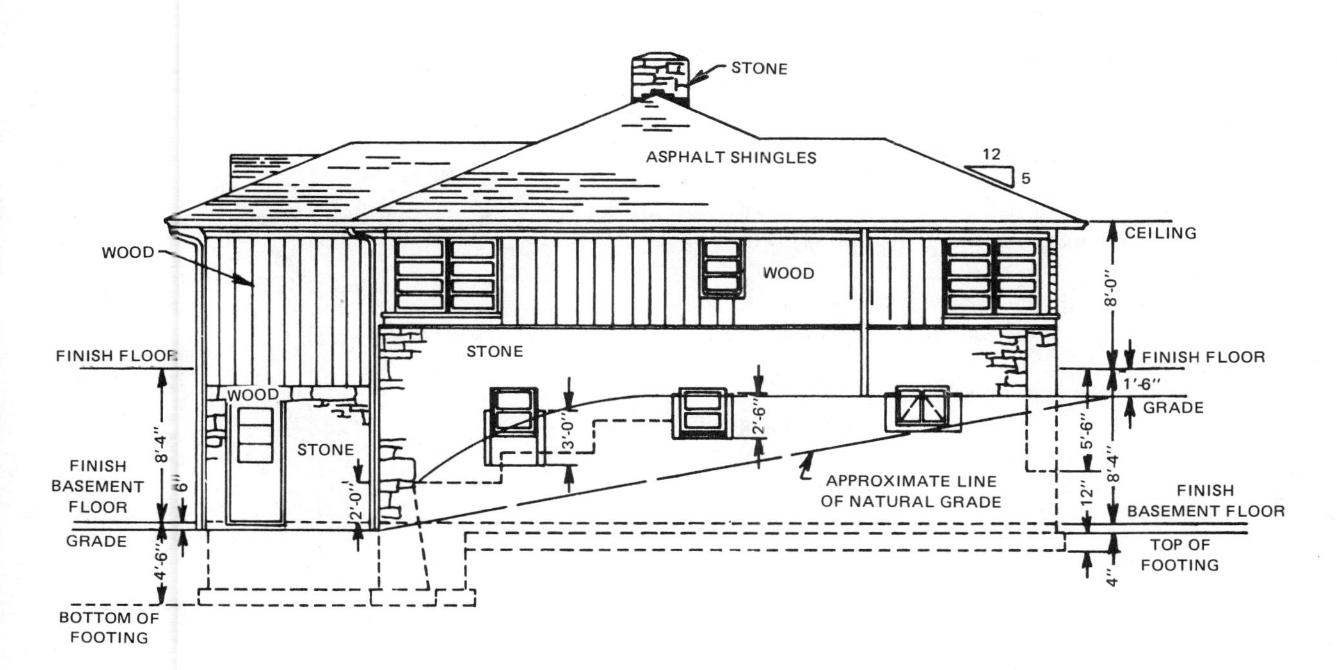

Fig. 33-4 Typical elevation view of house

Floor plans and elevations are usually drawn to the same scale. Windows and doors on the plan view are the same size and the same distance from the corners of the building on both views. By checking and comparing the plan view with the elevation view, the mason should be able to visualize approximately how the building will appear when it is constructed.

It is important to remember that the major difference between a plan view and an elevation view is that the plan view is always shown in a horizontal plane and the elevation view is shown from a vertical position. The information concerning measurements should be the same for each. Another important difference between the plan view and the elevation view concerns windows and doors. Measurements of windows and doors are given on the plan view but only the appearance of the window and door and the type are pictured on the elevation views.

Details not shown on the elevation view can usually be found by using the notes or references on the elevation plan which refer to another sheet on the drawings where they may be drawn to a larger scale.

SECTIONAL DRAWINGS

Features of construction, such as stairs, chimneys, wall thickness, placement of wall ties, and flashing installation cannot be shown clearly by an elevation view. *Sectional drawings* are views of specific portions of the structure. To understand the view given by a sectional drawing, compare it with a grapefruit cut in half. The exposed area of the fruit after cutting shows the inside of the fruit vertically from top to bottom. This is the same idea which the architect is trying to project to the builder through a sectional drawing.

There are three different definitions of a section view: a *longitudinal sectional view,* which is defined as a vertical cut through the long dimension of a building or roof; a *transverse sectional view,* which is a vertical cut through the short dimension of a building or room; and a *cross sectional view,* showing the composition of the wall, floor, or roof. All sectional views show how the separate parts of the structure are to be assembled or incorporated into the total structure.

A typical sectional view of a wall shows specific information and measurements, Figure 33-5. As an elevation view shows the outside of a structure, the sectional view allows the viewer to look inside the wall and see items that cannot be seen on other views of the plans.

Sectional plans are also used in the construction of fireplaces, Figure 33-6. Information such as the height of the damper, location of the ash dump, height of the mantel, and materials used in the hearth are shown in detail.

DETAILED DRAWING

As the name implies, a *detailed drawing* is an enlargement of a drawing on a smaller scale such as a sectional or elevation view. The detailed drawing, Figure 33-7, is used when the working drawing cannot show the desired information clearly.

Some of the many items generally shown in detail are front entrances, specific wall sections, complicated bond patterns in brick or stone, millwork, fireplace sections, and window and door installation details such as lock arrangements or sash mechanisms.

Since these details of construction are larger than the normal drawing, the mason must check the scale under each detailed drawing, as they may change on the same page of the plans. Details may vary from a scale of 1 1/2″ = 1′-0″ up to actual-size drawings.

Shop Drawing

In recent years, many manufacturers have developed building products which require specific instructions for their installation. Since installation instructions for specific brands of the same item may vary, architects sometimes require that a shop drawing be supplied by the company. A *shop drawing* is a drawing provided by a manufacturer to explain the installation of a product. Many times, this is the case with doors and windows, Figure 33-8. After the architect approves the shop drawing, a copy is sent to the contractor on the job. Another is given to the particular supervisor to whom this drawing pertains, who uses it as a working drawing.

REVISIONS TO THE WORKING DRAWING

Sometimes after the specifications have been written and the plans are drawn for a project, changes are made by the owner or architect for various reasons. These changes, known as *revisions,* must accompany the plans so that the contractor is informed of them before work is started. Revisions should be held to a minimum, but are a part of most plans.

As an example of a necessary revision, consider a door to the main entrance of a house. The size of the masonry opening is given on the floor plan, but the architect discovers after the floor plan is drawn that a door in the size specified is not available. To change the door size so that one may be obtained, a revision to the plans must be made. If the work has already been laid out by the mason, a *change order* is then issued by the architect.

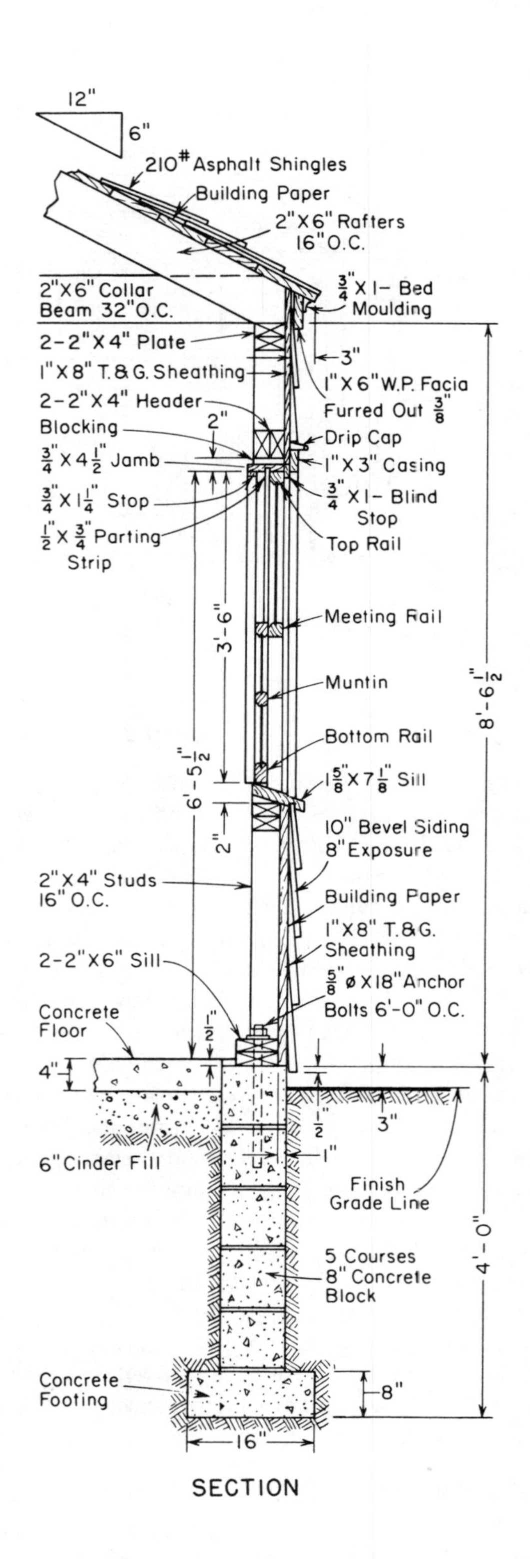

Fig. 33-5 Sectional view of wall

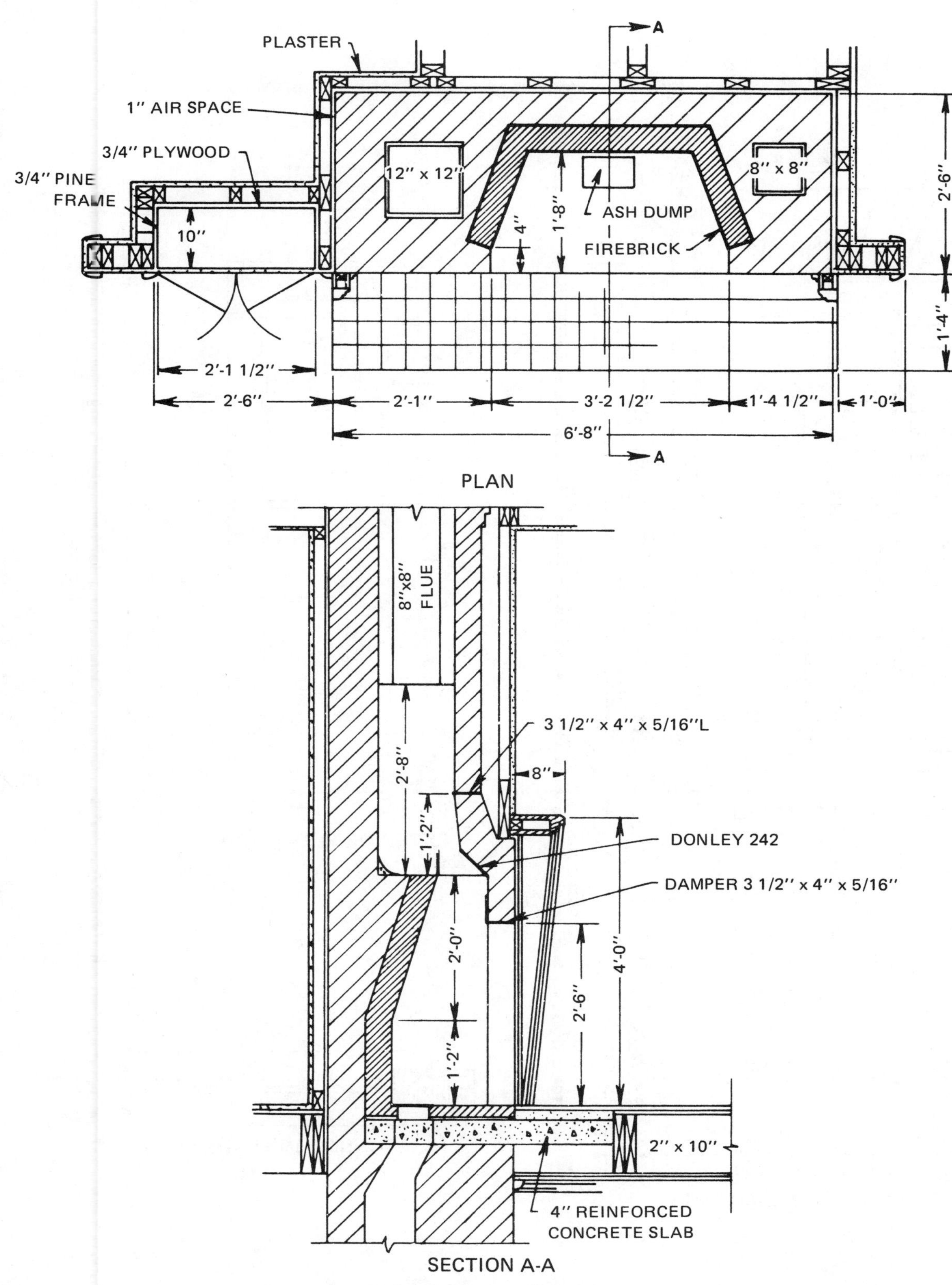

Fig. 33-6 Sectional plan for fireplace

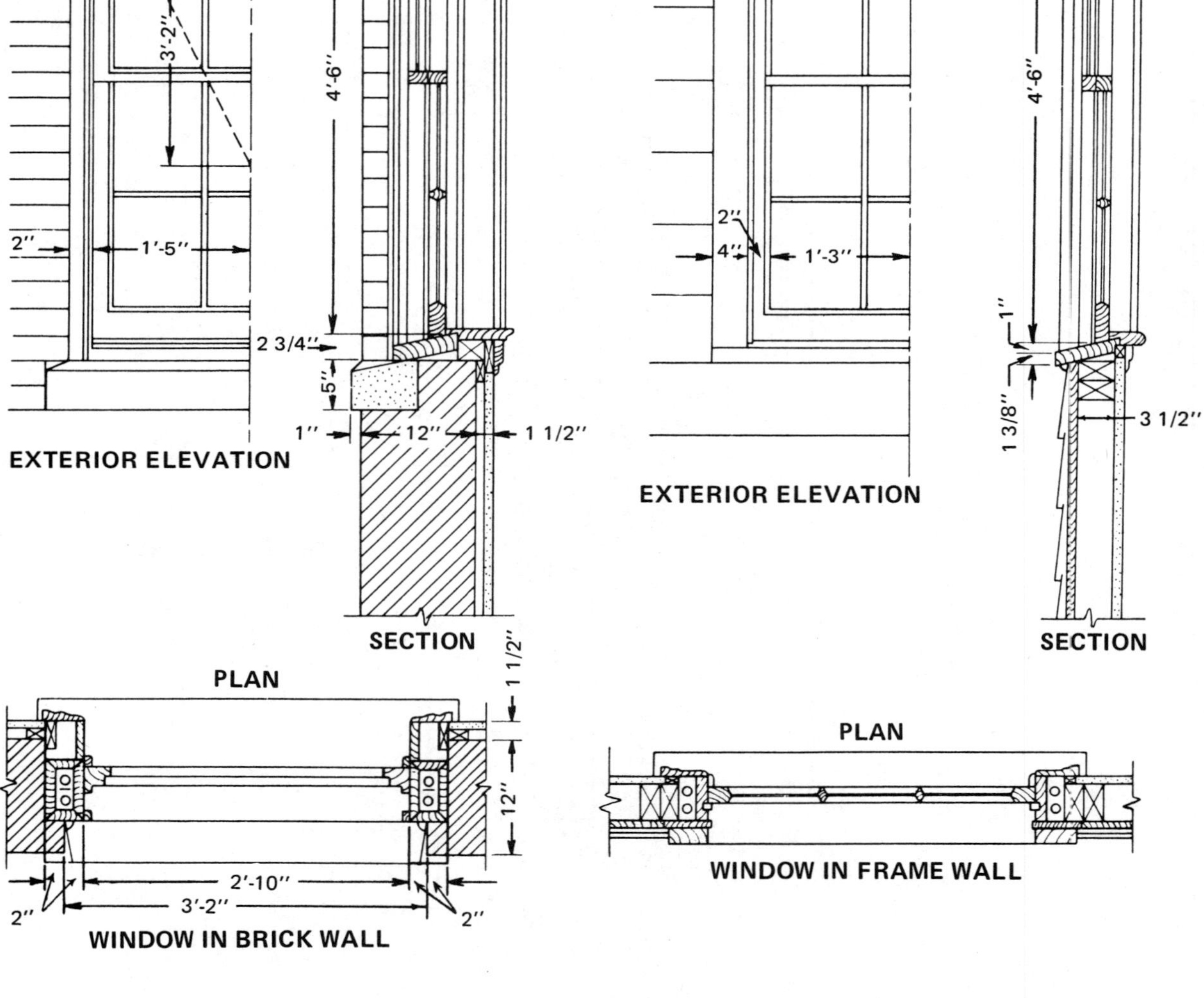

Fig. 33-7 Detailed drawing of windows in masonry and frame construction

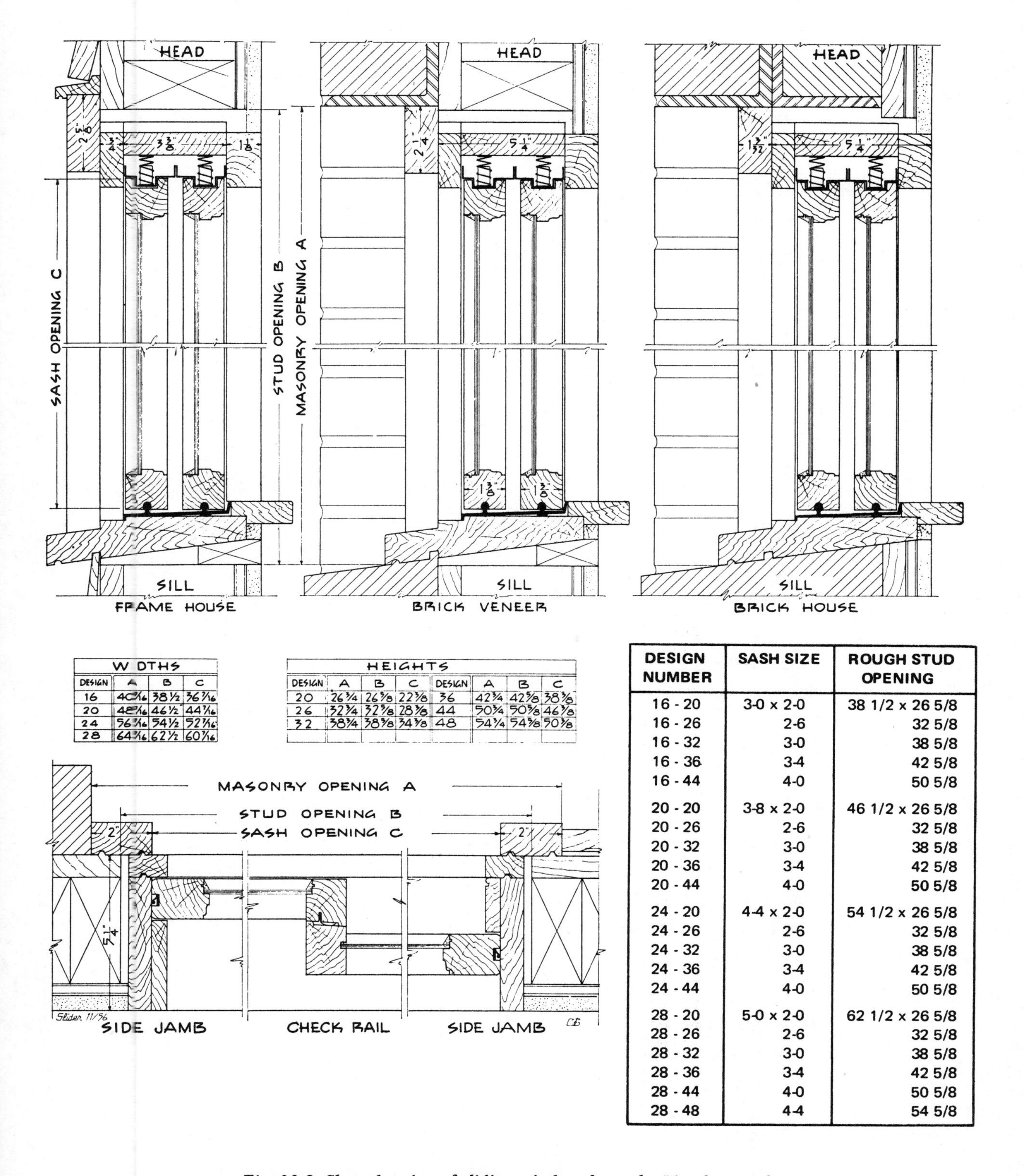

WIDTHS			
DESIGN	A	B	C
16	40 [illegible]/16	38 1/2	36 7/16
20	48 [illegible]/16	46 1/2	44 7/16
24	56 [illegible]/16	54 1/2	52 7/16
28	64 [illegible]/16	62 1/2	60 7/16

HEIGHTS							
DESIGN	A	B	C	DESIGN	A	B	C
20	26 3/4	26 5/8	22 3/8	36	42 3/4	42 5/8	38 3/8
26	32 3/4	32 5/8	28 3/8	44	50 3/4	50 5/8	46 3/8
32	38 3/4	38 5/8	34 3/8	48	54 3/4	54 5/8	50 3/8

DESIGN NUMBER	SASH SIZE	ROUGH STUD OPENING
16 - 20	3-0 x 2-0	38 1/2 x 26 5/8
16 - 26	2-6	32 5/8
16 - 32	3-0	38 5/8
16 - 36	3-4	42 5/8
16 - 44	4-0	50 5/8
20 - 20	3-8 x 2-0	46 1/2 x 26 5/8
20 - 26	2-6	32 5/8
20 - 32	3-0	38 5/8
20 - 36	3-4	42 5/8
20 - 44	4-0	50 5/8
24 - 20	4-4 x 2-0	54 1/2 x 26 5/8
24 - 26	2-6	32 5/8
24 - 32	3-0	38 5/8
24 - 36	3-4	42 5/8
24 - 44	4-0	50 5/8
28 - 20	5-0 x 2-0	62 1/2 x 26 5/8
28 - 26	2-6	32 5/8
28 - 32	3-0	38 5/8
28 - 36	3-4	42 5/8
28 - 44	4-0	50 5/8
28 - 48	4-4	54 5/8

Fig. 33-8 Shop drawing of sliding window from the Woodco catalog

CARE OF THE WORKING DRAWING

Proper care must be taken of the drawings if they are to last the entire job. The drawings should not be allowed to become wet and they should not be left in the sun for any length of time, as they will bleach out and be unreadable. The plans should be kept away from mortar or concrete mixtures and caution should be exercised when handling the drawings in the welding area. The plans should be collected at the close of the workday and returned to the job office or masons' shanty. Plans are expensive to duplicate and should be treated with great care.

ACHIEVEMENT REVIEW

A. Indicate in which of the following views the items below would be found: plot plan, floor and foundation plan, elevation view, sectional plan, or detailed drawings. If an item is shown in more than one plan, select the plan which gives the best view.

Sample:

Chimney base ____foundation plan____

1. Masonry opening for windows ________
2. Height of chimney above roof ________
3. Location of driveway ________
4. Layout of partition wall ________
5. Electrical layout ________
6. Flashing layout ________
7. Pier to support steel beam for first floor ________
8. Placement of wall ties in masonry wall ________
9. Height of fireplace mantel ________
10. Special front entrance ________
11. Outside wall showing installation of doors and windows ________

B. Answer each of the following questions.

1. What does the term *bench mark* mean?
2. What general information is shown on a plot plan?
3. What type of information is shown on a typical floor plan for a house?
4. How many elevation drawings are shown on a typical house plan?
5. What is meant by the term *shop drawings*?;
6. What does the term *revision* mean in a set of plans?
7. What is a longitudinal section view?
8. What is a transverse sectional view?
9. What happens to a set of plans if they are left out in bright sunlight for a long period of time?
10. Which of the working drawings discussed could also be called a site plan?

Unit 34
Dimensions and Scales

OBJECTIVES

After studying this unit, the student will be able to

- describe two methods of dimensioning when laying out masonry walls.
- define scale in relation to working drawings.
- identify the various scales used on a set of plans.
- explain the use and limitations of the mason's rule when scaling a drawing to determine a measurement.

Dimensions and scales are the architect's way of stating critical measurements to the persons carrying out the plans. It is essential for the mason to understand scales and the various methods of dimensioning if a project is to be laid out and constructed correctly. A plan without accurate measurements is a useless picture of the proposed structure. When studying working drawings, notice that most dimensions (more commonly called measurements) appear on the floor plans. Dimensions indicate the size of all important parts of the structure, drawn to a specified, predetermined scale.

The mason should not assume that all figures on a drawing are correct. All individual measurements on each wall in the plan should be rechecked. This is done by adding all of the separate dimensions and checking this figure with the overall dimension. If there is any difference in the two figures, the supervisor should be consulted. The mason should never change the original drawing without first consulting the supervisor.

DIMENSIONING

Methods

There are two general methods of dimensioning commonly used. One consists of measuring the overall dimension of the structure. The other is to measure the dimensions just to the outside of the unfinished wall. An overall dimension is taken from one extreme point to the other extreme point of the object or wall, Figure 34-1.

Both methods of measuring dimensions are shown in Figure 34-2. The dimension shown is 22′–0″ to the exterior face of the sheathing or plywood. This indicates that the sheathing is flush with the foundation wall. The dimension to the outside face of the wood studs is 21′–10 1/2″.

Some architects prefer to set the outside studs even with the outside face of the foundation wall. The sheathing would, in this case, overlap on the face top of the foundation wall. This method simplifies the procedure since the overall dimensions and the framing measurements are the same. The mason must take careful note of points of dimension location before laying out any work.

Dimensions as Stated on Plans

Dimensions are shown on plans in feet and inches, with a hyphen (-) between. For example, various measurements may be stated as 5′- 4″, 1′- 4″, 6′- 6″, or 10′ - 5 1/2″. Even measurements are shown with a hyphen and a 0 with inch marks (″) following the foot measurement, such as 4′- 0″. This method reduces the possibility of any misunderstanding concerning the measurement.

The exception to this practice is dimensions which are recognized standards in the modular system of

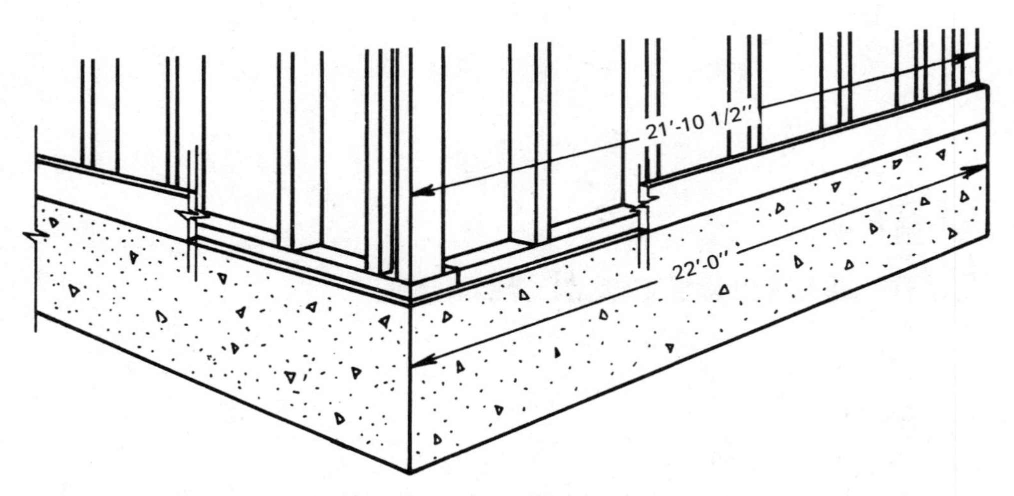

Fig. 34-1 Framed wall construction on concrete foundation. Notice that the overall measurement of the foundation is 22′ – 0″.

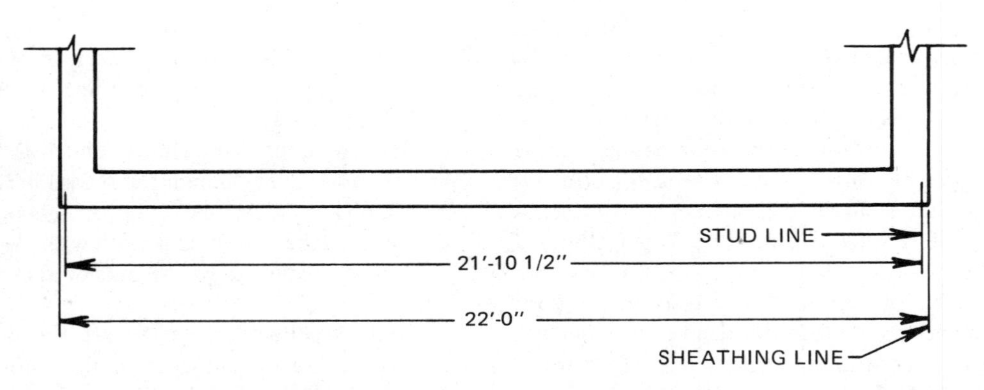

Fig. 34-2 Comparison of the dimension measured to the face of the studs and the dimension measured to the face of the sheathing. The difference between the two dimensions is the thickness of the sheathing.

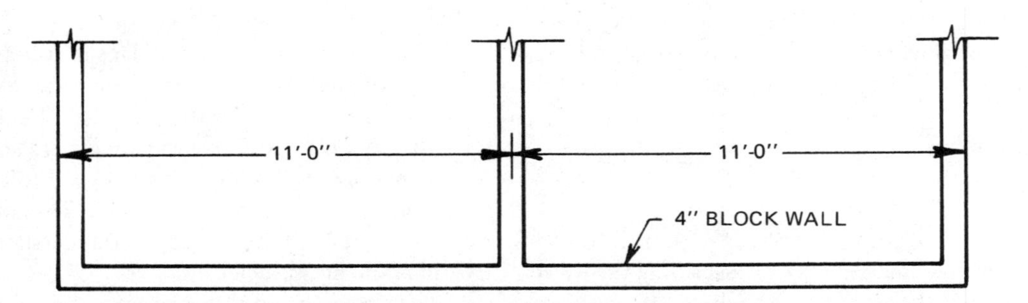

Fig. 34-3 An accurate method of dimensioning a partition wall involves measuring from the outside face of the wall to the center of the partition wall.

construction work, such as 16" center-to-center measurement of framing studs. This measurement could be shown simply as 16" O.C.

Generally, the majority of dimensions needed by the mason are located on floor plans. However, sectional views of walls should be examined especially carefully.

Practical Application

Dimensioning Partition Walls. Masonry partition walls are used to divide a certain amount of space into various rooms. They may or may not be load bearing. If they are not load bearing, they are known as *partitions.*

There are two general methods of laying out masonry partition walls. The first is to measure from the outside of the exterior wall to the center of the partition wall, Figure 34-3.

The second method is to measure from the outside of the exterior masonry wall to the inside face of the partition wall, Figure 34-4.

Dimensioning Openings in Masonry Walls. In dimensioning openings in a masonry wall, the measurements may be taken from either the center of the opening, Figure 34-5, or the side of the opening, Figure 34-6. Taking the measurement from the center of the opening is considered more accurate.

Points to Remember

When studying dimensions, the student should remember the following.

- Overall measurements are taken from one extreme end of the wall or object to the other extreme.
- Dimensions from the outside of the studding to the outside of the studding in framework must include sheathing thickness to be considered an overall measurement.
- The dimensions of masonry partition walls can be taken from the outside face of the exterior wall to the center of the partition wall, or from the outside face of the exterior wall to the inside face of the partition wall.

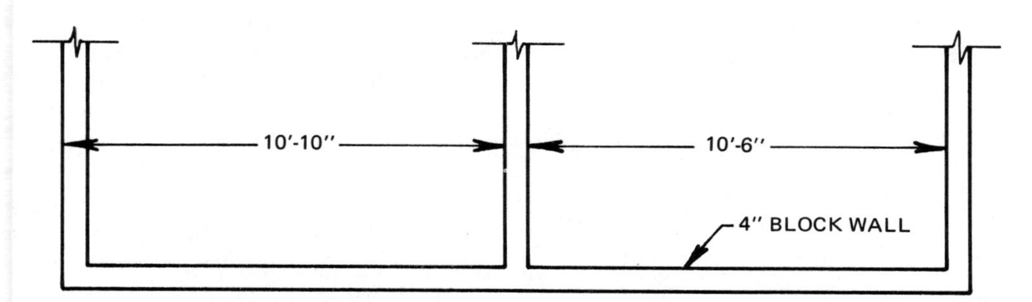

Fig. 34-4 Dimensioning a partition wall by measuring from the outside face of the main wall to the inside face of the partition wall

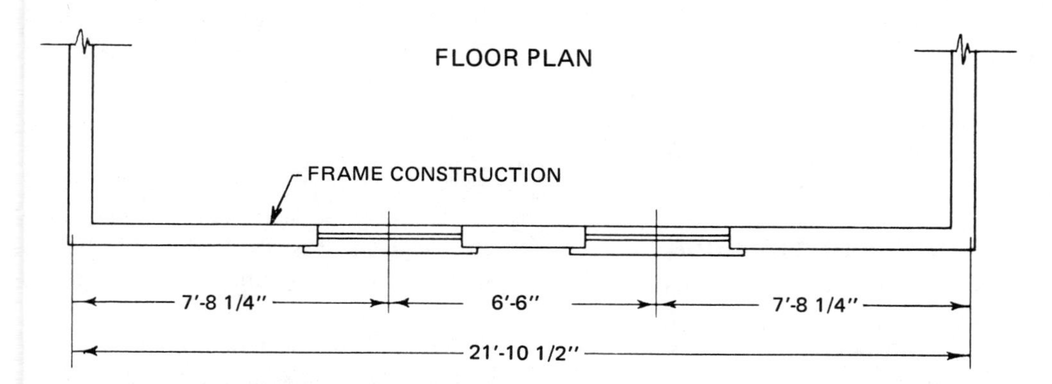

Fig. 34-5 Measuring from the outside face of wood studs to the center of window openings. This is customary for framing work.

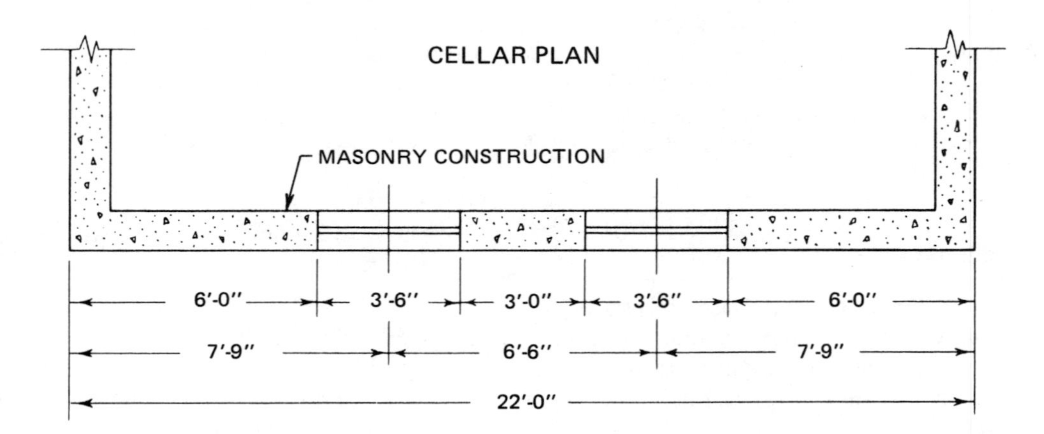

Fig. 34-6 Measuring from the outside face of the masonry wall to the edges of openings

- All dimensions are shown with the foot measurement first, followed by a hyphen and a 0 or the remaining number of inches (for example, 4'- 0" or 4'- 6").
- Although it is general practice to take measurements from the outside wall to the edge of openings, it is considered more accurate to measure to the center of openings then to the edge of the jambs. Always double-check all measurements of openings before proceeding with the work.
- The overall dimensions and all dimensions in between must always be checked against each other to be sure they are the same length.
- Dimensions should be shown on drawings according to their importance. The overall dimension is always given first, with the dimensions to center of openings next. The sizes of the openings are then shown. All of the smaller dimensions are shown inside the complete overall measurement.
- Dimensions serve two specific purposes; one, to indicate the location of a construction feature, and two, to indicate the size of this object or feature.

SCALES AND THE WORKING DRAWING

It stands to reason that working drawings cannot usually be shown in the actual size of the specific projects. The obvious solution is to reduce the plans, or draw the project to *scale.* On such drawings, all dimensions are reduced proportionately to reflect a certain fixed ratio between the size of the drawing and the actual finished structure. The size of the scale is determined by how complicated or important the information is for the viewer to understand.

Where to Find the Scale on a Drawing

To obtain dimensions from plans, the scale to which the plan is drawn must first be determined. The scale is usually shown on the drawings in the title plate. The scale is usually found in the lower right-hand corner or under a particular detail or sectional view. The *title plate* gives necessary information about the drawing not given in the drawing itself or in accompanying notes. It is possible that the scale will change several times on the same sheet. When studying working drawings, be sure to check all views for possible changes in scale.

Floor plans for various floors in a set of drawings are drawn to the same scale.

Large commercial drawings usually have an index of the various sheets contained in a set of plans and the page number where they can be found. The index is on the first page following the title sheet. After locating the correct sheet, determine the scale before making any measurements.

Other Scales Used in Construction Drawings

There are 10 basic scales used in dimensioning plans and drawings besides the full scale. These scales are known as the architect's scales, Figure 34-7.

Commonly Used Scales

The most commonly used scale is 1/4" = 1'- 0". This indicates to the mason that every 1/4" on the plans represents 1"- 0" of actual structure area. For example, in this scale, 1″ represents 4′- 0″ on the job; 5″ represents 20′- 0″. The 1/4″ scale is usually used on housing or small commercial drawings. This is done because all of the information can be drawn on a small sheet of paper and still be seen clearly. Buildings constructed from a 1/4″ scale are actually 48 times larger than the actual drawing when constructed.

On larger structures, the scale of 1/8″ = 1′- 0″ is most often used. Scales any smaller than 1/8″ would be very difficult for the worker to read and the possibility of error would be greatly increased. It should not be necessary for the mason to use a magnifying glass to read the plans. Buildings constructed to the 1/8″ scale are actually 96 times larger than the actual drawing.

The smaller the scale is, the more careful the mason must be when dimensioning. A mistake of 1/16″ on the 1/8″ scale would account for 6″ on the job. A line 2 1/2″ long drawn to the 1/8″ scale would actually measure 20′- 0″ on the job.

THE USE OF RULES IN DIMENSIONING DRAWINGS

The Architect's Scale

The *architect's scale* is a special instrument that the architect uses to dimension drawings which contain the above mentioned scales. A triangular-shaped architect's scale is used in the architect's office or by drafting students when preparing a drawing. This instrument is usually not practical on the job because of its length. A straight scale, Figure 34-8, is made that will easily fit into the shirt pocket. This type of scale can usually be obtained from any reliable store which stocks drafting supplies.

How to Read the Architect's Scale

Notice the scale in Figure 34-9. The left and right edges of the scale are laid out so that they contain two different scales with a common denominator. The 1/8″ scale (left side) is marked off with short lines, each representing 1′- 0″. The scale is read from left to right. The 1/4″ scale (right side) is marked with longer lines, each representing 1′- 0″. This scale is read from right to left. Each long line represents 1′- 0″ on the 1/4″ scale or 2′- 0″ on the 1/8″ scale. Each short line represents 1′- 0″ on the 1/8″ scale and 6″ on the 1/4″ scale.

Designated Scale	Designated Scale
3/32″ = 1′-0″	1/2″ = 1′-0″
1/8″ = 1′-0″	3/4″ = 1′-0″
3/16″ = 1′-0″	1″ = 1′-0″
1/4″ = 1′-0″	1 1/2″ = 1′-0″
3/8″ = 1′-0″	3″ = 1′-0″

Note: At times, a full scale (1″ = 1″) is used.

Fig. 34-7 The architect's scales

Each scale has a short section at the end with several lines. The fine lines represent inches or fractions of inches when the measurement exceeds an even foot dimension, such as 10′- 4″. The 10′- 0″ measurement is read on the scale and the remaining 4″ are found in the finely calibrated lines.

In the 3/4″ scale, notice that there are 12, 1/16″ marks in 3/4″. Since 3/4″ equals 1′- 0″ according to the scale, each of the marks represents 1″ since there are 12 of them. Therefore, 1/16″ in a 3/4″ scale always represents 1″ on the job.

In the 1″ and 1/2″ scales, every 1/8″ represents 1″ since there are 12, 1/8s in 1 1/2″.

To determine what fraction is equal to 1″ on the job, divide the fraction which equals 1′- 0″ on the job by 12, since there are 12″ per foot. With a scale of 1/2″ = 1′, divide the 1/2″ by 12; 1/24″ on this scale equals 1″ on the job.

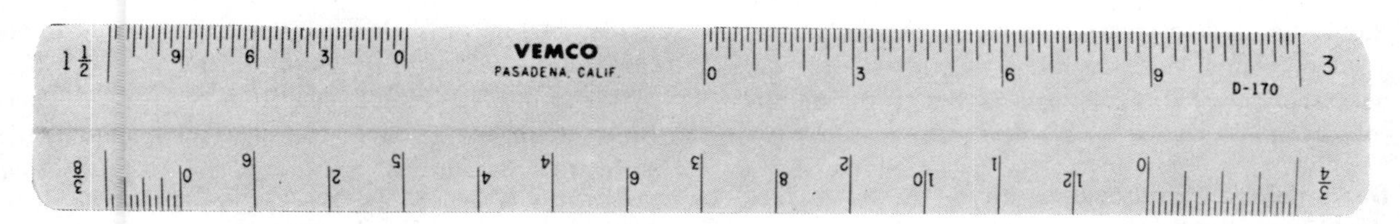

Fig. 34-8 Architect's scale rule. This rule contains the eight most often used scales in construction and fits easily into the worker's pocket. Four of the scales are shown in the illustration; the other four are on the reverse side.

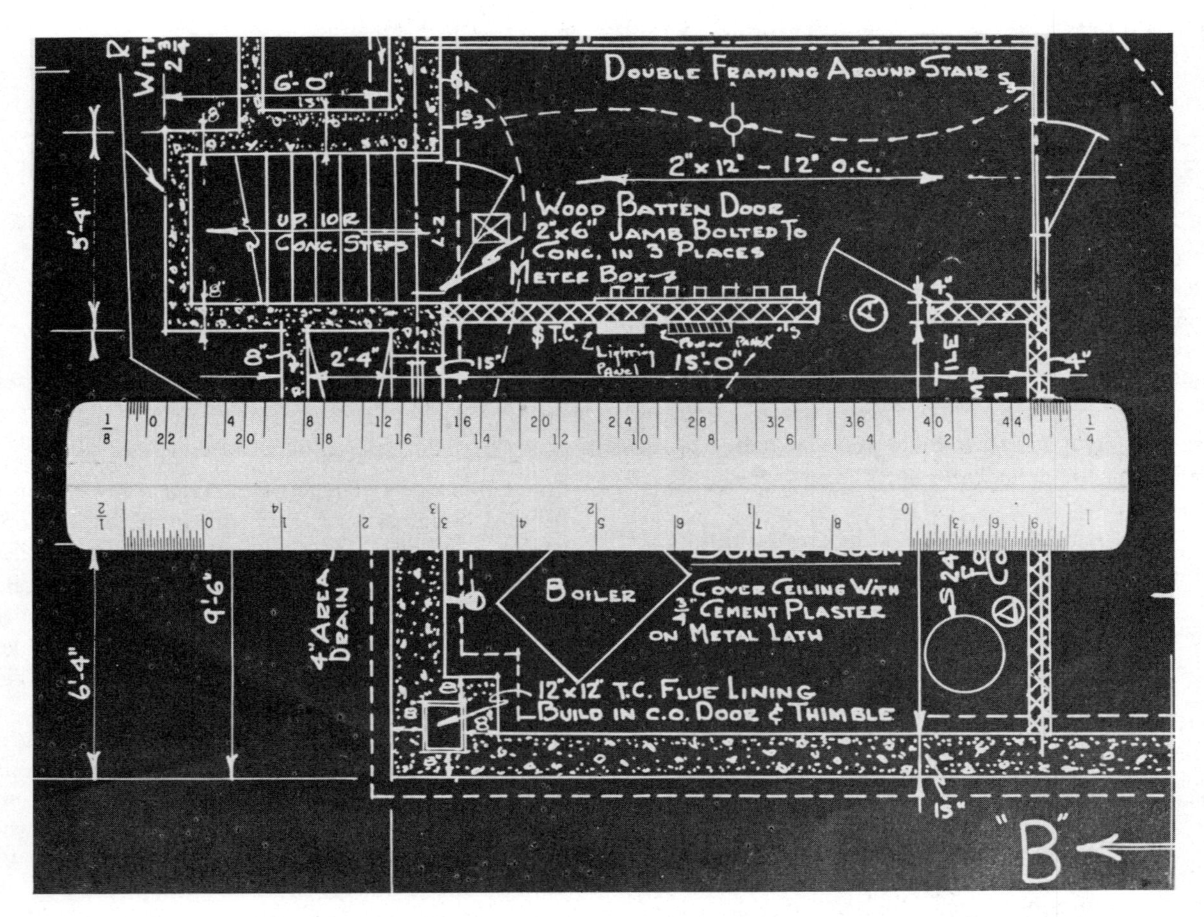

Fig. 34-9 Determining a measurement with the architect's scale rule. Using a 1/4-inch scale, notice that the length of the boiler room is 15 feet, just as the plan indicates.

If the scale used is 3/8" = 1' - 0", 3/8" is divided by 12 to determine the amount equal to 1", which would be 1/32" on the scale.

The Mason's Rule

Since the mason's rule is divided into sixteenths, it cannot be used easily for all scales. Where the scale is such that 1/16" equals even inches on the job, the rule can be used with no problem. In the scale 3/8" = 1'- 0", 1/16 equals 2" because 1/32" is equal to 1- 0" on the job. If possible, the architect's scale rule should be used to scale an unknown dimension from a set of plans. The possibility of error is decreased with the use of this rule, as opposed to the mason's rule.

Determining a Measurement with the Mason's Folding Rule

Scaling an unknown dimension from a plan with a folding rule should be done only as a last resort. The small divisions that are shown on the architect's rule are not shown on the mason's folding rule, Figure 34-10. Also it may be difficult to remember the various scales discussed and apply them correctly when measuring with the folding rule. On the architect's scale rule, the mason merely matches the scale rule with the item on the plans.

There are times, however, when the mason's rule is very useful in working with plans and drawings. For example, in cases where the mason must determine a missing dimension from a drawing, or sketch a particular feature to scale for clarification of an important point, the mason's rule is invaluable.

Blueprints are reproduced from original drawings by a process which may cause a small amount of shrinkage. Because of this, and for various other reasons, it is always good practice for the mason to double-check dimensions and scales. However, if an error is found, the supervisor should be consulted before any changes are made.

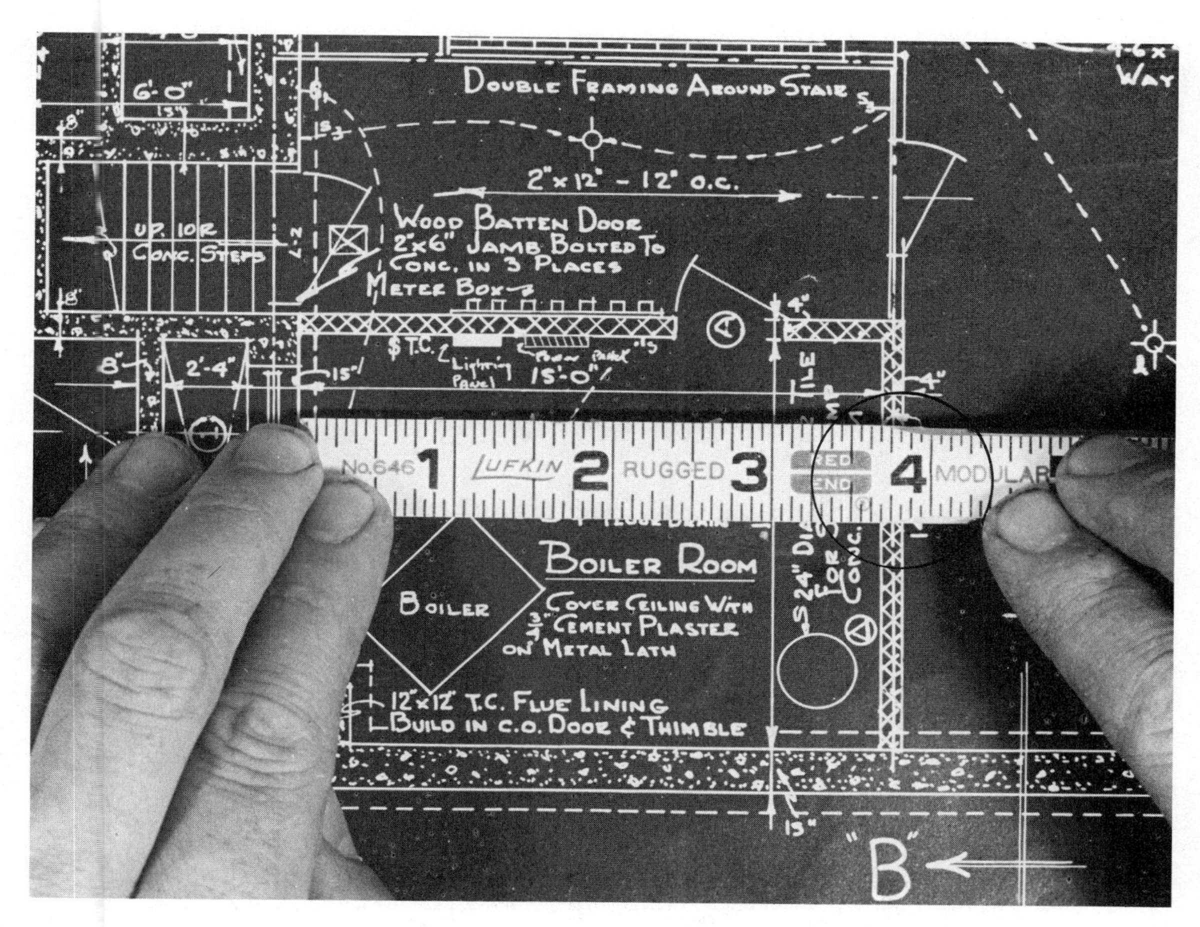

Fig. 34-10 Determining a measurement with the mason's folding rule. The plan, drawn to the 1/4-inch scale, shows that the length of the boiler room is 15 feet. Using the mason's rule, the measurement falls short of the correct measurements by 1/16 inch. (Compare figure 34-9, in which the measurement was very accurate.)

ACHIEVEMENT REVIEW

A. Select the best answer from the choices offered to complete each statement. List your choice by letter identification.

1. An overall dimension is taken from
   a. one extreme point on a drawing to the other extreme point.
   b. center-to-center of the opening.
   c. the outside face of the wall to the center of the opening.
   d. the total length of all outside walls.

2. The most accurate method of laying out a partition wall is from the
   a. exterior face of the wall to the inside face of the partition wall.
   b. interior face of the exterior wall to the inside face of the partition wall.
   c. outside face of the exterior wall to the center of the partition wall.
   d. center of the outside wall to the center of the partition wall.

3. A measurement of 5 feet and 4 inches is shown on a plan as
   a. 5′- 0″- 4″.
   b. 5′- 4″.
   c. 5- 4.
   d. 5″- 4′.

4. The most commonly used scale in house construction is
   a. 1/8" = 1'- 0". c. 1/2" = 1' - 0".
   b. 1/4" = 1'- 0". d. 3/4" = 1'- 0".
5. The most commonly used scale for large commercial buildings is
   a. 1/4" = 1'- 0". c. 1/2" = 1'- 0".
   b. 3/8" = 1'- 0". d. 1/8" = 1'- 0".
6. The specified scale for a drawing is usually found
   a. above the view. c. under the view.
   b. to the side of the view. d. in the middle of the view.
7. Not counting the full scale, the total number of scales found on an architect's scale rule is
   a. 10. c. 6.
   b. 8. d. 4.

B. Answer each of the following questions. Express your answer in inches.

1. In a 1/4" scale, what length of line in a plan would represent an actual measurement of 20'- 0"?
2. In a 1/4" scale, what length of line in a plan would represent an actual measurement of 40'- 6"?
3. In a 1/8" scale, what length of line in a plan would represent an actual measurement of 50'- 0"?
4. In a 1/8" scale, what length of line in a plan would represent an actual measurement of 30'- 6"?
5. In a 1/2" scale, what length of line in a plan would represent an actual measurement of 15'- 0"?
6. In a 3/4" scale, what length of line in a plan would represent an actual measurement of 10'- 0"?

C. The following lines drawn to a 1/4" scale in a drawing represent what lengths in an actual building?

1. 3/16"
2. 1/8"
3. 7/8"
4. 4 1/2"
5. 10"
6. 15 1/8"

Unit 35
The Mason and Metrics

OBJECTIVES

After studying this unit, the student will be able to

- list the basic units of measurement in the metric system.
- list the prefixes used to express units of measurement in metrics.
- discuss the positive aspects of use of the metric system.

There is much discussion today concerning the possible change from the traditional system of measurement used in the United States today, known as the English system of measurement, to the metric system. Various segments of American industry are already using the metric system to some extent. The majority of industrialized nations in the world today use the metric system as their sole system of measurement. Many times, measurements on building materials from other countries are expressed in metrics. For these reasons, masons should become acquainted with the basics of the metric system.

Prefixes are used to show the relationship of all linear measurements to the metre. The following are prefixes used and their values.

kilo (thousand) = 1000
hecto (hundred) = 100
deka (ten) = 10
deci (one tenth) = 0.1
centi (one hundredth) = 0.01
milli (one thousandth) = 0.001

THE METRIC SYSTEM AS A DECIMAL SYSTEM

The metric system is a decimal system based on the unit ten. The base units are metre (linear measure), the gram (weight measure), and the litre (volume measure). The relationship between the base units and the other units is a power of ten. This relationship makes the metric system easier to use than the English system.

Although the metric system is different from the customary system, the concept is not entirely new. Our number system is a decimal system. Since the founding of the United States, a decimal system has been used to measure money. All currency denominations in our money system are based on the unit ten. Notice the similarity of the following units in our money system and the metric units of measure.

one thousand dollars = $1000.00
one hundred dollars = $100.00
ten dollars = $10.00
dollar = $1.00
dime = $0.10
cent = $0.01
mill = $0.001

COMPARING THE ENGLISH AND METRIC SYSTEMS

Figure 35-1 shows the basic units of measure in the English and metric systems.

MEASURING IN METRICS

Linear Measure

In the metric system, the standard unit of measurement for length is the *metre*. The standard metric units of linear measurement are shown in Figure 35-2. The value of each unit in relation to the metre is shown, along with the English system equivalent.

Length			
Metric		**English**	
1000 millimetres	= 1 metre	12 inches	= 1 foot
100 centimetres	= 1 metre	3 feet	= 1 yard
1000 metres	= 1 kilometre	36 inches	= 1 yard
		5280 feet	= 1 mile
Weight			
Metric		**English**	
1000 grams	= 1 kilogram	438 grains	= 1 ounce
1000 kilograms	= 1 metric ton	16 ounces	= 1 pound
		2000 pounds	= 1 short ton
Volume			
Metric		**English**	
1000 millilitres	= 1 litre	2 cups	= 1 pint
		2 pints	= 1 quart
		4 quarts	= 1 gallon
		8 pints	= 1 gallon

Fig. 35-1 Commonly used units of measure in the English and metric systems of measurement

METRIC SYSTEM		ENGLISH SYSTEM
Common Linear Units (With Symbols)	Value of Unit in Relation to Metre	Value in English Units
1 millimetre (mm)	0.001 metre	0.039 inch
1 centimetre (cm)	0.01 metre	0.394 inch
1 decimetre (dm)	0.1 metre	3.937 inches
1 metre	1 metre	39.37 inches
1 dekametre (dam)	10 metres	32.80 feet
1 hectometre (hm)	100 metres	328.08 feet
1 kilometre (km)	1000 metres	0.6214 mile

Fig. 35-2 Linear measure in the metric system

More than one prefix should not be used when expressing one metric measurement. For example, a measurement would not be expressed as 3 decimetres, 4 centimetres, 6 millimetres (3 dm 4 cm 6 mm). The 3 dm is equivalent to 300 mm and the 4 cm is equivalent to 40 mm. The measurement is then equal to 300 mm + 40 mm + 6 mm. The measurement is expressed as 346 mm or 0.346 m.

Weight Measure

Metric weight measurements are computed in grams, Figure 35-3.

Volume Measure

The litre, which measures a little more than the quart, is used as the basic unit of volume measurement, Figure 35-4.

PRACTICAL APPLICATION: USING METRICS

The metric rule is sometimes used for measuring materials, Figure 35-5. The smallest division on the metric rule is the millimetre. There are 1000 millimetres in every metre.

All dimensions in the masonry foundation plan in Figure 35-6 are expressed in metres. The total distance around the foundation or the sum of the length of all the masonry walls (called the *perimeter*), is found by adding the measurements of all the walls together.

To find the perimeter, add all lengths together.

25 m + 2 m + 6 m + 10 m + 31 m + 8 m =
82 m (total length in metres)

A brick wall is to be built. The surface area in square metres must be found before the total number of bricks for the job can be calculated.

1 milligram	=	0.001 gram
1 centigram	=	0.01 gram
1 decigram	=	0.1 gram
1 gram	=	1 gram
1 dekagram	=	10 grams
1 hectogram	=	100 grams
1 kilogram	=	1000 grams

Fig. 35-3 Metric weight measurements expressed in grams

1 millilitre	=	0.001 litre
1 centilitre	=	0.01 litre
1 decilitre	=	0.1 litre
1 litre	=	1 litre
1 dekalitre	=	10 litres
1 hectolitre	=	100 litres
1 kilolitre	=	1000 litres

Fig. 35-4 Metric volume measurements expressed in litres

Fig. 35-5 Comparing the mason's 6-foot rule (top) to the metric meter rule (bottom). The 2 rules are shown on the concrete block to illustrate the different expressions of measurement.

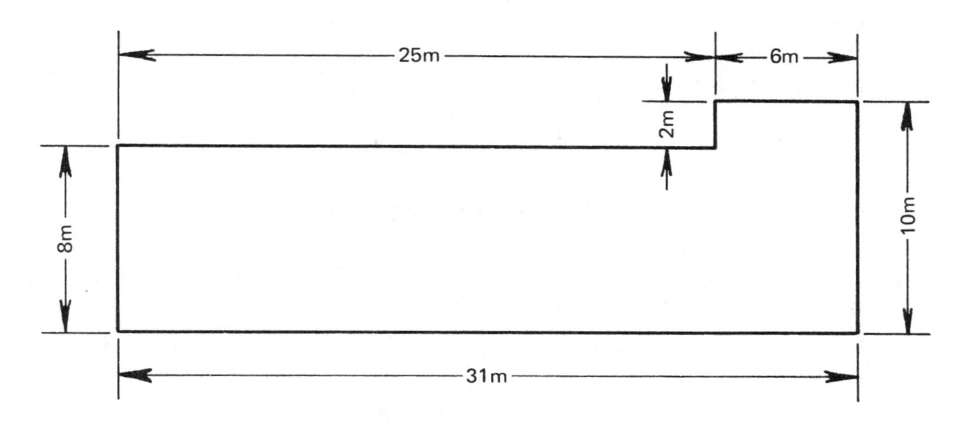

Fig. 35-6 Plan for masonry foundation

To find the surface area of the wall, multiply length by width.

$$16 \text{ m} \times 4 \text{ m} = 64 \text{ m}^2$$

Already, the metric system of measurement is being used in some areas in the United States. Athletic events are measured in metres rather than in yards or feet. When astronauts broadcast from the moon, they told a worldwide audience how far in metres their rockets landed from lunar hills. Many types of automobiles require metric tools. Lengths of camera film have long been measured in millimetres. The metric system has been used for years in many businesses, mechanical trades, and scientific fields.

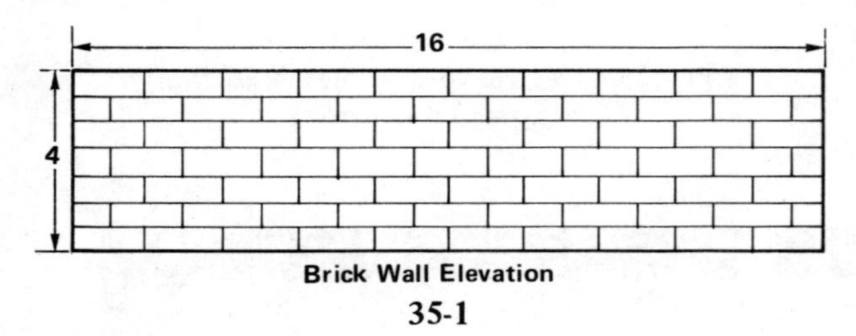

Brick Wall Elevation

35-1

As the need for one system of measurement for all nations becomes more and more apparent, it is likely that more pressure will be brought upon American industries and businesses to convert to the metric system of measurement. For this reason, students should familiarize themselves with this system.

ACHIEVEMENT REVIEW

A. Select the best answer from the choices offered to complete the statement or answer the question. List your choice by letter identification.

1. The standard unit of measure for length in the metric system is the
   a. gram. c. litre.
   b. metre. d. kilometre.
2. The standard unit of measure for weight in the metric system is the
   a. metre. c. litre.
   b. gram. d. degree Celsius.
3. The standard unit of measure for volume in the metric system is the
   a. degree Celsius. c. gram.
   b. metre. d. litre.
4. Which of the following prefixes indicates one thousandth in the metric system?
   a. Deci c. Milli
   b. Centi d. Kilo
5. Which of the following represents a kilogram?
   a. 100 grams c. 500 grams
   b. 10 grams d. 1000 grams
6. Which of the following equals one hectolitre?
   a. 10 litres c. 0.1 litre
   b. 100 litres d. 0.01 litre
7. Comparing the metric system with the English system of measurement, the metre is equal to which of the following?
   a. 12" c. 29.51"
   b. 16" d. 39.37"

B. Solve each of the following problems. Express your answer in metric measurements.

8. Find the surface area of the face of a brick wall that is 40 m long and 6 m high.
9. A foundation measuring 25 m long x 8 m wide is to be built. What is the total number of metres around the foundation?
10. A stone patio measuring 9 m x 4 m is to be laid. What is the surface area of the patio?
11. A mason orders 65 cubic metres of concrete for a job. Only 58 cubic metres are used on the job. How many cubic metres remain?

SUMMARY, SECTION 9

- Specifications are the written or printed instructions which accompany building plans and describe information not already shown on the plans.
- To construct a building, building plans and specifications must be studied together before any work is done.
- Specification points of instruction should be listed in the order the work is to be accomplished.
- When there is a conflict between specifications and plans, the specifications take precedence.
- Points of good workmanship should be a part of all masonry specifications.
- Lines of different thicknesses and design are used on building plans to denote various meanings to the viewer.
- Symbols are used to save space on plans and to identify various materials to be used in the construction of the building.
- Symbols for the same materials are shown differently in elevations and section views.
- Schedules are lists contained in the building plans which describe such specific items as sizes of windows and doors and finishes for floors and ceilings.
- Plans and specifications should be protected on the job to ensure that they last the duration of the job.
- Generally, a set of working drawings consists of a plot plan, foundation and floor plans, elevation view, sectional view, and detailed drawing.
- The plot plan shows the location of the structure on the lot and important features such as the sewer, driveway, and elevation of land.
- A floor plan is a horizontal view of the floor level of a structure, showing important information such as the layout of windows, doors, partitions, and stairways.
- An elevation drawing is a vertical view as one would see it while standing and facing the structure.
- A sectional drawing shows how the separate parts of the structure are to be assembled or incorporated into the total structure. It is a vertical slice through the structure showing all objects that would normally be hidden from view.
- Details on a working drawing show objects on a larger scale than they appear on other drawings. They provide a better understanding in the construction of the building.
- All of the working drawings for one project are directly related to one another and should be studied carefully to assure complete understanding of the job to be done.
- All dimensions of the length of a wall should be added together and the total figure checked against the overall dimensions to be sure they agree.
- The most accurate method of laying out masonry partitions or openings is to measure from the edge of the exterior face to the center of the opening or wall.
- Dimensions serve two basic purposes: to indicate the location of the object, and to indicate the size of the object.
- Drawing to scale involves reducing all views of a structure to a predetermined, fixed ratio that will allow the drawing to fit on a relatively small sheet of paper.

- The two most frequently used scales are the 1/4" (for home and small commercial construction) and the 1/8" (for larger commercial buildings).
- There are 10 basic scales used by architects in dimensioning plans.
- Plans should be scaled with a mason's rule only as a last resort.
- The supervisor should be consulted before any measurements on a plan are changed.
- The metric system of measurement is used more and more frequently in areas of industry, science, and business.
- In the metric system, the metre is the basic unit of measurement of length, as the gram is for weight and the litre is for volume.
- The metric system is a decimal system, much like the monetary system in the United States.
- The prime objection to changing to the metric system in the United States is the enormous cost of relabeling and re-education of the public.
- Realizing the possibility of future adoption of the metric system of measurement in the masonry trade, it is important that the mason become familiar with the basics of the metric system.

SUMMARY ACHIEVEMENT REVIEW, SECTION 9

Complete each of the following statements referring to material found in Section 9.

1. A printed or written description of work to be done which accompanies building plans is called the ______________.
2. A typical set of general specifications has a standard number of sections or divisions totaling ______________.
3. When a question arises on the job as to whether the specifications or plans are correct, the ______________ are considered correct.
4. A heavy, continuous line used to indicate visible outlines or edges of objects is called a (an) ______________.
5. A wavy line indicating that parts of a structure have been left out or that the full length of the object has not been drawn on building plans is called a (an) ______________.
6. A line composed of short dashes used to show that there are parts hidden from view on a plan is called a (an) ______________.
7. A list or drawing separate from the plans which generally describes windows, doors, and wall finishes is called a (an) ______________.
8. The building plans which the workers read on the job are more commonly called the ______________.
9. The part of the plans which shows the structure situated on the lot with service utilities and the contour of the land is called the ______________.
10. The drawing giving a vertical view of the wall or object as it should look when completed is called the ______________.
11. That part of the building plans which shows the layout of windows, doors, and openings with dimensions is called the ______________.

12. The drawing in which a part of the building or object has been cut away, giving a vertical view and exposing the interior of the structure or object, is called the ________________.

13. In building plans, drawings made to a larger scale to show important information are called ________________.

14. Changes that are made after the plans have been drawn are called ________________.

15. A dimension measured from one extreme point to another extreme point is called a (an) ________________.

16. The most accurate method of laying out a partition wall is to measure from the exterior face of the outside wall to the ________________.

17. A measurement of 5′- 0″ should be shown on a set of plans as ________________.

18. The scale usually used for small buildings and homes is ________________.

19. The most commonly used scale for commercial buildings is ________________.

20. There are ________________ scales shown on the architect's scale rule.

21. A set of plans should not be scaled with a mason's folding rule because________________ ________________.

22. A term which is attached to the beginning of a word to show the relationship of the measurement to another basic measurement is called a ________________.

23. In the metric system, length is computed in ________________.

24. In the metric system, weight is computed in ________________.

25. In the metric system, volume is computed in ________________.

26. The measurement which is 0.001 metre is known as a ________________.

Section 10 Recommended Working Practices

Unit 36 Safety on the Job

OBJECTIVES

After studying this unit, the student will be able to

- explain why good safety practices on the job are essential.
- describe the correct dress and safety measures to be utilized when on the job.
- list some of the more common hazards presented on jobsites and how to avoid them.

Nothing should be more important to masons than their own personal safety and the prevention of accidents on the job. Besides the physical handicap which results from injury, the worker may be deterred from work for a period of time. Loss of time costs the mason and the employer money.

Many accidents could be avoided if common sense were practiced and shortcuts were not taken. The quickest way to complete a job is not always the best way. Other reasons for injuries on the job include lack of interest and attention to the job at hand, tiredness, and careless, slipshod methods. Sometimes, there is a tendency for workers to become careless on the job after they have learned the basics of the job very well. It is imperative that the mason guard against this attitude. Safety requirements should be explained to the mason by the contractor when starting work. It is then the responsibility of the mason to observe and obey these regulations.

Besides the physical nature and the mental attitude of the worker, injury and accidents may result from unsafe tools and equipment and improper clothing and safety devices. Construction work is one of the most hazardous of all occupations because one must work with power tools, on scaffolds, and under and around moving equipment. For this reason, the federal law OSHA (Occupational Safety and Health Act) was passed to guarantee the worker a safe working environment. It should be remembered that OSHA only sets minimum standards and more strict state requirements may be in effect in many areas of the country. It also should be remembered that the OSHA law was passed because of the high accident rate due to careless methods and lack of regard for the safety of the worker. It is to everyone's benefit to practice good safety rules at all times.

THE IMPORTANCE OF PROPER DRESS

Clothing and Shoes

Clothing should be kept in good repair at all times when working, Figure 36-1. Torn coveralls or work pants may snag on materials or tools, thereby causing a fall. Shirt sleeves should be either buttoned or rolled up and shirttails tucked into the pants. Since pant cuffs may also catch on scaffolding or ladders when climbing, the mason should turn the cuffs down or wear pants without cuffs.

Short pants cut off above the knee may be comfortable in warm weather but offer no protection to the legs when working around protruding objects such as wall ties and wire. Also, pieces of masonry units and welding sparks may also cause injury to exposed legs. On many construction sites, pants cut above the knee are prohibited by the supervisor or safety engineer.

The mason should be careful to protect the feet as they may be struck and crushed from heavy weights. Steel-toed shoes which are capable of bearing heavy blows are available and are a recommended safety measure. Many masons resist wearing steel-toed shoes because of the extra weight, and because workers sometimes forget they are wearing them and injure themselves by kicking an object and jamming the toes against the hard surface of the shoe.

Working shoes should be equipped with sturdy heels so that the feet are properly supported and to reduce the chance of a turned ankle. Excessively thin soles on shoes should be avoided, as they can cause sore feet and fatigue.

Many firms offer safety programs which provide safety shoes to employees at a reduced price.

Protective Headgear

Protective headgear constructed of a sturdy material such as fiberglass are known as *hard hats*. Hard hats are required on all construction jobsites by federal and state law. Lightweight plastic hats, commonly known as *bump hats*, afford little protection and do not conform to regulations. Approved OSHA hard hats are recommended for use on all construction jobsites.

The hard hat should be raised slightly off the worker's head so that if a blow is encountered, the hard hat will cushion the effect, acting much as a shock absorber does in a car, Figure 36-2. It is extremely dangerous practice to remove the lacing from inside the hat, as the hat would be in direct contact with the skull and could cause serious, if not fatal, injury.

Eye Protection

The masonry saw presents a hazard, as particles may fly off during the cutting process and into the eyes. Eye protection in the form of goggles or shatterproof glasses should, therefore, always be worn when cutting masonry materials, Figure 36-3. Although goggles may be sprayed with water and partially obstruct sight when operating a wet saw, flying chips present a greater danger and should be worn when any saw is in operation. They also help considerably to prevent grit and dirt from entering the

Fig. 36-1 Mason in safety gear operating a saw. Notice the hard hat, safety glasses, rubber gloves, and coveralls.

Fig. 36-2 Mason wearing proper safety headgear. Notice that the hard hat is set slightly off the head.

eyes on windy days. If dirt, grit, or other foreign matter enters the eye and cannot be removed easily, the mason should see an eye doctor as soon as possible.

Working Gloves

While restricting the sense of feeling to some degree, work gloves protect the hands from rough materials or when cutting masonry work. Work is more efficient in cold weather when gloves are worn. Wear a glove light enough so that it is not clumsy and difficult to flex the fingers.

Hair

It should be remembered that exceptionally long hair presents a very dangerous hazard if it is allowed to hang loose when working around running machinery and protruding objects. Tuck the hair well under the hard hat or tie it back to prevent possible injury.

ARRANGING MATERIALS SAFELY IN THE WORK AREA

Materials in the general working area and on the stockpile should be stacked safely. Masonry units which are not stacked properly and bound are very likely to upset, possibly causing injury.

The accepted method on most jobs is to reverse the direction of every other course when stacking materials so that they are secure, Figure 36-4. The mason should attempt to keep the pile neat and vertically in line to eliminate the possibility of snagging the clothing. The pile should not be so high that the mason does not have easy access to the material. Materials stacked too high on the pile not only pose a safety hazard, but may drastically curtail the mason's production.

Mortar pans should be placed back from walls approximately 2′ to allow proper working space. Pans should have approximately 5′ between them lengthwise to allow stacking of masonry units between each pan. Tools should be stored underneath the mortar pans out of the paths of workers.

Workers should never run on the job or jump over trenches or stacks of materials. In cold weather, be especially cautious of walkways and runways into buildings as they may be covered with ice and may be very slippery.

SAFETY NETS

One of the most important life-saving devices to be developed in recent years is the safety net, Figure 36-5. During the two years that the 110-story Sears Tower was being built in Chicago, Illinois, nine construction workers fell into elevator shafts to what

Fig. 36-3 Mason operating saw with proper eye protection

Fig. 36-4 Stacked materials can be secured by reversing every other course on the pile.

could have been death or serious injury. But in each case, the worker was caught in a safety net hung across the shaft. Most of them returned to work immediately.

Construction workers in general feel much safer when working from heights if they know a net is in place underneath. Since it is costly to erect the nets, they usually are used only on big jobs.

The original nets were made of manila rope, but now synthetic fibers are used as they are stronger, lighter in weight, and more resistant to wear and weather elements. One of the most popular types is made from polypropylene and has more than twice the strength of manila rope.

USING TOOLS SAFELY

Tools are the masons' means of earning a living. They must be kept in good repair if the mason is to be productive and a safe worker. Poor tools can cause injuries and they become potential weapons. Worn and defective tools should be replaced or repaired immediately.

Hammers and chisels must be kept sharp and tempered to the proper degree of hardness for the materials being cut. The best procedure is to take the tools to a local blacksmith, if one is available, so that the steel can be heated and tempered when being sharpened. Excessive grinding of a cutting edge results in a defective tool. A sharp tool is a safe tool.

Mushroomed or Burred Chisels

After prolonged use, the head of a chisel becomes flared out from striking with a steel hammer, Figure 36-6. In these cases, the chisel has become *mushroomed* or *burred.* This is a dangerous condition, since the metal curled over the edges eventually comes off; and invariably, the pieces of steel become lodged in the body. These steel pieces are very difficult to remove and may cause injury and loss of time at work.

To prevent this from happening, grind off the chisel head on a grinding wheel to remove the burred metal. While doing this, cool the tool periodically with water to prevent it from becoming too hot and, therefore, losing the temper of the steel

Caution: Heavy gloves should always be worn when cutting with or grinding a chisel to protect the hands from possible blows and particles of steel.

Electrical Equipment

Grounding Electrical Tools. All electrical tools should be grounded. Grounded plugs have three prongs. The third prong acts as a grounding device for the electricity feeding into the piece of equipment. If it were not present, the electricity might pass

Fig. 36-5 Steel framework building protected by safety nets

Fig. 36-6 Burred chisel heads, a potential cause of injury, should be avoided.

through the worker's body in the event of an electrical problem. Most new plugs on tools act as grounding devices.

Electrical Extension Cords. A simple act such as the unplugging of an electrical extension cord can be dangerous if not done correctly, Figures 36-7A and 36-7B. Heavy-duty cords are also costly to replace if damaged through careless handling and neglect. The mason should never stand in water while unplugging a cord or yank the cord carelessly. When there is an electrical problem on the job, a trained electrician should always be called.

Using Ladders Safely

The ladder is an essential piece of equipment for the mason. Practical safety rules should be practiced whenever using ladders.

Be sure that the ladder is in a good workable condition, with no split rungs or sides. If a cracked rung is found, remove the defective rung with a hammer to warn others of the hazard. Report the unsafe condition immediately so that repairs can be made.

Do not erect a ladder in a place where a passing vehicle or passerby might collide with it. Always place the ladder on a solid, nonskid surface. When using a ladder to reach another walking surface, be certain that the ladder extends at least 3′ above the point where you expect to step off of it. Be sure that the ladder is firmly attached to a strong support. The areas on which the mason is stepping should be free from clutter and should offer some type of grip for the hands. Rung clamps must always be engaged before ascending the ladder, or the top section of the ladder will slip, possibly causing injury.

Aluminum ladders, while stronger, are lighter in weight. For this reason, they skid more easily than heavy wood ladders. Aluminum, an excellent conductor of electricity, must not touch or lay against electrical wire.

If tools are to be carried while climbing a ladder, shove the plumb rule through the handle of the tool bag and sling it over the shoulder. This frees both hands for climbing.

At no time should the mason stand on the highest rung of the ladder to work. This practice is extremely dangerous. Where there is a danger of slipping on concrete floors, another worker should hold the bottom of the ladder steady. Nonskid feet are also available for ladders.

Ragged-Edged Mortar Pans

The rough treatment given mortar pans, such as being beaten to remove mortar, soon takes its toll on the pans, in the form of sharp, torn metal edges. This can be very dangerous since the mason's hand is placed below the top edge when mortar is taken from the pan with the trowel.

When mortar pans are in this condition, they should be taken out of use and repaired immediately. If they are beyond repair, they should be replaced. Metal pans can be welded, but plastic ones must eventually be discarded.

Removing Nails from Lumber

When wooden braces which contain nails are removed, either remove the nails at once or bend them over into the wood. Never throw lumber with nails which have the sharp end facing up on the ground or

Fig. 36-7A Incorrect method of unplugging electrical cord. Notice the mason's grip on the wire and the downward pull. This is a potential work hazard.

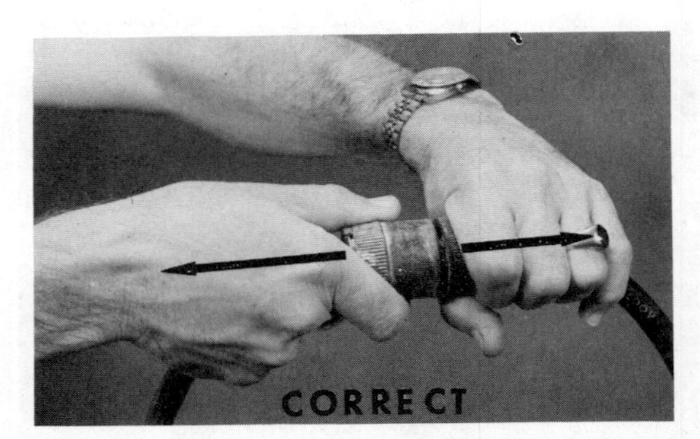

Fig. 36-7B Correct method of unplugging cord. The mason is gripping the plug firmly and pulling the plug in a straight, outward direction.

floor. A common accident on jobsites is stepping on nails, but this can be prevented through simple safety practices. If a nail punctures the skin, see a doctor immediately for treatment of the wound.

The sharp end of nails should never be exposed underneath scaffolding, since a person passing underneath must bend over and, therefore, will not see them. Remove or bend the nails over.

Overhead Objects

The danger derived from objects being dropped from overhead cannot be overemphasized. This danger may come from two major sources. First, workers located at a high level pose a threat as they may drop tools or materials on the workers below. Protective measures can be taken to prevent this, such as wooden planking laid on framework or heavy wire mesh. The best course is to avoid working directly underneath other workers, but this is not always possible. Always be on the alert for falling objects when working in this type of situation.

The second source of danger from overhead objects originates on high-rises or tall buildings, where cranes are necessary for lifting and placing various materials on the structure. Workers must exercise extreme caution in these situations. For example, sometimes steel bar joists which support floors must be fastened to a crane by means of a steel hook and cable so that they can be moved. It is entirely possible that the joists may become loosened from the grip as they are being transported to the desired level, endangering the workers below.

> **Caution:** Under no circumstances should any workers be located underneath cranes when they are transporting heavy equipment. As a rule, the best practice is to suspend work in the immediate area while the materials are being moved by the crane.

EXERCISING CARE ON SCAFFOLDING

Remember the general rules concerning safety on scaffolding, and always apply these rules when on the job.

- Inspect every scaffold to be sure it is built correctly before working on it. It should be set on a firm base, preferably a wooden plank (called a *mud sill*).
- Spaces or gaps left near the wall are potential hazards through which the mason could fall and be injured. A substantial thickness of plywood or boards should cover the hole and be nailed down on both sides.
- When more than 1 section of scaffolding is used, the danger becomes greater because of the increased height. Attach all braces and tighten thumbscrews. Lap boards far enough over so that they are secure. Guardrails and toe boards should be in place when the height demands them.
- Tie scaffolding to the structure if more than 2 sections are used or if it is more than 12′ in height. Use a strong wire that can be cut off later. Fill the holes with mortar.
- Tie ladders to the scaffold so that they won't become dislodged.
- Place stacks of materials over the scaffolding frames.
- Use good, strong planking for the flooring of the scaffold.
- Practical joking cannot be tolerated when working on scaffolding.

LIFTING SAFELY

Many masons suffer unnecessary, painful injuries such as back strain and hernias due to poor lifting practices. Good lifting practices are primarily the application of good common sense, Figure 36-8. The following are suggestions to be followed when lifting.

- Remove all scraps of materials from the working area and clear the area to which the object is to be moved. Be sure that the area is free from holes and uneven footing.
- Wear steel-toed safety shoes if possible.
- Get extra help for large or heavy objects. Never attempt to lift a weight when it is questionable as to whether you are able to lift it or not.
- Study the object to determine the best way to pick up and grip the object.
- Place the feet close to the object, and apply a good grip.
- Bend the knees and keep the back straight when lifting.
- Keep the load close to the body.

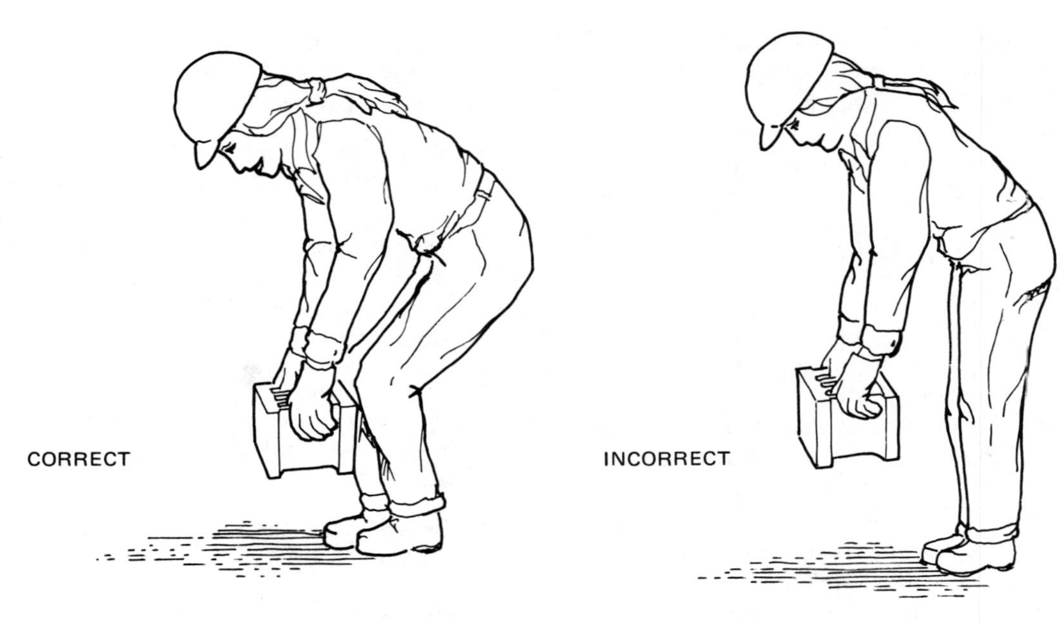

Fig. 36-8 Correct and incorrect methods of lifting. The worker on the right is lifting with her back, not her arms. The worker on the left bends her legs as she lifts the block, therefore reducing unnecessary strain.

WHEN AN INJURY OCCURS

Any injury serious enough to require medical attention should be reported to the supervisor or instructor immediately. There are two important reasons for this; quick action and first aid may stop the injury from becoming more serious, and workers compensation insurance requires that injuries be reported as soon as possible after they occur. Failure to report accidents may result in the loss of rights or benefits.

As a rule, the worker should not attempt to treat the injured person if the injury is serious. However, if a person is bleeding severely, immediate action must be taken. It should be remembered, though, that improper treatment may do more harm than good.

So that work is safe, profitable, and satisfying, be constantly alert to hazards on construction work and practice good safety rules at all times.

ACHIEVEMENT REVIEW

Answer the following questions with a short statement.

1. Describe a properly dressed mason ready for work.
2. State two reasons why good work shoes are important to the mason while working.
3. Give a reason why it is not desirable to wear work pants with cuffs while working.
4. Although the wearing of steel-toed shoes is recommended, there is a danger if care is not exercised. What is this danger?
5. Removing the lacing from inside a hard hat can be very dangerous. Why is this true?
6. Describe two situations when the wearing of safety glasses or goggles is required.
7. What is a *burred head* on a chisel and what action should be taken to correct this situation?
8. Leaving a brick hammer laying in the work path can cause a painful bruise of the leg. Describe how this may happen.

9. Describe the proper method of unplugging an electrical extension cord.
10. What is the purpose of a third prong on the end of the cord that is plugged into the receptacle?
11. The mason is preparing to climb a ladder set on a concrete floor. The ladder does not have nonskid feet. What is the proper procedure to follow to prevent an accident?
12. Describe the safe method of stacking bricks. Use a sketch or drawing to explain more fully.
13. What is the proper procedure to follow when erecting the first sections of scaffolding to properly secure it?
14. Accidents must be reported to the supervisor as soon as possible. Why is this necessary?

Unit 37
Recommended Practices and Tips

OBJECTIVES

After studying this unit, the student will be able to

- tie a girth knot.
- apply various tips to actual working situations.

Throughout this text, masonry practices and established methods of performing tasks and skills have been discussed in detail. However, there are certain tips in any trade or business which can be used to make the job easier and quicker. These tips can be extremely helpful to the apprentice mason. It is not possible, of course, to relate all of the tips and time-saving techniques which masons have used over the years, but some of the more important ones involving the use and care of tools and supplies are summarized here. Trade tips have developed over the years as easier, more effective ways of doing various jobs. Tips should never be applied at the sacrifice of safety.

TIPS ON THE CARE AND USE OF TOOLS

Cleaning the Trowel

Good mason's trowels are made from very high-quality steel. If they are allowed to remain wet for a long period of time, they will rust. Once a trowel is rusted or pitted, it is not effective in lifting mortar. The trowel should not be cleaned with water unless it is going to be dried immediately.

To clean fresh mortar from the trowel, the mason should work it into a sand pile or in loose floor dust, Figure 37-1. The mortar absorbs the dust which causes the mortar to dry rapidly. The blade of a hammer or the edge of a brick can then be used to scrape off the dry mortar. Steel wool also works well. However it is cleaned, the trowel should always be stored completely dry and clean. If the trowel is being stored for a long period of time, apply a light coat of oil to prevent rusting.

The Plumb Rule

Preserving the Plumb Rule. Wooden plumb rules should be rubbed down with boiled linseed oil, available at any hardware or paint store, Figure 37-2. This should be done periodically to preserve the wood and prevent mortar from sticking to the surface. Be careful not to get the linseed oil on the glass area, as it causes a buildup of oily scum which obscures the bubbles in the vials. If this happens, use a small piece of fine steel wool to clean the glass area.

All metal levels should be kept as free from mortar as possible as it builds up more quickly and easily on metal than on wood. Steel wool works well at removing the stains from a metal level after the mortar has been removed.

Fig. 37-1 Cleaning trowel with dry sand to prevent rusting

The Plumb Rule as a Measuring Rod. The standard 48″ plumb rule can be used as a measuring rod when laying out the linear bond on concrete block. (This applies only to standard concrete block which are 16″ long including the mortar joint.) Since the plumb rule is 48″ in length, it equals the length of 3 concrete blocks, including mortar joints.

Lay the plumb rule on the base and mark each end with a crayon or pencil. Move the plumb rule forward 1 full length and mark again. Continue this motion across the wall until the bond is completely marked. If a long wall is to be marked for bond, it is easier to use a steel tape rather than the plumb rule or mason's rule to mark the block off in divisions of 16″.

Lubricating the Folding Rule

The mason's folding rule should be lubricated periodically. If not, it will deteriorate from dust and mortar in the joints. Various types of lubricants can be used, but paste wax is best. Oil has a tendency to turn black and may leak from the joints into the mason's pocket. Any lubricant which has salt in it should never be used as it corrodes and deteriorates the joint.

Wax is applied by applying a small amount of wax on a small screwdriver, penknife, or stick and filling the joint in the rule, Figure 37-3. The wax should not be cleaned out with a cloth but left as it is. This method is effective for a considerable length of time.

Preserving Wheelbarrow Tires

Allowing a wheelbarrow to remain loaded for a long period of time causes unnecessary strain on the tire. To preserve this expensive tire, lean the wheelbarrow back, raising the front end. A brick should be placed underneath the steel frame in front of the tire, Figure 37-4. This relieves much of the stress. A wheelbarrow should never be allowed to sit overnight while loaded.

Using Adjustable Braces

For years, metal door frames were braced with wooden braces. This involved a great amount of lumber and the use of nails over the edge of the frame to hold it in place. The nails themselves can be a safety hazard. A good method of bracing metal frames in an interior masonry wall is with the use of adjustable metal braces, Figure 37-5. The braces are fastened to the overhead structural members and adjusted by sliding the brace on the frame to the desired position. They are then secured with a thumbscrew. Once in place, steel braces are very difficult to knock loose.

Fig. 37-2 Rubbing down the plumb rule with a cloth saturated with linseed oil

Recovering Tools Dropped Inside the Masonry Wall

On occasion, the mason may drop or knock a tool from the top of the wall and it lands inside a wall. One method of recovering the tool is to measure down the outside of the wall to the point at which the tool has dropped and cut into the wall with a chisel. This method, however, is not only costly to the contractor, but time consuming, since the masonry must be rebuilt.

If the object is not blocked off with reinforcement wire or wall ties, a simpler solution is to lower a small but powerful magnet on a length of line until contact is made with the tool. The tool can then be carefully pulled out of the wall.

When buying a magnet, test it by picking up a brick hammer or trowel, as these are the two tools for which it will be used most often. Most hardware stores stock inexpensive magnets of this type. Use of the magnet could save the expense of buying another tool.

If a magnet cannot be used, a wire hook can be made from a length of reinforcement wire.

Lengthening the Mason's Brush

At times, when using a scaffold to complete a wall, the work is too wet to brush immediately. This can be caused either by the masonry units themselves being wet or by damp weather conditions. Rather than detaining the job to wait for the wall to dry, the scaffolding can be dismantled and the wall can then be brushed.

This can be accomplished by use of a stepladder or by erecting 1 section of scaffolding and laying a board over it. Both of these methods are time consuming and costly. The use of a very tall ladder can also be

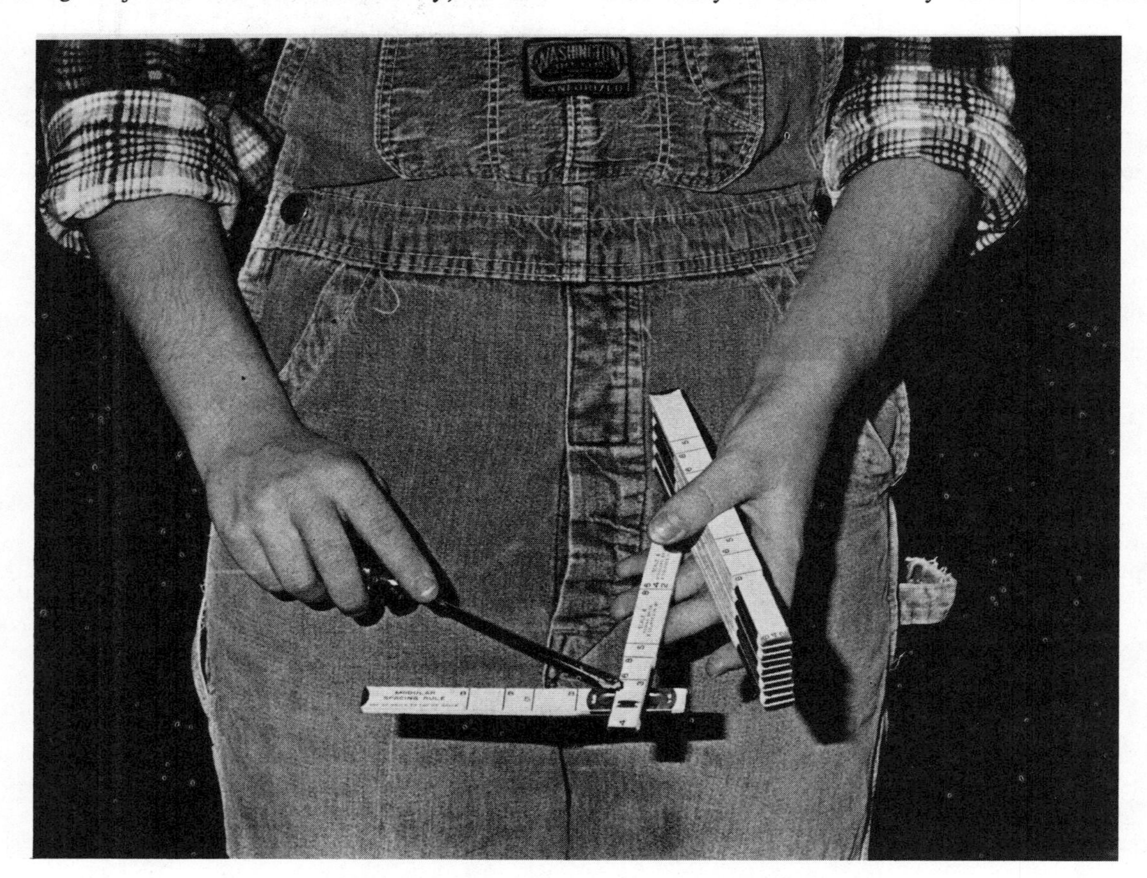

Fig. 37-3 Lubricating joints of folding rule with paste wax

Fig. 37-4 Using a brick to relieve stress on wheelbarrow tire

Fig. 37-5 Using metal braces to secure door frame

Fig. 37-6 The top section of a wall can be reached easily by attaching a brush to a pole which is long enough to reach the height.

dangerous. A simpler method, which has been used for many years, is to tie a long-handled brush (commonly called a *stove brush*) to a pole or wood strip so that the tall heights can be reached easily, Figure 37-6.

USING SUPPLIES MORE EFFECTIVELY

Detaining the Drying of Mortar

Mortar should always be mixed with as much water as possible, but still be useable for the materials to be laid. This decreases tempering. The setting time of mortar can be slowed down several ways. One precaution is to always wet the pan or board before placing mortar in or on it. This is especially necessary in hot, dry weather, as the evaporation rate is highest at this time.

Many times, masons prefer to temper their own mortar, especially when they are working on scaffolding, where it would be difficult for another person to reach. Tempering mortar is usually easier if the proper-size container and tool are chosen. Small containers, such as a 1-gal can or an old coffee can, are handy since they are easily handled and can be stored underneath the mortar stand. Small garden hoes are usually better tempering tools than the larger hoes used in mortar boxes. The hoe should always be cleaned with water immediately after use to prevent a buildup of mortar.

Once mortar has been placed in the pan it can be kept moist by pouring a small amount of water around the edges of the pan. The rate of evaporation from the mortar is decreased, so the mortar does not set as quickly. Use good judgment when adding water to mortar. Too much water destroys the proper consistency.

Using Nylon Line

Nylon line is used almost exclusively as a guide for laying masonry units. One of the most important features of nylon line is that it stretches without breaking when pulled tightly. Nylon line also withstands great amounts of water without deteriorating.

When very long pieces of nylon line are being used on the line and it breaks, it is sometimes best, for the sake of economy, to tie line together instead of discarding it. When two pieces of line are to be tied together, they should be tied so that the knot falls in a window or door space. In this way, the line will not interfere with the correct spacing of the units.

Sealing Nylon Line. To prevent raveling of the ends of nylon line, the ends may be burned with a match.

> **Caution:** When sealing line, do not touch the hot end of the line until it has cooled.

Tying a Girth Hitch. The girth hitch is a simple knot used to fasten the line to a nail or line pin. To tie a girth hitch, use the following procedure. Figure 37-7 illustrates each step.

Step 1. Form a loop in the line, allowing the sealed end to rest on the fingers of the right hand.

Step 2. Loop the line over the thumb and forefinger of the left hand. Hold the bottom of the triangular loop with the thumb and forefinger of the right hand.

Step 3. Turn the left hand downward, holding the thumb and forefinger together. This forms a loop around these two fingers.

Step 4. Pull downward firmly with the right thumb and forefinger until the two loops have al-

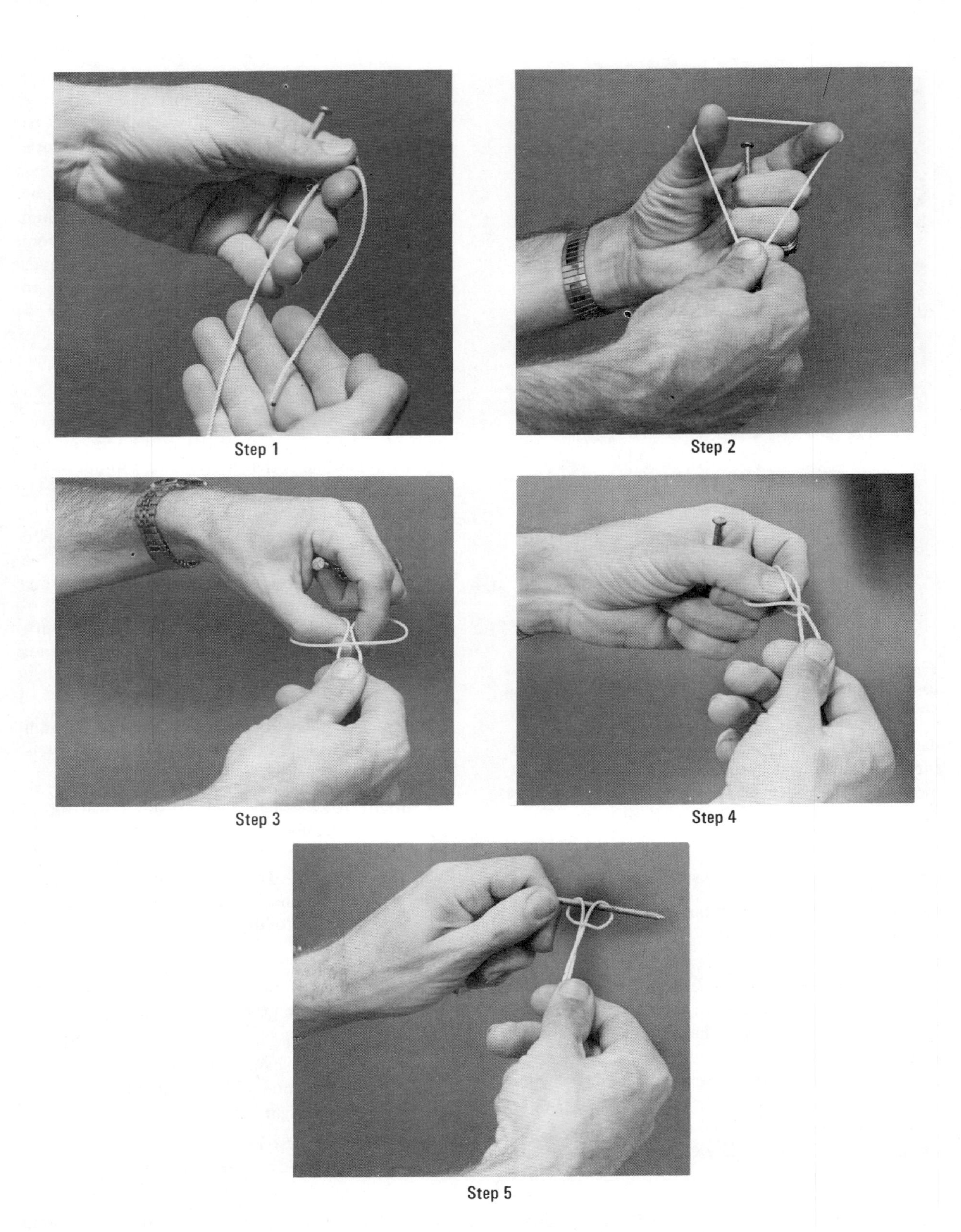

Fig. 37-7 Steps in tying a girth hitch

most reached each other. Do not pull the loops completely together as the nail must be inserted in the partially closed loop.

Step 5. Push the nail through the two loops and draw the loop tightly against the nail.

The girth hitch is also especially useful for securing a ball of line after a piece has been cut. This prevents the rest of the ball of line from unwinding.

Marking the Line. A line that has been pulled tight across a path can present a hazard to anyone working or walking around it. This is particularly true when workers with their heads bent slightly are moving equipment. A piece of material which is lightweight and which will not cause the line to sag should be fastened to the line. Aluminum foil dropped over the line works well. Be sure to squeeze the foil tightly so that it does not come off. The brightness of the foil serves as a visual warning to anyone approaching the line.

Making a Substitute Trig

On most masonry jobs, a metal trig given to the mason by building material manufacturers is used. There are times, however, when the line is being raised for another course and the trig is not available, or has been lost. So that the job is not delayed, a substitute trig can be made from a number of items such as a match cover, gum wrapper, tape, or a loop of line. The makeshift trig works effectively until it can be replaced with another metal clip-on trig.

Using Line on Corners. When pulling a line from a newly built brick corner, there is always the danger of pulling too hard and pulling the corner bricks loose, or ruining the entire corner. To prevent this from happening, insert the nail end of the line several courses of bricks down from the top of the corner, Figure 37-8. Pull the line over the corner and wrap it around a brick laid dry in a header position 1 full turn. Project the header brick about 2" out past the face of the wall and turn it on an approximate 45° angle. The line can now be pulled tightly by the person on the other end.

Adjust the brick until the line is even with the face of the wall. Place another brick on top to prevent the brick from moving. In this way, the force of the pulling of the line is transferred to the turned brick rather than the corner. This is an especially good measure to use when a line is being pulled against a small corner that has not completely set.

Bricks

When cutting bricks to fit against a jamb of a window or door, it is a good practice to make all necessary cuts to build to the top before actually laying any units. Cutting bricks one at a time for each individual course is very time consuming. Cuts made should be

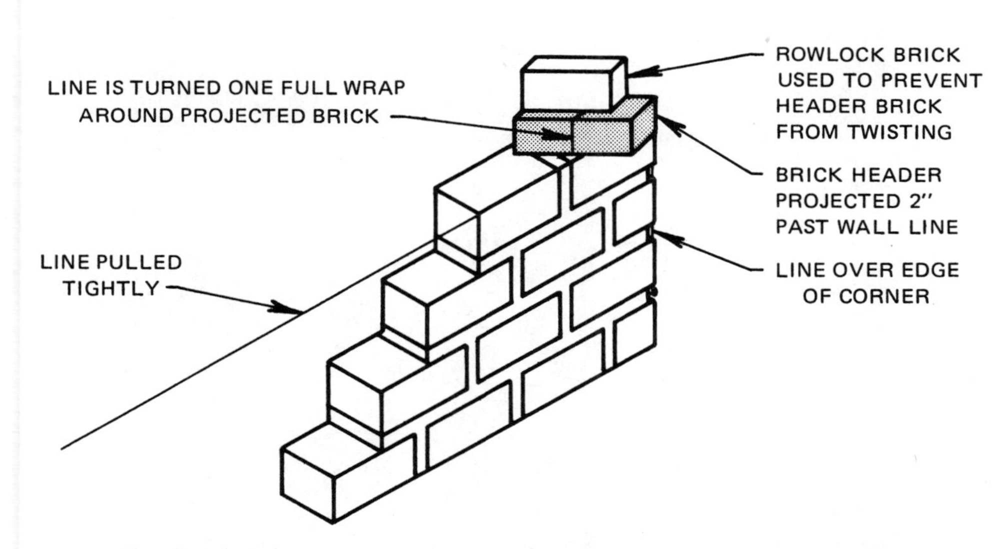

Fig. 37-8 The proper use of line can relieve stress on corner construction

the same size. Any difference in the wall space should be compensated for in the mortar head joints rather than in the size of the cuts.

Sand

Walkways leading into buildings sometimes become very slippery when wet or frozen. A generous sprinkling of sand on the area is a very economical method of stabilizing an otherwise poor footing by giving substantial traction. Calcium chloride is also used to melt ice.

Sand is very useful when sprinkled on concrete floors in the immediate area of masonry wall construction. The sand helps prevent the mortar from sticking to the floor, and reduces the amount of cleaning necessary at the completion of the job. A scraper with a long handle, such as the type used to chip ice from sidewalks, can be used to scrape the floor; the waste is then easily removed.

Sand is also sometimes used to cover brick windowsills after they have been installed to reduce the amount of work needed when cleaning with acid.

More convenient and efficient ways of working in the masonry trade are being developed constantly. However, practical methods of working should never be replaced by dangerous shortcuts. Never sacrifice safety to accomplish a job more quickly.

Glossary

absorption – the weight of water a masonry unit will absorb when immersed in either cold or boiling water for a stated length of time. This weight is expressed as a percentage of the weight of the dry unit.

adhesion – the characteristic of mortar which enables it to cling to a masonry unit

admixture – a substance or chemical added to mortar to achieve a special condition such as workability, air entrainment, water retention, acceleration of setting time, or retardation of setting time

adobe brick – a brick made in early times from clay and placed in a mold in the sun to dry and cure

aggregate – a material such as sand or gravel which is mixed with portland cement to form a concrete block

anchor bolts – steel bolts installed in a masonry wall to secure a structural member, such as wooden or steel plates

apprentice – a person who works under the direction of skilled craftspersons to learn a trade

architect's scale – ten different scales placed on a rule used to measure dimensions of drawings and plans; scale ranges from 3/32″ to 3″

ASTM (American Society for Testing and Materials) – an organization that sets standards for building materials

autoclave – a special metal cylindrical chamber used to rapidly cure concrete block; heat is approximately 360° F

autogenous healing – the ability of mortar with a high lime content to reknot or heal itself if the masonry cracks. Moisture is necessary for this healing to occur.

back filling – the process of filling dirt around a foundation after it has been built to reach grade level

backing up – the process of laying the inside masonry units of a composite wall after the exterior has been laid

banding – the placing of metal straps on cubes of up to 500 bricks by machine as the bricks leave the tunnel kiln

bat – the trade term for broken brick

batch – the combination of materials used to produce concrete block

batter – the recession of masonry in successive courses on a gradual incline; opposite of corbeling

beam – a piece of timber, steel, or other material which is supported at each end by walls, columns, or posts and is used to support heavy flooring or weight over an opening

bearing plate – a steel plate laid in mortar on a masonry wall to distribute the weight of beams which are placed over it

bearing wall – a masonry wall which supports a load other than its own weight

bed joint – a horizontal bed of mortar on which a brick is laid

beehive kiln – a semipermanent circular structure constructed of brick and wrapped with steel bands to control the expansion of the kiln when the bricks are being burned

blowout – the swelling or rupturing of a cavity wall caused by too much pressure from pouring liquid grout in a wall

blueprints or building plans – detailed drawings of a structure showing measurements and the various views which are necessary to build the structure. The term *blueprint* commonly refers to the reproduction of plans with white lines on a blue background. Persons in the masonry trade commonly call all plans blueprints.

bond – the process of (1) tying together various parts of a masonry wall by lapping units one over another or by connecting with metal ties; (2) the pattern formed by exposed face of the units; (3) the adhesion between mortar or grout and masonry units or reinforcement devices

bond beam – a course of masonry units having steel reinforcing rods inserted and filled solidly with mortar; may serve as a lintel or reinforcement beam to strengthen the wall

box anchor – a square metal tie used to tie two masonry units together

breaking joints – any arrangement of masonry units which prevents continuous vertical joints from occurring in succeeding courses

brick – a solid unit of clay or shale which has been burned in a kiln; usually rectangular in shape

brick kiln – a brick structure used to burn brick in a controlled heat

brick nogging – brick masonry that is built between wood posts or framing

British system of measurement – system of measurement used in the United States and Britain

building code – the legal requirement established by different governing agencies covering minimum construction practices

burning the joint – tooling mortar joints after they are too hard, shown by black marks. The black marks are caused by a reaction between the steel and the mortar joints.

burr – the ragged edge of metal on a chisel head resulting from the strike of a hammer

burred – the condition of a steel chisel when it is struck too frequently with a hammer; the chisel becomes frayed and ragged

buttering – the process of applying mortar on a masonry unit with a trowel to form a joint

buttress – projecting pilaster built against another wall to give the wall greater strength and stability; usually has a battered or sloping top

calcium chloride – a chemical used to accelerate the setting time of mortar

carrying the trig – the process of setting the trig brick in mortar on the wall

caulking or calking – a material used to seal cracks around such things as windows, doors, and floor lines; the material usually comes in a tube or a cartridge and is applied with a caulking gun

cavity wall – a wall consisting of 2 tiers of wythes of masonry units separated by a continuous air space not less than 2" wide. The space may be retained as insulation or filled with grout and steel reinforcements.

chase – a continuous recess built into a wall to receive pipes, wires, or heating ducts

checkpoints – mason's folding rule marks placed every 16" on the footing or base; used as a reference when laying masonry units

cinder block – a concrete block in which cinders are used as the aggregate

cinder brick – a brick made from cement and cinders

cleanout door – a metal frame and door built into a chimney which allows entrance to the chimney so that soot or ashes may be removed

clinker – a very hard, burned brick whose shape is distorted or bloated due to nearly complete burning in the kiln

clock method – the method of cutting and picking up mortar from a pan by using numbers on the face of an imaginary clock as a guide

closure brick – the last brick that is laid in the wall, usually located in the center of the wall

CMU – the abbreviation for Concrete Masonry Units

collar joint – the vertical joint between 2 tiers of masonry

common or American bond – an arrangement of bricks in stretcher position with a header course on the fifth, sixth, or seventh course of bricks

composite wall – any bonded wall consisting of wythes or tiers of different masonry units, such as a brick and concrete wall

compressive strength – the measured amount of downward pressure (load) that can be applied to a masonry material before it ruptures or breaks

concrete block – a hollow or solid block made from portland cement and aggregate

contractor – the person(s) who undertakes a job to construct a structure under a contract or agreement

control or expansion joint – a vertical joint built into masonry work in places where the forces acting on the wall cause pressure, possibly creating cracks. The joint is filled with a material which will will give flexibility, thereby relieving some of the pressure.

coping – the masonry covering laid on top of a wall. Coping is usually projected from both sides of the wall to provide a protective covering as well as an ornamental design.

corbeling – the projection of masonry units to form a shelf or ledge

crew – a group of persons who work together in constructing a building or project under the direction of a foreman

cross joint – the mortar head joint applied across the long side of a header brick

crowding the line – causing the line to swing or bounce with the brick or the fingers when laying brick; also called putting the thumb in the line

cubic foot measurement – the method of measuring masonry materials that are 1′- 0″ long, 1′- 0″ deep, and 1′- 0″ wide

cupping the mortar – a method of cutting and rolling mortar onto a trowel in a cupping motion from the mortar board

curing – the process of seasoning concrete block by using steam and open-air drying

curtain wall – a wall independent of the wall below; supported by structural steel or concrete frame of building

deadman – a post or prop to which the line on a wall is attached

details – in masonry, specific drawings of elements of construction such as lintel layout, flashing details, and installation of bolts in the wall. These are shown on a larger scale to simplify necessary procedures.

double jointing – the process of applying mortar to both ends of the closure brick as well as to ends of the bricks surrounding the area where the closure brick will be placed

dovetail anchor – a metal tie in the shape of a dove's tail which is used to tie masonry units to concrete walls

draft – the circulation of air which causes smoke or fumes to move through the chimney

drawing – the process of unloading the brick from the kiln after cooling

drawn to scale – plan drawn to a fixed ratio of measurement between the drawing and the actual completed structure. For example, a 1′- 0″ measurement on the structure usually equals a 1/4″ measurement of the plan.

drip strip – a strip of wood usually measuring 3″ x 3/4″ laid in the middle of a cavity wall to catch mortar droppings

drowning the mortar – the trade expression describing a condition of mortar resulting from too much water added to the mix

dry bonding – the laying out of bricks without mortar to establish the bond for the wall

dry-press process – a process in which a mixture of 90% clay and 10% water is formed into bricks in steel molds under high pressure; used for clays with very low plasticity

Dutch corner – the method of starting off a corner by using a 6″ piece of brick

ears (of a block) – the ends of a stretcher block where the head joint is formed; found only on stretcher block which do not have square ends

economy brick – a brick with nominal dimensions of 4″ x 4″ x 8″

efflorescence – a whitish deposit on the exposed surface of masonry units caused by the leaching of soluble salts from within the wall

elasticity – the ability of mortar to stretch and give

elevation – the vertical view of a structure or feature of a structure showing external, upright parts

engineered brick – a brick with nominal dimensions of 3 1/5″ x 4″ x 8″

English bond – a bond in which there are alternate courses of headers and stretchers

English corner – the method of starting a corner by laying a 2″ piece of brick off the return corner brick

face brick – a brick used in the front or face side of a wall; usually a better grade of brick

face shell bedding – a method of applying mortar to only the outside webs of a concrete block

face wall – a masonry wall which is tooled and exposed to view

fat mortar – a mortar which contains a high percentage of cementitious materials; sticky and hard to dislodge from the trowel

firebrick – a brick made from a highly fire-resistant clay found at a greater depth in the ground
flashing – (1) a special process used to add color to bricks by reducing the temperature in the kiln (reducing the amount of oxygen burning); (2) a type of metal such as lead, aluminum, or tin which is laid in the bed joint of a chimney at roof level to prevent the entrance of moisture and/or provide water drainage
flash set – the premature setting or hardening of cement in a mix due to excessive heat
Flemish bond – the brick bond in which a stretcher and a header alternate on the same course. The headers on every other course should be centered over the stretcher below.
flue – a built-in passageway to carry gases or fumes from a chimney
flue lining – the fired clay or fireproof terra-cotta pipe used to line the flue of a chimney. They may be round or rectangular in shape.
flue ring – another name for thimble. See thimble.
flush joint – a mortar joint which is even with the unit and in which there is no tooling done
fluted block – concrete block made with projected vertical ribs on face of block; used for textured walls
foot boards (also called blocking or hopping boards) – wooden planks laid on top of bricks or concrete block to allow the mason a higher reach when laying masonry units
footing – the broadened base of a structure which supports the weight of the structure; usually concrete
frog – a depression in the bed surface of a brick; used for design and as an aid for locking the mortar in place
full header – a header course that consists completely of headers, except for the corner starting piece
furring – the method of finishing the interior face of a masonry wall to provide space for insulation, prevent moisture transmittance, or to provide a place for application of some type of building material
furrowing – a slight indentation made in the mortar bed joint with the point of a trowel to prepare for laying brick
garden wall bond – the bond in which there are 3 stretchers alternating with a header on the same course
general contractor – the main or prime contractor on a job. They have the responsibility of coordinating all of the subcontractors' work to complete the structure according to the terms of the contract.
gingerbread work – the trade term given very decorative masonry work, especially that which is located on the cornice trim of homes
girth hitch – a form of knot used in masonry work to fasten a nail to a line
gram – the standard unit of weight measure in the metric system. One gram is equal to 1/28 oz in the English system of measurement
green brick – brick in its soft, pliable state before burning in the kiln
green staining – a reaction on masonry work caused by the salts contained in vanadium
grout – (1) a very thin mortar which is poured between two walls for reinforcement; (2) a liquid concrete that is poured in the center of a reinforced masonry wall. Consists of a portland cement, lime, and aggregates
guardrails – metal rails which fit onto the scaffolding to prevent the workers from falling. Guardrails are required by OSHA federal regulations on any scaffolding exceeding 10′ in height
guild – an organization of masons formed to guard and protect brickmaking processes and trade secrets
hacking – the procedure of stacking bricks on a pile with the use of brick tongs or carriers
half lap (of unit) – the lapping of one masonry unit halfway over another
hard brick – a brick with a very dense composition and a very low water absorption rate
hard to the line – the term which describes a masonry unit when it is pushing against the line
header – a masonry unit that is laid over two individual walls, thereby tying them together
header block – a special concrete block made to receive a brick header, also called a shoe block

header high – the trade term for the height of the wall where the headers are laid
head joint – the mortar that is placed vertically between the ends of the bricks
high-suction brick – a porous brick that absorbs moisture rapidly
hog in the wall – the trade term describing a masonry wall which is built to the same height on both ends but which has one end of the wall containing more couses than the other. This is usually caused by one mason laying bigger mortar joints under the bricks than are laid by another mason
humored – the trade term indicating a gentle adjustment of masonry work so that differences are not evident
hydrated or slaked lime – the remaining materials after quicklime has been treated with water. This is the type of lime used in masonry mortar.
hydrochloric acid – see muriatic acid.
initial set – the initial drying of mortar which bonds it to the masonry unit
jack-over-jack – the laying of one masonry unit over another without any overlapping; not a good practice unless a stack bond is being formed
jamb – vertical side of an opening, such as the side of a window or door
joints – in masonry work, the mortar between the units
joist – a type of beam used to support floor or ceiling loads, which in turn is supported by a larger beam, girder, or bearing wall
journeyman – a skilled worker who has served as an apprentice in a trade or profession and is now fully recognized as competent in that trade
jumbo or oversized brick – the term used by masons and brick manufacturers to describe a brick larger in size than the standard brick
kiln run bricks – bricks that meet no special standards as far as color, texture, or design are concerned
lead – the part of a wall which is built as a guide for attaching a line to lay the rest of the wall
lean mortar – a mortar which is lacking in cementitious materials and does not stick to the trowel or masonry unit
lintel – a horizontal member or beam placed over a wall or opening to carry or support the weight of masonry work
litre – the standard unit of measure in the metric system of capacity (more commonly called volume). A litre is primarily used to measure liquids and is equal to 1.0567 qt in the British system of measure.
load bearing – the term referring to a wall or other masonry work which supports a load
loam – a type of soil usually composed of clay, silt, sand, or other organic matter
London pattern – a pattern of trowel having a diamond-shaped heel
masonry – a material such as concrete block, bricks, or stone bonded together with mortar to form a wall or structure
masonry cement – prepared cement which contains all of the necessary cementitious ingredients, including any admixtures, so that only sand and water need to be added to form mortar. Masonry cement should meet standards for Type N mortar.
metre – the standard unit of length measure in the metric system. The metre is equal to 39.37″ in the British system.
Metric system – the decimal system of weights and measures in which the metre, gram, and litre are the basic units of measure. The metric system is used in most of the industrial nations of the world
modular masonry – a masonry construction in which the size of the building material is based upon the modular unit of 4″
module – the unit of measure used as a standard for construction; a module is 4″
mud – the trade term for masonry mortar or plaster
muriatic acid – an acid used to clean masonry work by mixing a percentage of water to acid. The acid solution works to break down stains and particles on the work.
nailing block – a wooden block inserted into a masonry wall to which the frame is secured by the use of nails

nipples – steel pins used to connect 2 sections of steel scaffolding

nominal – a dimension greater than a specified masonry dimension because of an amount allowed for the thickness of a mortar joint not exceeding 1/2″

non-load bearing – a term referring to a wall or other masonry work which supports only its own weight

Norman – a brick with nominal dimensions of 2 2/3″ x 4″ x 12″

OSHA (Occupational Safety and Health Act Public Law 91-596) – the law passed by Congress in 1970 establishing safety requirements for all places of employment in the United States

overall dimension – a measurement from one extreme point to another to obtain the total distance of the structure, such as a foundation wall

overhand work – the laying of a masonry wall from the inside, with all work done from only one side of the wall

parging – the application of a thin coat of mortar to the back of a wall to waterproof the wall

partition – an interior wall, usually 1 story or less in height, used to separate one room from another

peg end – than end of the line which contains the nail

Philadelphia pattern – a pattern of trowel having a square-shaped heel

pier – an unattached vertical column of masonry work which is not bonded to a masonry wall. The horizontal length of a pier should not be over four times its thickness.

pilaster – a wall portion projecting from wall faces and serving as a vertical column and/or beam

pin end – that end of the line which contains the line pin and to which the line is pulled

plot plan – a plan showing the complete piece of property and important information such as boundary lines, driveways, contour lines, septic systems, and wells

plugging chisel – a chisel with a tapered blade used for removing mortar from joints

plumb bond – the alignment of all head joints in a vertical position

portland cement – the fine, grayish powder formed by burning limestone, clay, or shale and then grinding the resulting clinkers. The result is a cement which hardens under water and which is used as a base for all mortar. Portland cement is a grade of cement, not a brand.

poultice – a paste made with a solvent and an inert material which is used to remove objectionable stains on brickwork

prefixes – abbreviations used to identify the varying degrees of measure in the metric system; the prefixes indicate the relationship of the units to the metre

presetting – the application of heat at a moderately low temperature (about 72°F) to set cement in concrete block

proprietary compound – a chemical compound protected by a patent or copyright; used to clean masonry work

puddling – the process of settling or consolidating grout in a masonry reinforced wall with a stick to prevent the formation of voids in reinforced sections of the wall

pug mill – a machine which extrudes clay in the form of a ribbon which is then cut into brick lengths

pumice – a light, porous volcanic rock used in a finely crushed powdered form as an aggregate of concrete block

putlog – a piece of wood or steel left in the masonry wall with one end resting in a hole and the other end laying on solid ground or on a suitable base. The scaffold frame rests on the putlog, making a strong base on which to start the scaffolding.

quicklime – a white powder which remains after limestone has been burned at a high temperature in a kiln

racking – the process of laying back each course of masonry work so that it is shorter than the course below it

raggle – a groove in a joint or masonry unit to receive roofing or flashing at the roof line

range number – the identifying number or letter assigned to each product of a brick manufacturer to designate the color, texture, or blend of the brick

ranging the corner – the process of aligning one corner with another by using a line stretched from one point to another

reinforced masonry – a type of masonry work consisting of 2 tiers of masonry units with reinforcements of steel and grout in the center for extra strength
reinforcement rod – a steel rod or bar placed in walls and concrete work to lend strength
return – the turn and continuation of a masonry wall in another direction from which it was started
reveal – that portion of a masonry jamb or recess which is visible from the face of the wall back to the frame; it is placed between the jambs, such as a brick window jamb
rip block – a concrete block less than full size in height which is generally used as a starting piece
rock face block – a concrete block made in a mold which resembles a stone wall
rolling scaffolding – a scaffolding which is mounted on wheels; used on relatively small jobs
Roman – a brick with nominal dimensions of 2″ x 4″ x 12″
rough wall – a masonry wall which is not tooled; must still be plumb and level
round joint – the trade name for a half-circular indentation formed in a mortar joint by use of a convex jointer
rowlock (also spelled rolok) – a brick laid on its face edge so that the normal bedding area is facing the bed joint; commonly used on windowsills
rubbed joint – a flush mortar joint which has been rubbed flat with some type of material to receive paint or to simply appear flat
running bond or all stretcher bond – a bond in which all of the masonry units are laid in a stretcher position
rustic – a type of brickwork intended to resemble American colonial masonry or to present a roughly finished appearance
sailor – a brick that is laid in a vertical position with the largest side facing the front of the masonry wall
salmon brick – a brick which is relatively soft, pinkish, and underburned; not recommended for outside walls
salt and pepper brick – a brick that has black marks on the finish caused by iron spots in the clay
sandblasting – the process of spraying sand which is under great air pressure onto a masonry wall to remove old paint or stains that cannot be removed any other way
sand lime brick – a brick made from sand, lime, and cement
sand-struck brick – bricks that have a sand finish because the mold is lubricated with sand to keep the brick from sticking to the mold
sash block – a specially made concrete block which has a slot in the end to receive a metal window frame
scaffold high – the height to which masonry walls are built when scaffolding is needed to complete the structure; usually 4′ to 5′ high
scaffold sill – wooden scaffold frames resting on ground level; helps to prevent the scaffold from settling after heavy rains and distributes the weight of the scaffold and materials
schedule – a list added separately to plans which describes such items as windows, doors, floors, and wall finishes
scove kiln – an outdated type of brick kiln, rectangular in shape, with holes or tunnels at the bottom in which fires were built to supply the heat to burn the bricks
section view – a drawing which details a vertical slice through a structure, showing all important information such as the composition of the wall, the width and height of the wall, wall ties, collar joints, and, in many cases, the various heights of openings in the wall
set – the length of time required for mortar to pass from the soft state to the hard state that will sustain the weight of the masonry placed on it
set of the trowel – the angle of the rise of the trowel handle in relation to the blade when laying in a flat position
setting up – the action of the mortar hardening
sheathing – a facing material such as plywood or insulated siding nailed to the outside of a wood-studded wall

shiner – a brick that is laid in a horizontal position with the largest side facing the front of the wall
shop drawing – a drawing supplied by the manufacturer to explain the installation of a product
shoring – a bracing placed against a wall or under a beam to provide temporary support
shove joint – the procedure of buttering a mortar joint on a brick and shoving it into place against another brick
sieve test – a method of measuring sand by passing it through a specific size of mesh and grading it to ASTM specifications
sill high – the height at which windowsills are installed in a masonry wall
siltation test – a method of determining the amount of loam and impurities in sand by placing the sand in water and measuring the loam that settles on top of the sand
slack to the line – the term which describes a masonry unit laid too far from the line
slaking – the process of changing quicklime into hydrated lime through a chemical reaction by adding water
slushing (a wall) – the process of using a trowel to fill collar joints with mortar
snap header – a half brick laid in a wall to resemble a header
soft-mud process – a process in which units are formed in wooden molds and removed to be burned in the kiln; used for clays which contain too much water for the stiff-mud process
soldier – a brick laid in a vertical position with its longest, narrowest side out, facing the front of the wall
solid masonry wall – a wall built of masonry units laid with full mortar joints between them and with no type of framing present
spall – a small chip or piece of block or brick that has become loosened from the unit
specifications – the detailed written description of the work to be accomplished in a building. Specifications accompany the plans and describe such things as quality of materials used, workmanship, and methods of construction.
splitrock – a solid block made of a coarse stone aggregate in which the split face is used as the surface of the finished wall
spotting a brick – the process of laying a brick to establish a specific point without the aid of the line
stack – the trade term for a chimney which contains a flue for the passage of gases or smoke
stack bond – a bond in which masonry units are laid directly over one another without breaking the bond. All of the head joints should line up in a plumb vertical position
standard brick – the brick most often used in masonry, with nominal dimensions of 8″ x 4″ x 2 2/3″
stiff-mud process – the process in which bricks are made by extruding clay through a pug mill, cutting the column with wire cutters, and burning the units in a kiln
story rod – a wooden pole or rod upon which are indicated all of the masonry courses and openings, such as windows or doors, required in the job
stretcher – a brick laid in a horizontal position with the longest, narrowest side facing the front of the wall. This is the most common position in which bricks are laid.
stringing (mortar) – the process of spreading mortar with a trowel on a masonry wall
subcontractor – a person who performs work negotiated with the prime contractor on a job or project, such as a masonry contractor
suspended or swinging scaffolding – scaffolding with heavy-duty suspended frames on which planking fits
tailing the lead – the process of aligning the ends of the courses on a corner
T-bar tie – a heavy-gauge steel strap tie used to tie intersecting bearing walls to main walls. The tie bar is built into the opposite bed joints of intersecting walls.
tempering – the process of thinning masonry mortar with water to provide workability
tender – a laborer who tends or brings materials to the mason
texture – the arrangement of particles in masonry materials which accounts for the brick's appearance. The various effects created by tooling mortar joints are also considered part of the texture.
thimble – a round, terra-cotta or fired clay insert which fits into a chimney to receive the furnace pipe, forming a fireproof connection to the masonry wall

throwing a mortar joint – the trade term for the process whereby mortar is applied on a brick with a swiping motion to form a joint

tier – a single vertical thickness or row of masonry units; also called a wythe

tight away – a term used to signal to the mason on the pulling end of the line that it is all right to pull the line tight against the line pin

toe boards – wooden boards that are attached to the bottom outside edge of scaffolding to prevent objects from falling off. Toe boards are an OSHA requirement on all scaffolds exceeding 10′ in height.

tooling or striking (of joints) – the trade term which describes the finishing of mortar joints by passing a striking tool over the joints

toothing – a toothlike projection of one masonry unit over another which forms a temporary end for a structure

tower scaffolding – a steel-framed scaffolding erected in a towerlike position, with adjustable height

trig brick – a brick on which a trig is placed to prevent the line from sagging

tuckpointing – the process of replacing the old mortar in joints with new mortar

tunnel kiln – a long, rectangular brick structure in which the process of burning bricks is controlled by computers and is completely automatic. This kiln is the most modern of all kilns used for brick manufacturing.

veneered wall – a masonry wall with a facing which is attached, but not bonded, to the backing to to act as a load-bearing wall

veneer tie – a corrugated metal wall tie used to tie masonry units to a backing, usually of a frame construction, by nailing

void – an open space in an area, such as the space between two masonry walls

walling the brick – the process of stacking brick on the back section of a wall that has been previously built. This process eliminates the need for the mason to reach into the brick pile to pick up individual bricks when building the front wall.

wall plate – a horizontal wooden member anchored to a masonry wall, to which other structural elements may be fastened. It is usually attached to the top of a masonry wall.

wash – sloping application of mortar or cement which prevents the entrance of water into masonry work

water retentivity – a quality of mortar which prevents the rapid loss of water to masonry units through absorption; enables mortar to retain moisture

water struck brick – bricks which have a very smooth finish because the molds are lubricated with water to keep the bricks from sticking to the mold

web – the cross walls connecting the face shells of hollow concrete masonry units

weep holes – small holes left in mortar joints for drainage of masonry walls. Weep holes are usually spaced 24″ apart and are placed over the first course of flashing.

welding splatter – the metal beads splashed on the surface of the masonry wall which soak into the surface of the masonry unit

winning – the removal of raw material (clay or shale) from the ground

wire mesh or hardware cloth – metal wire, usually formed in 1/4″ squares, used to tie walls which are not load bearing to masonry walls

wire track – the trade term for metal wire joint reinforcement used to strengthen or tie 2 tiers of masonry units together to form one composite wall

Z tie – a metal tie used to tie 2 tiers of masonry units together; may have a drip crimp in the center

Acknowledgments

The author extends special thanks to the Brick Institute of America, which provided assistance and special materials throughout the preparation of this text.

Reviewers

Thomas Redmond, P.E., National Concrete Masonry Association
William Roark, Director of Manpower Development, Brick Institute of America and Staff
L.E. Barbrow, National Bureau of Standards
Kenneth A. Gutschick, National Lime Association

Contributions by Delmar Editorial Staff

Copy Editors – Karl Anderson, Mary Grauerholz
Editor – Barbara A. Christie

Other help was provided by:

William B. Smith, Jr., P.E.
Joseph Thomas, Executive Director, Mid-Atlantic BIA
R.W. Otterson, Director, BIA
Ronald G. Nickson, P.E., BIA
Alan H. Yorkdale, P.E., Director of Engineering and Research, BIA
Robert P. Anderson, P.E., Engineer, BIA
Dean C. Patterson, P.E., Engineer, BIA
Daniel C. Cammer, S.E.T., Engineer, BIA
Carl P. Bongiovanni
Howell King Cole
Dominic Massett
Dr. Donald Maley, Chairman of Industrial Education, University of Maryland
Dr. Kenneth Strough, Department of Industrial Education, University of Maryland
Lefty Kreh
Irving Swope
George Rocus
George Kuhn
Donald Kreh
Aubrey Markus Lyles
Charles Hancock
Thomas Shoemaker
Harold Staley
Sherwood Mackenzie
Cleon R. Stull, Sr.
Tedd L. Godbee
James Fitzgerald, Chief Estimator, Frederick Contractors
Henry Toennies, NCMA
Ray White, Occupational Safety and Health Administration
Charles Chamberlain III, Associated Builders and Contractors, Inc.
Frank J. Fisher

William Ring
Frederick County Board of Education, Frederick, Maryland
Robert W. Sheckles, Inc., Frederick, Maryland
Frederick Contractors, Inc., Frederick, Maryland
Masonry Institute, Inc., Washington, D.C.
Portland Cement Association, Chicago, Illinois
Safeway Steel Products, Milwaukee, Wisconsin
Students, Catoctin High School Building Trades Classes

The author would also like to thank his friends and fellow masons for countless ideas and encouragement in completing this book.

Contributions to Classroom Testing

The instructional material in this text was classroom tested at Catoctin High School in Thurmont, Maryland.

Contributions of Content and Illustration

The Brick Institute of America – Figures 1-3A, 1-3B, 1-4, 2-1, 2-7, 3-3, 3-4, 7-8, 7-9, 7-10, 7-11, 10-3, 10-4, 10-6, 21-2, 22-2, 22-3, 22-4, 22-6, 22-7, 22-8, 22-9, 23-1, 23-3, 25-5, 26-6, 27-4, 29-1, 30-1, 32-1, 32-2, 32-3, 32-4, 32-6, 32-7, 33-2, 33-4, 33-5, 34-1, 34-2, 34-5, 34-6

The Dur-O-Wal Company - Figures 21-7, 25-6

ADJUSTOMATIC Sectional Steel Scaffolding Automatic Devices - Figure 27-1

Bil-Jax Scaffolding Company, Inc. - Figures 27-3, 27-5

Patent Scaffolding Company, Division of Harsco Corporation - Figures 27-6, 27-7, 27-8, 27-9, 27-10, 28-2

The Scaffolding and Shoring Institute - Figure 28-1

WACO Scaffolding and Shoring Institute - Achievement Review, Part B

The Baltimore Brick Company - Figure 2-8

The National Concrete Masonry Association - Figures 4-1, 4-3, 14-6, 29-8

The Goldblatt Tool Company - Figures 6-5, 6-15, 8-1, 8-10, 8-15

The National Lime Association - Figures 9-1, 12-1

The Masonry Specialty Company - Figure 15-2

Woodco Company - Figure 33-7

Park Tool Company - Figure 8-12

DCA Educational Products - Figures 8-2, 8-3, 8-4, 8-6

All other photographs were taken by the author.

Index

11/81(1C405)